T0180208

Geosynchronous SAR: System and Signal Processing

Teng Long · Cheng Hu · Zegang Ding
Xichao Dong · Weiming Tian
Tao Zeng

Geosynchronous SAR: System and Signal Processing

Teng Long
Beijing Institute of Technology
Beijing
China

Xichao Dong
Beijing Institute of Technology
Beijing
China

Cheng Hu
Beijing Institute of Technology
Beijing
China

Weiming Tian
Beijing Institute of Technology
Beijing
China

Zegang Ding
Beijing Institute of Technology
Beijing
China

Tao Zeng
Beijing Institute of Technology
Beijing
China

ISBN 978-981-13-3928-8 ISBN 978-981-10-7254-3 (eBook)
https://doi.org/10.1007/978-981-10-7254-3

Softcover re-print of the Hardcover 1st edition 2018

Printed on acid-free paper

This Springer imprint is published by Springer Nature
The registered company is Springer Nature Singapore Pte Ltd.
The registered company address is: 152 Beach Road, #21-01/04 Gateway East, Singapore 189721, Singapore

Foreword

The geosynchronous SAR (GEO SAR) runs on a geosynchronous orbit at a height of 36,000 km. As the orbit height of GEO SAR increases, the limitation in traditional low Earth orbit (LEO) SAR is no longer valid, such as the quality factor. Therefore, GEO SAR can provide the large imaging swath (hundreds to thousands of kilometers) and high temporal resolution (hours to 1 day) remote sensing data. It is a good complementation for the current spaceborne SAR imaging system. Through the construction of the GEO SAR constellation, it can achieve the near-real-time continuous surveillance over terrestrial or marine area. It can realized the monitoring and forecast of disasters and hazards, such as flooding, earthquake, landslide, tsunami and oil spilling, etc.

However, in GEO SAR, the ultra-long propagation path is at the level of 40,000km, and results in the significant path loss, which poses very harsh requirements on the antenna size and the transmitting power. Therefore, since the continental GEO SAR system with an inclination of 50 deg was firstly proposed by Tomiyasu in 1980s, the research and development show a stagnant state as the limitations of the then engineering technology. With the rapid development and progress of technology in 21st century, GEO SAR becomes feasible and thus is a research focus again. The research groups from the US, Europe and China carry out mounts of work, including the GEO SAR novel concepts, signal acquisition mechanism, information processing approaches, as well as the corresponding equivalent validation experiment for GEO SAR. Especially in China, as a scientific satellite mission, GEO SAR has been listed in the development program of 'National Civilian Space Infrastructure'.

The authors and their research group have been enrolled in the related researches in GEO SAR since 2007, sponsored by several National Natural Science Foundation of China (NSFC) projects (Grant No.: 61032009, 61225005, 61471038, 61501032), such as the NSFC key project and the National Science Fund for Distinguished Young Scholars. The book states the latest progress based on these years' researches, such as the system design and performance analysis, accurate imaging algorithms in frequency domain, atmospheric effects and compensation,

GEO InSAR/D-InSAR techniques, and the equivalent experiment validation based on global navigation satellite system.

We are pleased to acknowledge the interactions and support of the colleagues from the China Academy of Space Technology. We are also very grateful to and thank Springer staff and, in particular, Praveen Anand, Jessie Guo, who frequently solicited by our questions, doubts, and requests, have always found a way out.

Beijing, China

Teng Long
Cheng Hu
Zegang Ding
Xichao Dong
Weiming Tian
Tao Zeng

Contents

1 Introduction 1
1.1 SAR Concept 1
1.2 SAR Modes 3
1.2.1 Stripmap SAR 4
1.2.2 ScanSAR Mode 4
1.2.3 Spotlight SAR Mode 4
1.2.4 Sliding Spotlight SAR Mode 6
1.2.5 TOPS Mode 6
1.2.6 Multi-channel Mode 7
1.2.7 Summary 7
1.3 Summary of Spaceborne SAR Development 9
1.4 GEO SAR Concept 12
1.5 State of Art 13
1.6 Special Issues and Challenge 20
1.7 Outline 21
References 21

2 GEO SAR System Analysis and Design 27
2.1 Characteristics Analysis of GEO SAR 28
2.1.1 Characteristics Differences Between LEO SAR and GEO SAR 28
2.1.2 Coverage and Revisiting 29
2.1.3 Motion Characteristics 30
2.1.4 Doppler Characteristics 37
2.1.5 Work Modes of GEO SAR 41
2.2 System Parameters Analysis and Design 42
2.2.1 Resolution 42
2.2.2 Power Budget 54
2.2.3 Ambiguity 54

2.3 Strategies of Attitude Steering 59
2.3.1 2D Attitude Steering in Space-Borne SAR 59
2.3.2 Optimized Resolution Attitude Control 66
2.3.3 Performance Comparison 69
2.4 Summary 74
References 75

3 Algorithms for GEO SAR Imaging Processing 77
3.1 Introduction 78
3.2 Echo Signal Model 80
3.2.1 Error Analysis of the "Stop-and-Go" Assumption 80
3.2.2 AccurateSlant Range Model in GEO SAR 84
3.2.3 Two-Dimensional Spectrum of Echo Signal 86
3.2.4 Spatially Variant Slant Range Model Coefficients 92
3.3 Time Domains Algorithm 94
3.3.1 Traditional BP Algorithm 94
3.3.2 Fast BP Algorithm 96
3.3.3 Computer Simulation 97
3.4 Frequency Domain Algorithm 99
3.4.1 Analysis of Difficulties 99
3.4.2 Derivation of Azimuth Compensation 105
3.4.3 Details of 2D NCSA Based on the Azimuth Compensation 109
3.5 Discussion 116
3.5.1 Computer Simulation 118
3.6 Summary 122
References 127

4 Analysis of Temporal-Spatial Variant Atmospheric Effects on GEO SAR 129
4.1 Introduction 129
4.1.1 Troposphere 130
4.1.2 Ionosphere 133
4.1.3 Summary 135
4.2 Tropospheric Influences 135
4.2.1 Signal Model Considering Time-Varying Troposphere 135
4.2.2 Theoretical Analysis of Influences on Focusing 137
4.2.3 Simulation 138
4.2.4 Influences on Focusing 140
4.3 Background Ionospheric Influences 142
4.3.1 Background Ionosphere Models 142
4.3.2 Time-Frequency Signal Model 144
4.3.3 Influences on Focusing 145
4.3.4 Effects on Range Focusing 147
4.3.5 Effects on Azimuth Focusing 149

4.3.6 Performance Analysis and the Changing TEC Boundaries . . . 150
4.3.7 Simulations . . . 154
4.4 Ionospheric Scintillation Influences . . . 160
4.4.1 Characteristics and Modelling Ionospheric Scintillation in GEO SAR . . . 160
4.4.2 GEO SAR Signal Model Considering Ionospheric Scintillation . . . 165
4.4.3 Theoretical Analysis . . . 168
4.4.4 Simulations . . . 175
4.5 Summary . . . 180
References . . . 186

5 Ionospheric Experiment Validation and Compensation . . . 189
5.1 Introduction . . . 190
5.2 Experiment Principle and Signal Model . . . 191
5.2.1 Background Ionosphere . . . 191
5.2.2 Ionospheric Scintillation . . . 192
5.2.3 GEO SAR Signal Models . . . 194
5.2.4 Experiment Overview . . . 194
5.3 Background Ionosphere Experiment and Compensation . . . 196
5.3.1 GPS Data Recording . . . 196
5.3.2 Data Pre-processing . . . 197
5.3.3 Experimental Data Processing . . . 199
5.3.4 Result and Discussion . . . 201
5.3.5 Compensation: Autofocus Methods . . . 203
5.4 Ionospheric Scintillation Monitoring Experiment . . . 204
5.4.1 Experimental Data Processing . . . 204
5.4.2 Discussion . . . 206
5.5 Ionospheric Scintillation Compensation . . . 208
5.5.1 Avoidance Based on Orbit Design . . . 208
5.5.2 Joint Amplitude-Phase Compensation Based on Minimum Entropy . . . 214
5.6 Summary . . . 226
References . . . 227

6 Geosynchronous InSAR and D-InSAR . . . 231
6.1 Interferometry Basics . . . 231
6.2 GEO InSAR Special Issues . . . 234
6.2.1 Un-parallel Repeated Tracks of GEO InSAR . . . 234
6.2.2 Squint Looking in GEO InSAR . . . 237
6.3 Optimal Data Acquisition and Height Retrieval . . . 239
6.3.1 OMRD Data Acquisition Method . . . 240
6.3.2 GEO InSAR Height Retrieval Model . . . 245
6.3.3 Simulation Verifications . . . 246

6.4 GEO InSAR and D-InSAR Processing . . . 252
6.4.1 SAR Interferometry . . . 252
6.4.2 Differential Interferometry . . . 253
6.4.3 Performance Analysis . . . 257
6.4.4 GEO InSAR Baseline Analysis . . . 262
6.4.5 Analysis of Measurement Accuracy . . . 264
6.5 Summary . . . 270
References . . . 271

7 Three Dimensional Deformation Retrieval in GEO D-InSAR . . . 273
7.1 Limitation of 1D Deformation Measurement . . . 273
7.2 State of Art of 3D Deformation Retrieval in Spaceborne SAR . . . 274
7.3 Multi-angle Measuring in GEO SAR: 3D Deformation Retrieval . . . 276
7.3.1 Method and Accuracy Analysis . . . 276
7.3.2 Optimal Multi-angle Data Selection . . . 278
7.3.3 Simulation and Discussion . . . 281
7.4 Summary . . . 289
References . . . 289

About the Authors

Teng Long is a Full Professor in the School of Information and Electronics (SIE), Beijing Institute of Technology (BIT), Beijing, China. He is Fellow of the Institute of Engineering and Technology (IET) and Fellow of the Chinese Institute of Electronics (CIE). His research interests are radar system, embedded real-time digital signal processing, and remote sensing signal processing. He has published two academic books and over 200 refereed journal and conference papers, and granted over 50 national invention patents of China. He has received one Second Prize of National Technology Invention Award, two First Prizes, and four Second Prizes of Ministerial Awards of Technological Invention and Progress, National Top Tier Talent Award, National High-level Personnel of Special Support Program (Ten Thousand Talent Plan), Yangtze River Scholar Distinguished Professor, and Distinguished Young Scholar Fund of the National Natural Science Foundation (NNSF).

Cheng Hu is a Full Professor in SIE, BIT, Beijing, China. He is the IET Fellow and IEEE Senior Member and Senior Member of CIE. His research interests are the geosynchronous synthetic aperture radar (GEO SAR), bistatic SAR, ground-based SAR, and forward scattering radar. He has published more than 120 refereed journal and conference papers, and granted 30 and filed 30 national invention patents of China. He was a recipient of the best paper awards of the IET International Radar Conference 2013 and the IEEE International Radar Conference 2011, and the best poster awards in Land Remote Sensing of DRAGON 3 Program in 2012.

Zegang Ding is a Full Professor in SIE, BIT, Beijing, China. His research interests are the system design and image formation for spaceborne SAR, and the information processing for airborne SAR. He has published more than 30 refereed journal and conference papers, and granted 20 and filed 25 national invention patents of China.

Xichao Dong is an Assistant Professor in SIE, BIT, Beijing, China. His research interests are GEO SAR and microwave remote sensing. He has published more than 40 refereed journal and conference papers. He was a recipient of the IEEE CIE International Radar Conference Excellent Paper Award in 2011.

Weiming Tian is a Lecturer in SIE, BIT, Beijing, China. His research interests are high-resolution SAR system and signal processing technology, and ground differential interferometry SAR. He has published more than 20 academic papers.

Tao Zeng is a Full Professor in SIE, BIT, Beijing, China. He is the IET Fellow. He has been mainly working on the bistatic and multistatic radar, and so on. He has published more than 100 refereed journal and conference papers, and granted 70 and filed 60 national invention patents of China. He has received one Second Prize of National Technology Invention Award and the Distinguished Young Scholar Fund of the NNSF.

Chapter 1
Introduction

Abstract Synthetic aperture radar (SAR) can achieve high-resolution imaging, which is independent of flight altitude. It can provide observation and monitoring in all weathers and all day. Now it is widely used in remote sensing, topography, oceanography, glaciology, geology, forestry, etc. It has been developed from single function and single mode to multi-functions and multiple modes. Now geosynchronous SAR (GEO SAR) is an important and fascinating branch which has great advantages and challenges. In this chapter, SAR basic concept and work modes are given as the foundation of the book. Then the development of spaceborne SAR is summarized. As the start of the book, the GEO SAR concept and state of art are proposed. Finally, special issues and challenges are discussed and the outline of the whole book is given.

1.1 SAR Concept

The Radar is an electronic device that uses electromagnetic waves to detect targets. Early radar systems used time delays to measure the distance between the radar and the target. It also used the antenna pointing and the Doppler shift to determine the direction of the target and the target velocity. In 1951, Carl Wiley of Goodyear Aerospace Corporation in the United States was the first to find out that side-looking radar can improve the azimuth resolution by utilizing the Doppler shift present in echoes. This landmark discovery marks the birth of Synthetic Aperture Radar (SAR) technology. Using this technique, a two-dimensional radar image of the Earth's surface can be obtained.

Compared with the traditional optical remote sensing, SAR is an active radar system that can penetrate clouds, smoke, fogs, etc., with all weather all day observation capability. It works in the microwave band, and uses the collected target echoes to produce a two dimensional image, unveiling new features of the Earth's surface, and hence presenting an important tool in the field of Earth remote sensing [1].

T. Long et al., *Geosynchronous SAR: System and Signal Processing*,
https://doi.org/10.1007/978-981-10-7254-3_1

SAR can perform a two-dimensional high-resolution imaging. Figure 1.1 shows the SAR working geometry. The aircraft with a mounted radar onboard moves along a straight trajectory at a constant velocity V in a particular direction. The radar transmits and receives signals in a side looking direction at a fixed repetition frequency. It has a horizontal beam width β defined as

$$\beta = \frac{\lambda}{D} \tag{1.1}$$

where D is the real aperture length, and λ is the wavelength.

If the amplitude and phase of the received signal are stored and coherently superimposed with previously received signals, an equivalent linear array will be formed as the radar moves [2]. The length of the equivalent linear array is limited by the radar position when the target enters the radar beam and the position where the target leaves the illuminating beam, and is defined as the length of the synthetic aperture L_s, and given as

$$L_s = \beta R \tag{1.2}$$

where R is the distance from the target to the platform trajectory.

After coherent processing, the equivalent beam width formed by the synthetic aperture is

$$\beta_s = \frac{\lambda}{2L_s} \tag{1.3}$$

where the factor 2 is due to the round-trip propagation of the signal.

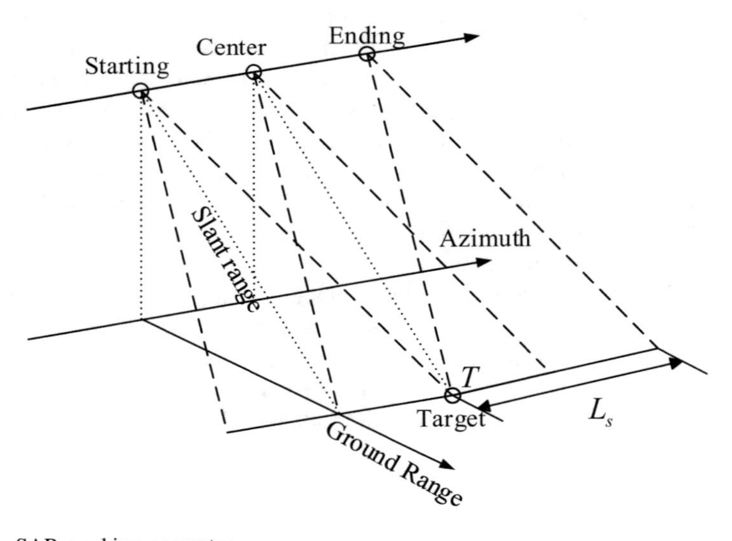

Fig. 1.1 SAR working geometry

Therefore, the azimuth resolution ρ_a is

$$\rho_a = \beta_s \times R = D/2 \tag{1.4}$$

On the other hand, high range resolution can be achieved by transmitting broadband signals (such as chirp signals) and using matched filtering techniques for pulse compression. The range resolution ρ_r is given as

$$\rho_r = \frac{c}{2B_d} \tag{1.5}$$

where c is the light velocity and B_d is the signal bandwidth.

The differences between SAR and traditional real aperture radar can be summarized as follows.

1. SAR azimuth resolution is independent of the slant range. It is equal to half the real aperture of the antenna. This is due to the fact that the synthetic aperture length is proportional to the target slant range. However, the azimuth resolution of a real aperture radar is inversely proportional to the range when the beamwidth is constant, i.e. the farther the distance is, the coarser the azimuth resolution is.
2. The azimuth resolution of a SAR system is independent of the wavelength. This is because the length of the synthetic aperture L_s is proportional to the wavelength. In the real aperture radar, longer wavelength induces wider beamwidth, and hence produces coarser azimuth resolution.

1.2 SAR Modes

Different SAR modes can satisfy various requirements in terms of system resolution and swath width. Basic SAR modes include stripmap mode, ScanSAR mode and spotlight SAR mode [2]. The sliding spotlight SAR mode was proposed as an evolution of the spotlight SAR to increase the azimuth swath width. Likewise, in order to overcome the scalloping effect in ScanSAR, the Terrain Observation by Progressive Scans (TOPS) mode was proposed [3]. However, the aforementioned SAR modes do not overcome the limitation of quality factor, and do not provide high-resolution and wide-swath imaging simultaneously. Alternatively, multi-channel mode was proposed by adding more transmitting and receiving channels. In fact, the quality factor can be improved by a number of equivalent channels [4–6], but at the cost of more complicated and expensive systems. In addition, the imbalance between channels is rather difficult. Therefore, the multi-channel mode has not been widely used in spaceborne SAR except for some experimental modes.

1.2.1 Stripmap SAR

Stripmap SAR is a standard SAR mode. In this mode, the beam pointing remains side-looking as the radar moves, resulting in consecutive strip images. It can provide a relative large azimuth swath providing a long working period of time. However, the azimuth resolution is limited by the azimuth antenna size (the maximum possible azimuth resolution equals to half the azimuth antenna aperture) while the range swath is limited by the range antenna size (Fig. 1.2).

1.2.2 ScanSAR Mode

ScanSAR is a common wide-swath mode and can be thought of as the extension of stripmap SAR mode. It transmits and receives echoes using different beam positions (or beam pointing), enabling the generation of wide-swath images by image combination. In ScanSAR, the acquired data at a given beam position is called a 'burst'. For the N-burst ScanSAR, the integration time of each sub-strip decreases by N times, i.e. the azimuth resolution drops N times while the swath width increases N times. However, inherent scalloping effects exist in ScanSAR, and need to be mitigated using multi-look processing, but at the cost of coarser azimuth resolution (Fig. 1.3).

1.2.3 Spotlight SAR Mode

Compared to the stripmap SAR, spotlight SAR can provide higher azimuth resolution. In this mode, the beam points to a certain fixed scene which has the effect of increasing the synthetic aperture and hence the coherent integration time resulting

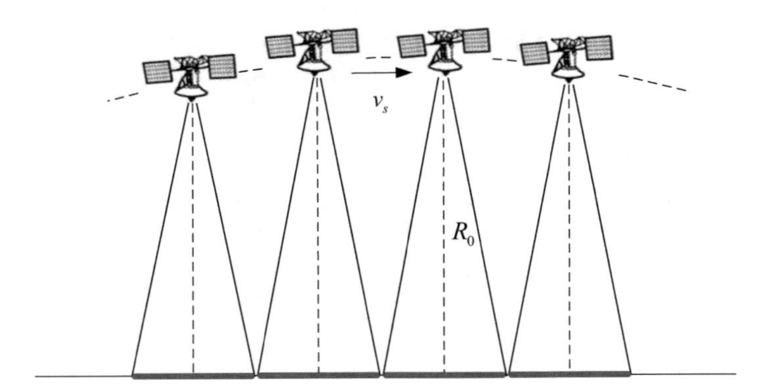

Fig. 1.2 Sketch map of stripmap SAR mode

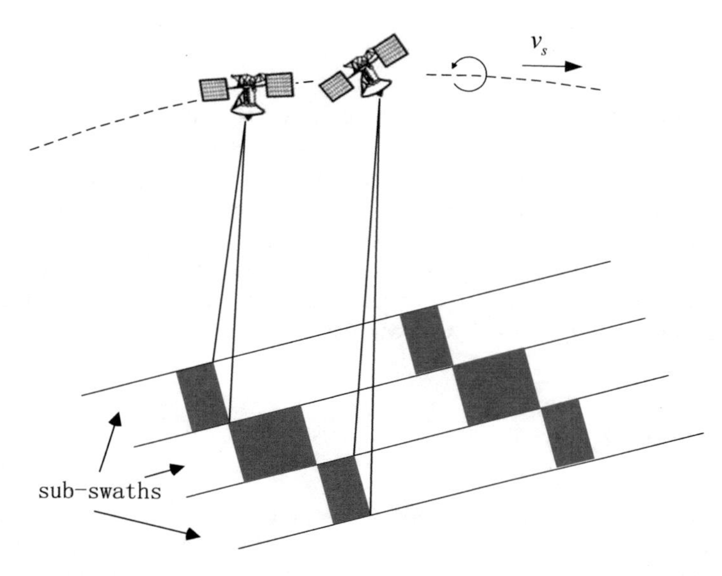

Fig. 1.3 Sketch map of ScanSAR mode

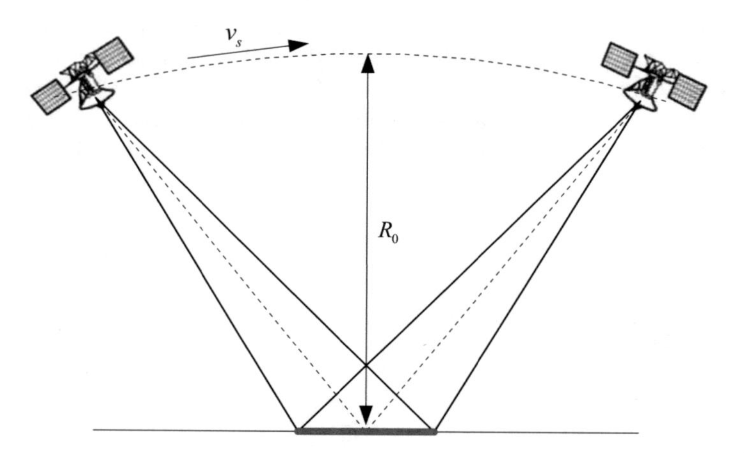

Fig. 1.4 Sketch map of spotlight SAR mode

in a better azimuth resolution. However, the imaging width in azimuth is small and limited to the area being illuminated by the beam. Furthermore, the antenna pointing needs to be mechanically or electronically monitored in azimuth (Fig. 1.4).

1.2.4 Sliding Spotlight SAR Mode

Sliding spotlight SAR can be thought of as the combination of stripmap SAR and spotlight SAR. Its beam points to a certain virtual center which is always far from the imaging scene. The position of the virtual center can determine the velocity of the beam footprint. Therefore, resulting in an increase of the coherent integration time and an improvement of the azimuth resolution compared with stripmap SAR. Furthermore, the sliding spotlight SAR azimuth resolution is a tradeoff between stripmap SAR and spotlight SAR; while, its azimuth swath width increases, and is located between spotlight SAR and stripmap SAR (Fig. 1.5).

1.2.5 TOPS Mode

TOPS is the abbreviation for Terrain Observation by Progressive Scans. It was mainly designed to remove the serious scalloping effects in ScanSAR. In this mode, the antenna not only swings to illuminate the different sub-swaths, but also images each target completely with the same whole antenna pattern by scanning from back to front in azimuth direction. This can solve the non-uniform imaging problem by employing only partial illuminating in ScanSAR. Thus, it can obtain images with high radiation accuracy and wide swath. In TOPS, the beam control method in a sub swath is totally contrary to the sliding spotlight SAR. The virtual rotating center is above the platform and the beam scans the ground rapidly from back to front. It is therefore equivalent, in some sense, to inverse sliding spotlight SAR (Fig. 1.6).

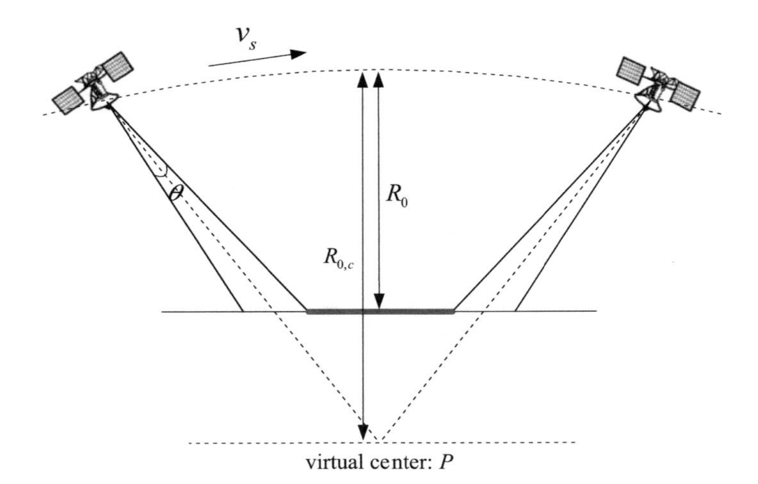

Fig. 1.5 Sketch map of sliding spotlight SAR mode

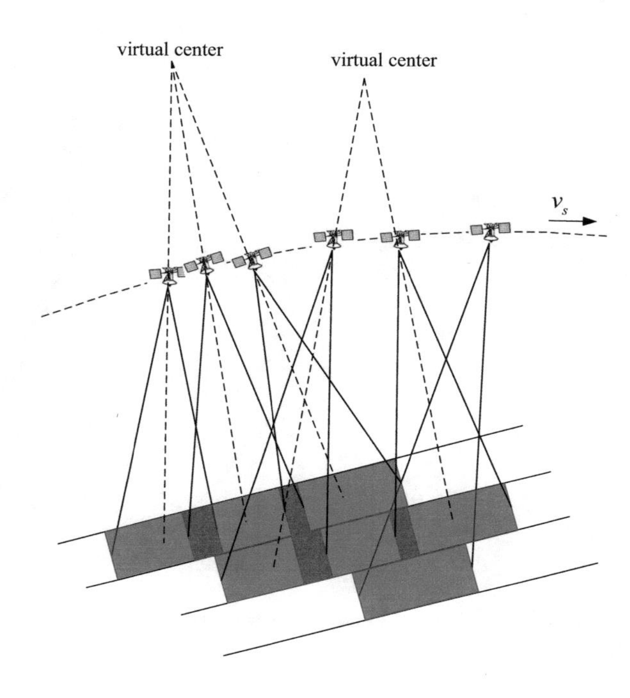

Fig. 1.6 Sketch map of TOPS mode

1.2.6 Multi-channel Mode

Multi-channel mode can achieve high azimuth resolution and wide range swath simultaneously. Its antenna consists of several subarrays whose phase centers are uniformly distributed along the platform trajectory. Subarrays form a synthesized transmitting beam and collect echoes respectively and independently. Thus for each pulse, echoes can be acquired at different positions along the trajectory, which can increase the spatial sampling rate, and improve the equivalent PRF, leading to a higher azimuth bandwidth and a better azimuth resolution. For an N-channel system, azimuth resolution can be improved by N times. However, amplitude and phase imbalances among channels may exist, resulting in performance degradation. In addition, the possible non-uniform sampling will result in a more complicated imaging processing (Fig. 1.7).

1.2.7 Summary

Generally, SAR is a pulse radar system. For different SAR modes, echoes located within two adjacent pulses should satisfy

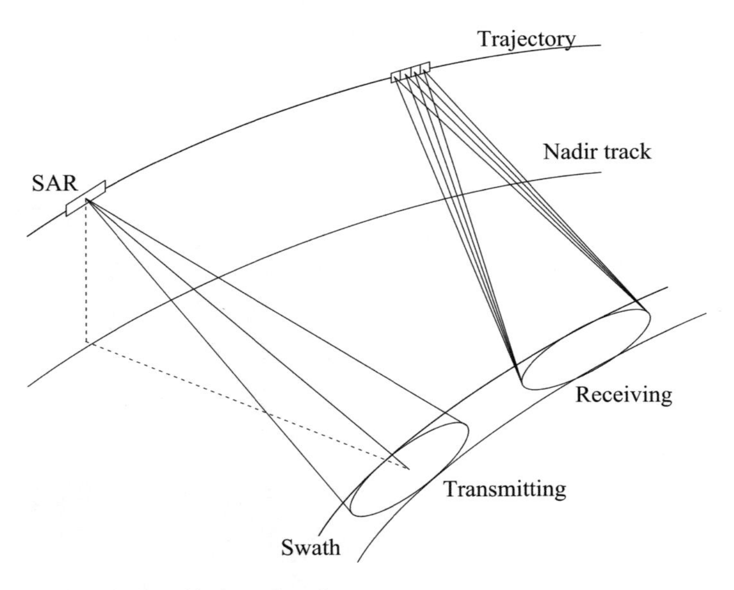

Fig. 1.7 Sketch map of multi-channel mode

$$W_g < \frac{c}{2PRF \sin \varphi} \tag{1.6}$$

where, W_g is the swath width, PRF is the pulse repetition frequency and φ is the incidence angle.

The azimuth resolution is related to the echoes Doppler band B_D (the larger the Doppler band is, the better the azimuth resolution is). But it is limited by the Nyquist sampling theorem, as follows

$$\rho_a \approx \frac{v_g}{B_D} \geq \frac{v_g}{PRF} \tag{1.7}$$

where v_g is the velocity of beam footprint.

Therefore, according to (1.6) and (1.7), the relationship between the swath width and the azimuth resolution can be expressed as

$$\frac{W_g}{\rho_a} < \frac{c}{2v_g \sin \varphi} \tag{1.8}$$

The constrained relationship in (1.8) is independent of the system parameters and is constant when the acquisition geometry is fixed. This numerical value is defined as the quality factor [1]. In LEO SAR, it is commonly around 10 k. For example, the azimuth resolution of 3 m is required for a 30 km swath width.

Thus, for the spaceborne SAR, the quality factor is generally constant. For different SAR modes, it represents a tradeoff between the swath width and the azimuth resolution. For example, spotlight SAR increases the coherent integration time, achieving a very high azimuth resolution at the cost of a small swath width; while for ScanSAR, the imaging scene is improved by multiple bursts, but the azimuth resolution becomes worse.

The New Digital Beam Forming (DBF) technique can form several receiving channels in azimuth or range. It can decrease the PRF system requirement and overcome the quality factor limitation. However, system and signal processing complexities increase greatly.

1.3 Summary of Spaceborne SAR Development

In the early days, synthetic aperture radars were mostly developed on airborne platforms. The airborne SAR is also an effective verification platform for spaceborne SAR technology. With the progress of SAR theory and technology, spaceborne SAR has been developed rapidly. In 1978, the United States National Aeronautics and Space Administration (NASA) launched Seasat which was designed for remote sensing of the Earth's oceans [7]. It had on board the first spaceborne SAR in the world. This pioneer spaceborne SAR collected in a short period of 105 days, more marine information gathered by ships over the past 100 years [8], demonstrating the ability of spaceborne SAR to acquire surface information rapidly and stably.

Spaceborne SAR gradually gained the attention of scientific researchers from all over the world and showed a rapid development. Currently, countries and agencies that have already launched or are about to launch a spaceborne SAR include the United States, the European Space Agency, Russia, Germany, Japan, Canada, China, India, Israel, South Korea, Argentina, etc. [9]. Spaceborne SAR also evolves from single operation mode, single polarization mode to multi-mode operation, and multi-polarization, covering various bands ranging from X-band to P-band. Nowadays, spaceborne SAR is being developed towards high resolution, wide swath, high temporal resolution, high radiation resolution and high radiation accuracy. Some typical spaceborne SAR system parameters are detailed in the Table 1.1.

As can be seen from Table 1.1, the orbit altitudes of existing spaceborne SARs are below 1000 km, and the swath width ranges from tens to hundreds of kilometers depending on operation modes. Compared with airborne SAR, spaceborne SAR has two main advantages: rapid response and wide coverage, which opens a wide range of application domains, such as agriculture, forestry, geology, hydrology, sea ice, hazards monitoring, reconnoitering and surveillance, etc.

However, due to the quality factor limitation, LEO SAR cannot simultaneously realize high-resolution and wide-swath imaging. The larger the width is, the coarser the resolution is. From (1.8), we can see that the quality factor is related to the

Table 1.1 Typical system parameters of Spaceborne SAR [64]

Satellites	Orbital altitude/km	Frequency/GHz	Mode	Resolution		Swath width/km
				Azimuth/m	Range/m	
Seasat	800	1.28 (L)	Stripmap	25	25	100
JERS-1	568	1.27 (L)	Stripmap	18	18	75
RADARSAT-1	800	5.3 (C)	Standard	28	25	100
			Wide (1)	28	30–48	165
			Wide (2)	28	32–45	150
			Fine Resolution	9	9–11	45
			ScanSAR (N)	50	50	305
			ScanSAR (W)	100	100	510
			Extended (H)	28	19–22	75
			Extended (L)	28	28–63	170
ERS-1/2	785	5.3 (C)	Stripmap	30	26.3	100
ENVISAT/ASAR	800	5.33 (C)	Image Mode	28	28	100
			Wide Swath Mode	150	150	400
			Alternating	29	30	100
			Wave Mode	28	30	5
			Global Monitoring	950	980	>400
ALOS/PALSAR	700	1.27 (L)	Fine (FBS)	7–44		40–70
			Fine (FBD)	14–88		
			Direct downlink	14–88		
			ScanSAR	100		250–350
			Polarimetry	30		30
SAR-Lupe	500	9.6 (X)	Stripmap	5	5	60
			Spotlight	0.5	0.5	5.5
COSMO-SkyMed	620	9.6 (X)	Spotlight	<1		10
			HIMAGE (Stripmap)	3–15		40
			Wide Region (ScanSAR)	30		100
			Huge Region (ScanSAR)	100		200
			Ping Pong (Stripmap)	15		30
TerraSAR-X [65]	514	9.6 (X)	Spotlight HS	1	2	10
			Spotlight SL	1	2	10
			Experimental Spotlight	1	1	10
			Stripmap (SM)	3	3	30
			ScanSAR (SC)	16	16	100

(continued)

Table 1.1 (continued)

Satellites	Orbital altitude/km	Frequency/GHz	Mode	Resolution		Swath width/km
				Azimuth/m	Range/m	
RADARSAT-2	800	5.4 (C)	Standard polarization	28	25	25
			Fine polarization	9	11	25
			Multi-look fine	9	11	50
			Ultra-fine	3	3	20
Sentinel-1 [66]	693	5.4 (C)	Stripmap	5	5	80
			Interferometric Wide swath	20	5	250
			Extra Wide swath	40	20	400
			Wave	5	5	20
TerraSAR-X 2	505–533	9.6 (X)	SpotLight HS (Single)	1		15
			SpotLight HS (Dual)	1		10
			Spotlight VHS 0.5 m	0.5		10
			Spotlight VHS 0.5 m MAPS	0.5		20
			Spotlight VHS 0.25 m	0.25		5
			Stripmap SM 3 m	3		24
			Stripmap SM 1 m	1		10
			Stripmap SM 1 m MAPS	1		20
			Stripmap SM 2 m	2		10
			TOPS 30 m MAPS	30		400
			TOPS 12 m	12		100
			TOPS 5 m	5		50
BIOMASS P-band SAR	660	0.435 (P)	Stripmap	<50	<50	>100

platform velocity (dependent on orbit height) and the incident angle. As the orbit height increases, the platform velocity decreases, and the quality factor is improved. If the SAR is mounted on a geosynchronous orbit, it has the potential of overcoming the quality factor limitations and hence achieves high-resolution and wide-swath imaging simultaneously. Its swath can reach hundreds or even

Table 1.2 Revisit time of some typical spaceborne SAR or constellation

Satellite	Revisit time (day)
COSMO-SkyMed[a] [67]	5
Sentinel-1 [68]	6
TerraSAR-X [69]	11
ALOS-2 PALSAR [70]	14
Radarsat-2 [71]	24

[a]The revisit time of the COSMO-SkyMed constellation can achieve up to 12 h dependent on the agility of beam pointing

thousands of kilometers. In the meantime, the geosynchronous orbit satellite has a shorter revisit time, whose access time per day for China ranges from 4 to 14 h.

From an application aspect, the improvement of temporal resolution to day-level even hour-level is of great significance. For example, disaster monitoring and hazards management, including flooding, fires, landslides, hurricanes, and earthquakes, require a high temporal measurement frequency (1–2 days, even hours) to determine process dynamics and changes. Take the example of flooding, the changing water level should be monitored in a near-real-time way to be able to forecast the flooding situation. For other applications, near-real-time mapping of earth surface changes, caused by earthquakes, volcanic eruptions, landslides, or fires are needed. However, the current spaceborne SAR runs on the low Earth orbit (LEO) with an orbital height of no more than 1000 km. Its revisit period is several to tens of days for area of interest (as in Table 1.2), and can reach at least 1 day through orbital maneuvering. Therefore, the LEO SAR has a low temporal resolution and needs more time to deal with emergencies, and hence cannot provide the timely information for hazard or disaster relief work.

1.4 GEO SAR Concept

The geosynchronous SAR (GEO SAR) runs on a geosynchronous orbit at a height of 36,000 km. This geosynchronous orbit has a certain inclination and thus is not geostationary. The relative movement between GEO SAR and ground targets can be used to achieve two-dimensional SAR imaging. The GEO SAR nadir track will form a "figure 8" shape (see Fig. 1.8a). Compared with the traditional LEO SAR, GEO SAR has irreversible advantages of short revisit period (several hours), large observation coverage (hundreds to thousands of kilometers) etc.

Through rational design of the orbital parameters, a GEO SAR satellite can ensure that the region of interest is observed once or twice per day, and can even achieve continuous monitoring for specific areas (see Fig. 1.8b). All these advantages reflect the GEO SAR potential in disaster prediction, ocean surveillance, military monitoring, etc. Therefore, GEO SAR is gradually becoming a research focus around the world.

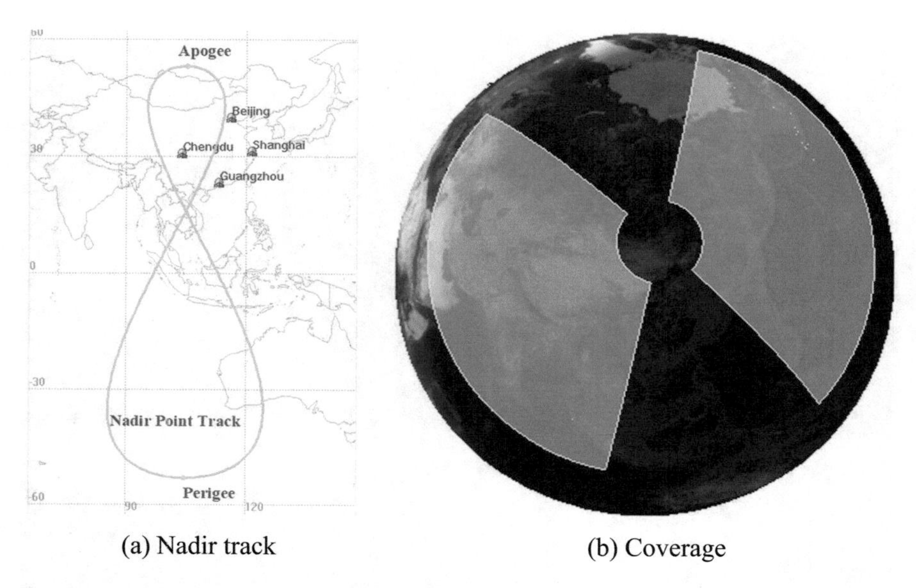

(a) Nadir track (b) Coverage

Fig. 1.8 Nadir track and coverage of GEO SAR

1.5 State of Art

The GEO SAR concept was first proposed in 1978 [10]. Initially, the GEO SAR was assumed to run on an elliptical geosynchronous orbit with the orbital inclination of 1°, eccentricity of 0.009 and argument of perigee of 90°, which produces a quasi-circular trajectory in Earth-center Earth-fix (ECEF) system. With radar frequency of 2.45 GHz and antenna diameter of 7.3 m, the average transmitter power is 8 kW in case that the normalized radar cross section is −20 dB. It would be able to observe the whole United States with a resolution of 100 m within 4 h. Four orbit configurations were mentioned, as shown in Fig. 1.9. In order to improve the relative motion, an orbit with an orbital inclination of 50° was proposed in 1983 [11]. No eccentricity was assumed, which results in satellite nadir track of "figure 8". It could map a 2.5×10^7 km^2 area in 6 h with 100 m resolution and 4-azlook using a 15 m-diam steerable antenna and 1312 W average power.

In 1987, Lesley conducted a study on the concept, feasibility, application and system design of GEO SAR [12]. Then, in 1997, Guttrich studied bistatic GEO SAR configuration for airborne or ground moving target indication (AMTI or GMTI), with GEO SAR as the transmitter and the unmanned aerial vehicle (UAV) or LEO SAR as the receiver [13]. The system could show significant advantages over conventional monostatic space-based systems. For example of AMTI detecting a 0.5 m^2 target with 12 dB signal to noise ratio (SNR) in L band, with transmitter power of 20 kw, transmit aperture of 100 m, the receive aperture was 36 m and 6 m × 1 m for satellite receiver with 1600 km altitude and UAV

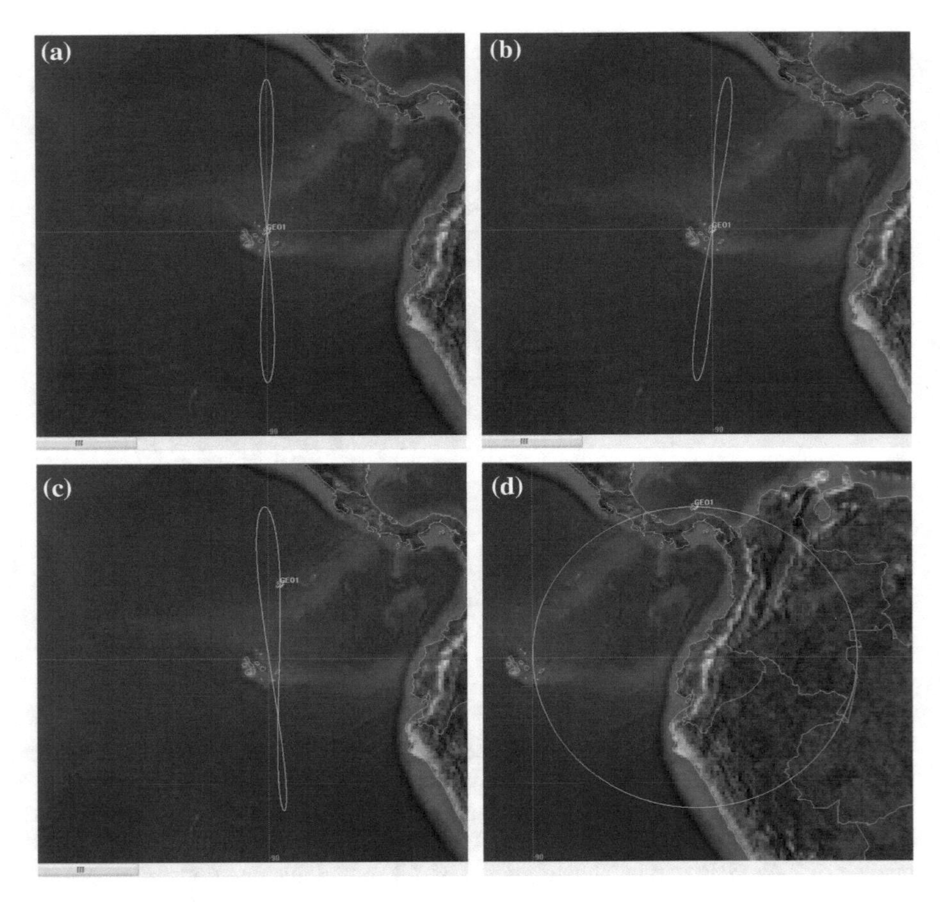

Fig. 1.9 Four types of orbit configurations: **a** inclined only; **b** a small eccentricity added; **c** a greater eccentricity added; **d** 90° of argument of perigee added

with 300 km range, respectively. For example of GMTI detecting the target with an added 0.1 g (1 g = 9.8 m/s^2) acceleration in S band, with transmitter power of 2 kw, transmit aperture of 25 m, the receive aperture was 10 m × 10 m and 1 m × 0.25 m for satellite receiver with 1600 km altitude and UAV with 200 km range, respectively.

Based on these studies, it has been recognized that the most important advantages of GEO SAR are the huge imaging coverage and short revisit period compared to LEO SAR. However, in GEO SAR, the synthetic aperture time, the required transmitting power and the antenna size are much larger than those of LEO SAR, which made the realization of a GEO SAR based imaging system difficult because of the existing technical limitations back then. As a result, research works on GEO SAR showed a stagnant state during the following decades.

With the rapid technological development and progress during the 21st century, researches on GEO SAR become active again. In 2001, a GEO SAR system with an orbit inclination of 50°–65° was discussed, covering North and South America [14–17]. This system has multiple working modes, with a squint angle range of −60° to 60° and covering a wide swath of 1000–5000 km. Using 20 kW of DC power and 30 m diameter antenna with ±7° electronic scanning in both azimuth and elevation, its short revisit period (12 h) can be used for disaster prediction and environmental monitoring, such as earthquakes, volcanoes, hurricanes, fires, floods, ecology and so on, and the sub-centimeter-level surface displacement measuring enable earthquake-related application. Still, GEO SAR has to face the technical difficulties induced by light, large size and electric-scan antenna.

In 2003, NASA proposed the Global Earthquake Satellite System (GESS) [18] in which 10 GEO SAR satellites were divided into 5 groups. It can provide a global consecutive coverage, as shown in Fig. 1.10. The 2 satellites in each group have the same orbits with a phase difference of 180°, which provide 12 h of revisit time and over 80% global coverage at any time. GESS can achieve SAR differential interferometry (D-InSAR) and significantly improve the capacity of global crustal deformation observation. It uses L-band antenna with a diameter of 30 m, which is made of inflatable membrane material. Its peak transmitting power is 60 kW, and the variable bandwidths are 10–80 MHz, ensuring a 20 m resolution at different ranges. GESS can obtain emergency images within 10 min and generates the three dimensional deformation measurement with a millimeter-level accuracy.

It was also pointed out by NASA that large aperture antenna is the core technology in the development of GEO SAR. The antenna should satisfy the following requirements: lightweight, low density antennas will be the first choice; durable, at

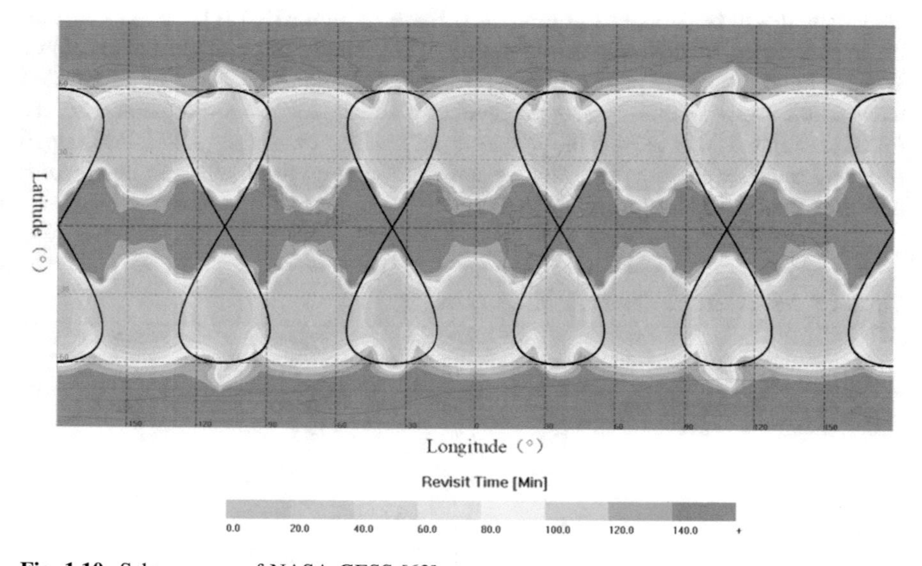

Fig. 1.10 Scheme map of NASA GESS [63]

least capable of sustainably working in space environment for 15 years; can work in L band; strong stability to ensure stable pointing accuracy of antenna; within 8–12 h, the pointing error should be within ±0.2°.

In Europe, researchers gradually proposed a series of near-zero inclination systems, which require extreme long synthetic aperture time (several hours) for compensating the long range attenuation. This GEO SAR configuration with near-zero inclination, moderate antenna size and low transmitted power generally has the advantages of low cost and low imaging difficulty, but the effects of clutter and atmospheric disturbances must be compensated.

In 1998, the parasitic GEO SAR system was firstly proposed [19], namely, a design of a bistatic passive SAR system using L-band broadcast signal receivers and satellite broadcast signal transmitters both placed in geostationary orbit. Imaging and interferometry processing can be performed using orbit perturbations. However, the system can only achieve high-resolution imaging of stable targets with a coherence time of more than half a day. When the target coherence time is between tens of milliseconds to half a day, the imaging resolution will decrease; when the target coherence time drops to only a few tens of milliseconds, the target will not be imaged.

In the next two decades, researchers from the United Kingdom (UK), Italy and Spain conducted a study on the system design and signal processing methods of the near-zero inclination GEO SAR scheme and proposed GeoSTARe (Geosynchronous SAR for Terrain and Atmosphere with short Revisit) concept [20]. It combines the continuous view capabilities with super-continental access of geostationary orbits with the all-day and all-weather imaging capabilities of SAR, leading to a novel and unique observation system with quasi-continuous revisit. Now GeoSTARe is under study by an international consortium of more than forty scientists, it will be proposed as next ESA Earth Explorer IX [21]. GeoSTARe can be implemented in different configurations [22]: single or multiple beams, C or L + X band, hosted payload on telecommunication satellite or dedicated stand-alone mini-satellite. For single beam or platform, the proposed parameters are listed in Table 1.3. It is scalable allowing for trading coverage, resolution, image time and revisit with cost [23]. The L-band configuration has the largest beamwidth and thus the largest footprint or swath.

Table 1.3 Parameters for single beam GeoSTARe [23]

	Unit	C-band	L-band	X-band
Ground range resolution	m	10	150	4.5
Azimuth Res—coarse (15 min)	m	340	1500	200
Azimuth Res—fine (7 h)	m	10	50	6.5
Swath (Az × Rg)	km	350 × 650	1500 × 3000	200 × 500
Averaged power	W	430	500	250
Antenna diameter	m	6	6	6
NESZ	dB	−24	−26	−20

Hobbs studied the concept from a systemic point of view: system design performance analysis and application prospects. Furthermore, coverage characteristics, orbit characteristics and their application in interferometry and soil moisture monitoring were analyzed in detail [24]. He compared and analyzed different GEO SAR schemes, and built a GEO SAR echo simulator to analyze the effect of long synthetic aperture time on imaging processing.

Hobbs et al. also proposed the GEO SAR constellation design [25] which employed 12 GEO SAR satellites. The constellation is divided into three groups. Each contains 4 GEO SAR satellites, corresponding to three different regions: America, Europe/Africa, and Asia/Oceania. Each group is arranged into two concentric circles, whose radius are 46.19 and 20 km for the outer and inner circles, respectively. The outer circle has eccentricity of 0.0005774 and inclination of 0.06616°, used to create the synthetic aperture, and the inner circle is used to decrease sidelobe energy, enhance SNR and thus to improve image quality. With total mass of 338 kg for single spacecraft, the required power is 100–280 W and 120–300 W in no eclipse case and in eclipse case, respectively. Considering antenna density and damage caused by micrometeoroids, inflatable antenna concept is decided. Further, Hobbs et al. studied the effects of various factors on GEO SAR system performance and indicated three important factors affecting GEO SAR imaging [26]: tidal, tropospheric and ionospheric perturbations. They pointed out that the ionospheric perturbation was the main problem in GEO SAR, and that the autofocus algorithm had important potentials for dealing with ionospheric disturbance.

Moreover, Guarnieri et al. focus on the quasi-geostationary GEO SAR. In 2009, [27] studied the influence of clutter and atmosphere on the imaging focusing in near-zero inclination GEO SAR system in Ku-band. It is pointed out that the Ku-band GEO SAR system can monitor the water vapor in a 6000 km^2 region at a resolution of approximately 300 m every 5 min. The orbital, resolution and Doppler characteristics of the near-zero inclination GEO SAR system are analyzed in detail [28, 29], as well as the SNR performance and design requirements. At the same time, Antoni pointed out that one potential application of this system is to effectively estimate the atmospheric phase, that is, the coarse resolution imaging can be performed by sub-aperture division, and then the atmospheric phase can be estimated through the stable point on the ground as Ref. [30]. They verified the effectiveness of the extraction and compensation method of atmospheric phase through ground-based SAR experiments.

Bistatic satellites configuration can be used to improve the APS estimation accuracy. In 2012, another GEO SAR constellation system, GEMINI, is proposed, which stands for Geosynchronous SAR for Earth Monitoring by Interferometry and Imaging [31]. The design uses communication satellites as main transmitting sources and one or more pairs of receivers installed at dedicated small satellites which run slowly in geosynchronous orbits. GEMINI can form synthetic apertures of tens of kilometers twice a day, and then perform imaging and interferometry processing. The system uses close receiver pairs to perform along-track interferometry processing in order to achieve the estimation of the atmospheric delay gradient. After reasonable compensation of atmospheric phase, not only the coherent accumulation of hours-scale signals can be realized, but also the analysis and early warning of surface deformation disaster can be completed effectively. At the same time, the meteorological data obtained can be used for weather forecasting.

In 2015, the GEO SAR constellation system based on Multi-input Multi-output (MIMO) were proposed to achieve finer resolution imaging in shorter integration time, which was named the ARGOS system (Advanced Radar Geosynchronous Observation System) [32, 33]. In ARGOS, a swarm of $N(= 6)$ iso-frequency mini-satellites are used to form a quasi-geostationary constellation. Each satellite can transmit and receive signals simultaneously, thus they form a lot of bistatic SAR systems, which produce $N(N + 1)/2$ equivalent phase centers (EPC). The system SNR enhanced by N^2 times, meanwhile the full aperture integration time is reduced to $2/(N(N + 1))$, namely, from 8 h to 30 min for the rapid high-resolution imaging. Therefore, high resolution and short revisit period are achieved at the same time.

In the implementation of ARGOS, the eccentricities, the ascending node times, and the center longitudes are tuned to configure the constellation. Two configurations are proposed including iso-elliptical swarm and concentric swarm. For iso-elliptical swarm, all satellites have identical orbit parameters except the center longitudes. Thus they have the identical Earth-related motion. While for concentric swarm, all satellites are configured as concentric orbits, i.e., they have the same center longitude but different eccentricity vectors. This configuration can avoid the non-uniform spectral coverage with time and improve the spectral coverage.

In addition to the research community in the US and Europe, another active research group is in China, focusing on GEO SAR system parameter analysis and design, imaging processing, the atmosphere effect analysis and compensation and so on. Systematic researches on GEO SAR have been carried out in Beijing Institute of Technology (BIT), including the system performance analysis and design, attitude control and imaging processing. The differences between GEO SAR and LEO SAR in terms of system performance analysis and design were analyzed in detail [34, 35], and the Earth rotation influence on Doppler frequency and resolution analysis was deduced [36]. Moreover, the influence of elliptical orbit and Earth rotation on GEO SAR attitude control were analyzed for the first time. Likewise, a small-angle attitude control method based on two-dimensional phase scanning was proposed [37]. Regarding atmospheric effects, the space-time-variant model was established [38, 39], and its influence on the long aperture time and large wide swath was analyzed. It was pointed out that the influence of the ionosphere in the L-band is very serious and corresponding compensation approaches were put forward to mitigate its effect [40, 41]. Some ionospheric experiments based on global navigation satellites system were carried out, verifying the theoretical analysis [42].

In terms of imaging processing, the errors introduced by “stop-and-go” in GEO SAR were discussed, and an accurate signal model as along with an improved spectrum analysis algorithm (SPECAN) were proposed to realize the precise focusing of small scene in GEO SAR [36]. Besides, an improved secondary range compression algorithm (SRC) was proposed based on the fact that the commonly used linear trajectory model and the Fresnel approximation assumption no longer hold if the GEO SAR satellite is positioned over the Equator [43] and the image is focused by compensating higher order phase. On the other hand, an improved range migration algorithm (RMA) was proposed to solve the problem of large migration range and strong space variance over the Equator [44]. Through adaptive compensation of Doppler parameters space variance, GEO SAR imaging of large scenes over the Equator was realized. Taking into account the curve trajectory and large imaging scenes, the two-dimensional spectrum of the GEO SAR signal was deduced using high order polynomial approximation and series reversion theory. In addition, the traditional chirp scaling algorithm (CS) [45] and nonlinear chirp scaling algorithm (NCS) ware improved [46], and used to realize the GEO SAR large scene imaging, and the process of GEO SAR imaging algorithm was analyzed [47]. The relevant research works mentioned above will be described in detail in Chap. 3 of this book.

For other groups, most works are also about the GEO SAR system characteristics analysis and imaging processing algorithms. The improved linear chirp scaling algorithm (CSA) considering the characteristics of elliptical orbit and the high-order Taylor series slant range model are proposed [38, 48]. The back projection algorithm (BPA) is also analyzed for the high-resolution imaging problems in GEO SAR [49].

The other configurations of GEO SAR are also studied. One aspect is bistatic configuration of a geostationary transmitter and airborne or UAV receiver, which is called “one-stationary BiSAR”. Synchronization method including space, time, frequency and phase synchronization was proposed [50]. A detailed systematical analysis on one-stationary BiSAR was given, including SNR, spatial resolution characteristics and mission design, pointing out that the most significant advantage is the improved performance provided by properly adjusting the receiver motion parameters [51]. Other studies cover resolution analysis, equivalent nadir point analysis and imaging algorithms [52–56].

Another aspect is the circular GEO SAR (GEOCSAR). In imaging processing, imaging model and simulation was studied [57], while spatial baseline decorrelation and corresponding critical baselines were analyzed [58]. Furthermore, three-dimensional surface deformation was provided using GEOCSAR [59]. In [60], effects of orbit errors on imaging and interferometric processing was analyzed, concluding that the error should be restricted within centimeter level for ensuring the image qualities, and an orbit error of 10 mm will cause a several millimeters of deformation error. Ionospheric and tropospheric effects on L-band GEOCSAR imaging were studied in detail [61, 62].

1.6 Special Issues and Challenge

The orbit height of 36,000 km brings outstanding performance to GEO SAR, but also produces theoretical and technical difficulties to its further development. First, the power constraints are harsh: the remote slant range (at the level of 40,000 km) in GEO SAR causes more than 300 dB of transmission attenuation, so the system must have a very high transmitting power and very large antenna aperture, which also put strict requirements on the platform carrying-capacity. However, with the development and progress of engineering and aerospace technology, large reflector antennas, high-power transmitters and higher carrying-capacity platforms are more likely to meet future GEO SAR requirements.

On this basis, the core difficulty of GEO SAR is the problem of information acquisition and recovery: synthetic aperture time reaching several minutes or even tens of minutes, echo delay reaching sub-second order of magnitude, observation swath covering thousands of kilometers, all these are increased by two orders of magnitude with respect to those in LEO SAR. Due to the huge variance of the slant range in the space-time dimension, the classical SAR system design and signal processing used in many assumptions have obvious errors as described in the following

(1) Information acquisition

- Firstly, the influence of satellite dynamics has changed. On the one hand, due to a longer synthetic aperture time, some slow-changing perturbations that are not considered will become relatively significant and cannot be ignored. On the other hand, the huge antenna changes the dynamics of the satellites, and the vibrations caused by the sway may remain for half an hour. Hence, GEO SAR cannot use the traditional yaw control method to counteract the rotation of the Earth as LEO SAR does. The related work is presented in Chap. 2 in details.
- Secondly, the impact of the atmospheric propagation effects needs to be re-considered. In fact, the spatial-temporal variant ionosphere and its propagation effects have to be studied. Further, the disturbances within the synthetic aperture time will cause considerable phase delay and other changes that must be dealt with. The details are presented in Chaps. 4 and 5.

(2) Signal processing

- The Fresnel approximation and the corresponding linear trajectory model, on which the LEO SAR imaging algorithm relies, are inaccurate in GEO SAR due to the long aperture time. Likewise, the echo delay, which is at the sub-second level, undermines the "stop-and-go" assumption. Also, the complex relative motion and stronger influence of space-variant characteristics are not negligible. The developed imaging algorithms are presented in Chap. 3 in details.

- In addition, many classical autofocus methods are no longer applicable since GEO SAR systems have temporal-spatial variant errors, and the effect of higher order terms is more pronounced. Furthermore, the short revisit time and the long access time for specified regions will induce significant advantages for GEO SAR interferometry and differential interferometry, as well as the multi-angle deformation retrieval which will be presented in Chaps. 6 and 7 in details.

1.7 Outline

This book focuses on the information acquisition and signal processing methods in GEO SAR, and proposes system design guidelines and accurate analysis methods suitable for GEO SAR. The book is organized as follows.

This chapter gives the background and state of art of GEO SAR. Chapter 2 presents the differences between GEO SAR and LEO SAR, along with the effects of Earth rotation and elliptical orbit. Besides, system performance and design are also presented. In Chap. 3, the imaging algorithms are discussed in detail. For the "stop-and-go" assumption, the equivalent straight trajectory model and the Fresnel approximation in LEO SAR are analyzed. The theoretical framework of GEO SAR imaging algorithms under the conditions of wide swath and curved trajectory is put forward. In Chaps. 4 and 5, the accurate GEO SAR model that considers the temporal-spatial variant atmosphere is established, and a quantitative analysis is given. At the same time, an experiment is carried out using navigation satellites to verify the ionospheric influence on GEO SAR, and corresponding compensation approaches are proposed. In Chaps. 6 and 7, the advantages and problems of GEO InSAR and D-InSAR are analyzed, and a data processing method is proposed. Furthermore, the three-dimensional deformation retrieval technique capability and performance are presented as an extended application of GEO InSAR and D-InSAR.

References

1. Curlander JC, Macdonough RN (1991) Synthetic aperture radar: systems and signal processing. Wiley, New York
2. Cumming IG, Wong FH (2005) Digital processing of synthetic aperture radar data. Artech House 1(2):3
3. Zan FD, Guarnieri AM (2006) TOPSAR: terrain observation by progressive scans. IEEE Trans Geosci Remote Sens 44(9):2352–2360. https://doi.org/10.1109/TGRS.2006.873853
4. Younis M, Fischer C, Wiesbeck W (2003) Digital beamforming in SAR systems. IEEE Trans Geosci Remote Sens 41(7):1735–1739. https://doi.org/10.1109/TGRS.2003.815662

5. Krieger G, Gebert N, Moreira A (2004) Unambiguous SAR signal reconstruction from nonuniform displaced phase center sampling. IEEE Geosci Remote Sens Lett 1(4):260–264. https://doi.org/10.1109/LGRS.2004.832700
6. Gebert N, Krieger G, Moreira A (2009) Digital beamforming on receive: techniques and optimization strategies for high-resolution wide-swath SAR imaging. IEEE Trans Aerosp Electron Syst 45(2):564–592. https://doi.org/10.1109/TAES.2009.5089542
7. Jordan RL (1980) The Seasat-A synthetic aperture radar system. IEEE J Oceanic Eng 5 (2):154–164. https://doi.org/10.1109/JOE.1980.1145451
8. Thompson T, Laderman (1976) A Seasat-A synthetic aperture radar: radar system implementation. In: Oceans, pp 247–251
9. Huang H, Zhang Y, Dong Z (2015) Spaceborne synthetic aperture radar interferometry noel technology. China Science Publisher (in Chinese)
10. Tomiyasu K (1978) Synthetic aperture radar in geosynchronous orbit. In: Antennas and propagation society international symposium, May 1978, pp 42–45. https://doi.org/10.1109/aps.1978.1147948
11. Tomiyasu K, Pacelli JL (1983) Synthetic aperture radar imaging from an inclined geosynchronous orbit. IEEE Trans Geosci Remote Sens GE-21 (3):324–329. https://doi.org/10.1109/tgrs.1983.350561
12. Murphy LM (1987) Synthetic aperture radar imaging from geosynchronous orbit-concept, feasibility and applications. In: Brighton international astronautical federation congress
13. Guttrich GL, Sievers WE, Tomljanovich NM (1997) Wide area surveillance concepts based on geosynchronous illumination and bistatic unmanned airborne vehicles or satellite reception. In: Proceedings of the 1997 IEEE national radar conference, 13–15 May 1997, pp 126–131. https://doi.org/10.1109/nrc.1997.588225
14. Madsen SN, Edelstein W, DiDomenico LD, LaBrecque J (2001) A geosynchronous synthetic aperture radar; for tectonic mapping, disaster management and measurements of vegetation and soil moisture. In: Geoscience and remote sensing symposium, 2001. IGARSS '01. IEEE 2001 international, 2001, vol 441, pp 447–449. https://doi.org/10.1109/igarss.2001.976185
15. Madsen SN, Chen C, Edelstein W (2002) Radar options for global earthquake monitoring. In: Geoscience and remote sensing symposium, 2002. IGARSS '02. IEEE international, 2002, vol 1483, pp 1483–1485. https://doi.org/10.1109/igarss.2002.1026156
16. Edelstein WN, Madsen SN, Moussessian A, Chen C (2005) Concepts and technologies for synthetic aperture radar from MEO and geosynchronous orbits, pp 195–203
17. Moussessian A, Chen C, Edelstein W, Madsen S, Rosen P (2005) System concepts and technologies for high orbit SAR. In: Microwave symposium digest, 2005 IEEE MTT-S international, 12–17 June 2005, 4 p. https://doi.org/10.1109/mwsym.2005.1517017
18. NASA J (2003) Global earthquake satellite system: a 20-year plan to enable earthquake prediction
19. Prati C, Rocca F, Giancola D, Guarnieri AM (1998) Passive geosynchronous SAR system reusing backscattered digital audio broadcasting signals. IEEE Trans Geosci Remote Sens 36 (6):1973–1976
20. Hobbs S, Guarnieri AM, Wadge G, Schulz D (2014) GeoSTARe initial mission design. In: Geoscience and remote sensing symposium (IGARSS), 2014 IEEE international, 13–18 July 2014, pp 92–95. https://doi.org/10.1109/igarss.2014.6946363
21. Rogers NC, Quegan S, Jun SuK, Papathanassiou KP (2014) Impacts of ionospheric scintillation on the BIOMASS P-Band satellite SAR. IEEE Trans Geosci Remote Sens 52 (3):1856–1868. https://doi.org/10.1109/TGRS.2013.2255880
22. Hobbs S, Convenevole C, Guarnieri AM, Wadge G (2016) Geostare system performance assessment methodology. In: 2016 IEEE international geoscience and remote sensing symposium (IGARSS), 10–15 July 2016, pp 1404–1407. https://doi.org/10.1109/igarss.2016.7729359
23. Wadge G, Guarnie AM, Hobbs SE, Schul D (2014) Potential atmospheric and terrestrial applications of a geosynchronous radar. In: 2014 IEEE geoscience and remote sensing symposium, 13–18 July 2014, pp 946–949. https://doi.org/10.1109/igarss.2014.6946582

24. Hobbs S, Mitchell C, Forte B, Holley R, Snapir B, Whittaker P (2014) System design for geosynchronous synthetic aperture radar missions. IEEE Trans Geosci Remote Sens 52 (12):7750–7763. https://doi.org/10.1109/TGRS.2014.2318171
25. Bruno D, Hobbs SE, Ottavianelli G (2006) Geosynchronous synthetic aperture radar: concept design, properties and possible applications. Acta Astronaut 59(1):149–156. https://doi.org/10.1016/j.actaastro.2006.02.005
26. Bruno D, Hobbs SE (2010) Radar imaging from geosynchronous orbit: temporal decorrelation aspects. IEEE Transac Geosci Remote Sens 48(7):2924–2929. https://doi.org/10.1109/TGRS.2010.2042062
27. Guarnieri AM, Rocca F, Ibars AB (2009) Impact of atmospheric water vapor on the design of a Ku band geosynchronous SAR system. In: 2009 IEEE international geoscience and remote sensing symposium, 12–17 July 2009, pp II-945–II-948. https://doi.org/10.1109/igarss.2009.5418254
28. Rodon JR, Broquetas A, Makhoul E, Guarnieri AM, Rocca F (2012) Results on spatial-temporal atmospheric phase screen retrieval from long-term GEOSAR acquisition. In: Geoscience and remote sensing symposium (IGARSS), 2012 IEEE international. IEEE, pp 3289–3292
29. Ruiz-Rodon J, Broquetas A, Makhoul E, Monti Guarnieri A, Rocca F (2014) Nearly zero inclination geosynchronous SAR mission analysis with long integration time for earth observation. IEEE Transac Geosci Remote Sens 52(10):6379–6391. https://doi.org/10.1109/TGRS.2013.2296357
30. Ruiz Rodon J, Broquetas A, Monti Guarnieri A, Rocca F (2013) Geosynchronous SAR focusing with atmospheric phase screen retrieval and compensation. IEEE Transac Geosci Remote Sens 51(8):4397–4404. https://doi.org/10.1109/TGRS.2013.2242202
31. Guarnieri AM, Tebaldini S, Rocca F, Broquetas A (2012) GEMINI: Geosynchronous SAR for earth monitoring by interferometry and imaging. In: Geoscience and remote sensing symposium, pp 210–213
32. Monti Guarnieri A, Broquetas A, Recchia A, Rocca F, Ruiz-Rodon J (2015) Advanced radar geosynchronous observation system: ARGOS. Geosci Remote Sens Lett IEEE 12(7):1406–1410. https://doi.org/10.1109/LGRS.2015.2404214
33. Guarnieri AM, Bombaci O, Catalano TF, Germani C, Koppel C, Rocca F, Wadge G (2015) ARGOS: a fractioned geosynchronous SAR. Acta Astronaut. https://doi.org/10.1016/j.actaastro.2015.11.022
34. Tian W, Hu C, Zeng T, Ding Z (2010) Several special issues in GEO SAR system. In: 2010 8th European conference on Synthetic aperture radar (EUSAR), 7–10 June 2010, pp 1–4
35. Dong X, Yangte G, Hu C, Zeng T, Chao D (2011) Effects of earth rotation on GEO SAR characteristics analysis. In: Proceedings of 2011 IEEE CIE international conference on radar, 24–27 Oct 2011, pp 34–37. https://doi.org/10.1109/cie-radar.2011.6159469
36. Cheng H, Teng L, Tao Z, Feifeng L, Zhipeng L (2011) The accurate focusing and resolution analysis method in geosynchronous SAR. IEEE Trans Geosci Remote Sens 49(10):3548–3563. https://doi.org/10.1109/TGRS.2011.2160402
37. Teng L, Xichao D, Cheng H, Zeng T (2011) A new method of zero-doppler centroid control in GEO SAR. Geosci Remote Sens Lett IEEE 8(3):512–516. https://doi.org/10.1109/LGRS.2010.2089969
38. Ye T, Cheng H, Xichao D, Tao Z, Teng L, Kuan L, Xinyu Z (2015) Theoretical analysis and verification of time variation of background ionosphere on geosynchronous SAR imaging. Geosci Remote Sens Lett IEEE 12(4):721–725. https://doi.org/10.1109/LGRS.2014.2360235
39. Hu C, Tian Y, Yang X, Zeng T, Long T, Dong X (2016) Background ionosphere effects on geosynchronous SAR focusing: theoretical analysis and verification based on the BeiDou navigation satellite system (BDS). IEEE J Sel Top Appl Earth Observations Remote Sens 9 (3):1143–1162. https://doi.org/10.1109/JSTARS.2015.2475283
40. Hu C, Li Y, Dong X, Ao D (2016) Avoiding the ionospheric scintillation interference on geosynchronous SAR by orbit optimization. IEEE Geosci Remote Sens Lett (99):1–5. https://doi.org/10.1109/lgrs.2016.2603230

41. Hu C, Li Y, Dong X, Wang R, Ao D (2016) Performance analysis of L-Band geosynchronous SAR imaging in the presence of ionospheric scintillation. IEEE Transac Geosci Remote Sens (99):1–14. https://doi.org/10.1109/tgrs.2016.2602939
42. Dong X, Hu C, Tian Y, Tian W, Li Y, Long T (2016) Experimental study of ionospheric impacts on geosynchronous SAR using GPS signals. IEEE J Sel Top Appl Earth Observations Remote Sens 9(6):2171–2183. https://doi.org/10.1109/JSTARS.2016.2537401
43. Hu C, Tian Y, Zeng T, Long T, Dong X (2016) Adaptive secondary range compression algorithm in geosynchronous SAR. IEEE J Sel Top Appl Earth Observations Remote Sens 9 (4):1397–1413. https://doi.org/10.1109/JSTARS.2015.2477317
44. Liu F, Hu C, Zeng T, Long T, Jin L (2010) A novel range migration algorithm of GEO SAR echo data. In: 2010 IEEE international geoscience and remote sensing symposium, 25–30 July 2010, pp 4656–4659. https://doi.org/10.1109/igarss.2010.5651127
45. Cheng H, Zhipeng L, Teng L (2012) An improved CS algorithm based on the curved trajectory in geosynchronous SAR. IEEE J Sel Top Appl Earth Observations Remote Sens 5 (3):795–808. https://doi.org/10.1109/JSTARS.2012.2188096
46. Hu C, Long T, Tian Y (2013) An improved nonlinear chirp scaling algorithm based on curved trajectory in geosynchronous SAR. Prog Electromagnet Res 135:481–513
47. Cheng H, Teng L, Zhipeng L, Tao Z, Ye T (2014) An improved frequency domain focusing method in geosynchronous SAR. IEEE Transac Geosci Remote Sens 52(9):5514–5528. https://doi.org/10.1109/TGRS.2013.2290133
48. Min B, Yi L, Zi. Jing T, Meng. Dao X, Ya. Chao L (2011) Imaging algorithm for GEO SAR based on series reversion. In: Proceedings of 2011 IEEE CIE international conference on radar, 24–27 Oct 2011, pp 1493–1496. https://doi.org/10.1109/cie-radar.2011.6159844
49. Li Z, Li C, Yu Z, Zhou J, Chen J (2011) Back projection algorithm for high resolution GEO-SAR image formation. In: 2011 IEEE international geoscience and remote sensing symposium, 24–29 July 2011, pp 336–339. https://doi.org/10.1109/igarss.2011.6048967
50. Ling L, Yinqing Z, Li J, Bing S (2008) Synchronization of geo spaceborne-airborne bistatic SAR. In: IGARSS 2008. IEEE international geoscience and remote sensing symposium, 7–11 July 2008, pp III-1209–III-1211. https://doi.org/10.1109/igarss.2008.4779574
51. Sun Z, Wu J, Pei J, Li Z, Huang Y, Yang J (2016) Inclined geosynchronous spaceborne-airborne bistatic SAR: performance analysis and mission design. IEEE Trans Geosci Remote Sens 54(1):343–357
52. Qiu X, Hu D, Ding C (2008) An improved NLCS algorithm with capability analysis for one-stationary BiSAR. IEEE Trans Geosci Remote Sens 46(10):3179–3186
53. Wang J, Wang Y, Ge J (2010) Research on the equivalent nadir point of bistatic SAR with geostationary illuminator and LEO or UAV receivers. In: Synthetic aperture radar, 2009. Asian-Pacific conference on Apsar 2009, pp 227–230
54. Wang Y, Wang J, Zhang J, Ge J (2011) Research on the resolution of bistatic SAR with geostationary illuminator and LEO receiver. In: 3rd international Asia-Pacific conference on synthetic aperture radar (APSAR), pp 1–4
55. Wang J, Wang Y, Zhang J, Ge J (2013) Resolution calculation and analysis in bistatic SAR with geostationary illuminator. IEEE Geosci Remote Sens Lett 10(1):194–198
56. Wu J, Li Z, Huang Y, Yang J (2014) Omega-K imaging algorithm for one-stationary bistatic SAR. IEEE Trans Aerosp Electron Syst 50(1):33–52
57. Qi L, Tan WX, Lin Y, Wang YP, Hong W, Wu YR (2012) SAR raw data 2-D imaging model and simulation of GEOCSAR. In: IEEE CIE international conference on radar, pp 833–836
58. Kou L, Wang X, Xiang M, Zhu M (2011) Spatial baseline decorrelation of geosynchronous circular SAR interferometry. In: IEEE CIE international conference on radar, pp 1578–1581
59. Kou L, Zhu M, Wang X, Xiang M (2012) Interferometric three-dimensional displacement measurement using geosynchronous circular SAR
60. L-L Kou, X-Q Wang, M-S Xiang, J-S Chong, M-H Zhu (2011) Effect of orbital errors on the geosynchronous circular synthetic aperture radar imaging and interferometric processing. J Zhejiang Univ Sci C 12(5):404–416. https://doi.org/10.1631/jzus.C1000170

61. Kou L, Xiang M, Wang X, Zhu M (2013) Tropospheric effects on L-band geosynchronous circular SAR imaging. IET Radar Sonar Navig 7(6):693–701
62. Kou L, Xiang M, Wang X (2014) Ionospheric effects on three-dimensional imaging of L-band geosynchronous circular synthetic aperture radar. Radar Sonar Navig IET 8(8):875–884
63. JPL (2003) Global earthquake satellite system: a 20-year plan to enable earthquake prediction
64. ESA (2017) Satellite missions database. https://directory.eoportal.org/web/eoportal/satellite-missions. Accessed 14 Nov. 2017
65. Pitz W, Miller D (2010) The TerraSAR-X satellite. IEEE Trans Geosci Remote Sens 48 (2):615–622. https://doi.org/10.1109/TGRS.2009.2037432
66. Snoeij P, Attema E, Davidson M, Duesmann B, Floury N, Levrini G, Rommen B, Rosich B (2010) Sentinel-1 radar mission: status and performance. IEEE Aerosp Electron Syst Mag 25 (8):32–39. https://doi.org/10.1109/MAES.2010.5552610
67. Migliaccio M, Nunziata F, Montuori A, Paes RL (2012) Single-look complex COSMO-SkyMed SAR data to observe metallic targets at sea. IEEE J Sel Top Appl Earth Observations Remote Sens 5(3):893–901. https://doi.org/10.1109/JSTARS.2012.2184271
68. Mateus P, Catalão J, Nico G (2017) Sentinel-1 interferometric SAR mapping of precipitable water vapor over a country-spanning area. IEEE Trans Geosci Remote Sens 55(5):2993–2999. https://doi.org/10.1109/TGRS.2017.2658342
69. Iglesias R, Monells D, Centolanza G, Mallorquí JJ, Fabregas X, Aguasca A (2012) Landslide monitoring with spotlight TerraSAR-X DATA. In: 2012 IEEE international geoscience and remote sensing symposium, 22–27 July 2012, pp 1298–1301. https://doi.org/10.1109/igarss.2012.6351300
70. Shimada M (2013) ALOS-2 science program. In: 2013 IEEE international geoscience and remote sensing symposium—IGARSS, 21–26 July 2013, pp 2400–2403. https://doi.org/10.1109/igarss.2013.6723303
71. Brule L, Baeggli H (2001) RADARSAT-2 mission update. In: IGARSS 2001. Scanning the present and resolving the future. Proceedings. IEEE 2001 international geoscience and remote sensing symposium (Cat. No.01CH37217), 9–13 July 2001, vol 2586, pp 2581–2583. https://doi.org/10.1109/igarss.2001.978095

Chapter 2
GEO SAR System Analysis and Design

Abstract GEO SAR is a new type of space-borne SAR, which runs on the geosynchronous orbit and has the advantages of short revisit time and wide coverage. However, due to the high geosynchronous orbit, the system characteristics and design methods of GEO SAR are quite different from those of the traditional low-earth-orbit SAR (LEO SAR), which are introduced in this chapter. Firstly, the system characteristics of GEO SAR are analyzed. The differences between GEO SAR and LEO SAR are introduced briefly, and then the coverage and revisit abilities are compared with the traditional LEO SAR. After that, a curvature circle motion model (CMM) is introduced to describe the curved satellite track, the Doppler characteristics are analyzed and the work modes are compared. Secondly, the parameter design methods of resolution, power and ambiguity are presented for GEO SAR, considering the high squint observation. Thirdly, two attitude steering strategies are introduced for GEO SAR. The total zero Doppler steering (TZDS) method aims to compensate the serious effects of Earth rotation and obtain a zero Doppler centroid, and the optimized resolution steering (ORS) method attempts to reduce the required steering angle and obtain an optimal resolution feature.

T. Long et al., *Geosynchronous SAR: System and Signal Processing*,
https://doi.org/10.1007/978-981-10-7254-3_2

2.1 Characteristics Analysis of GEO SAR

2.1.1 Characteristics Differences Between LEO SAR and GEO SAR

Space-borne synthetic aperture radar (SAR) is an important application branch of SAR. The current space-borne SARs onboard all run on the low-earth orbit (LEO), which is less than 1000 km. GEO SAR refers to the space-borne SAR running on the inclined Geosynchronous Satellite Orbit (IGSO) which is not geostationary orbit. Thus, there exists the relative motion between the satellite and the ground, which is essential for SAR imaging. Compared with LEO SAR, the largest advantage of GEO SAR is the short revisit time. With properly designed orbit parameters, GEO SAR is able to observe hot areas near China once or twice every orbit period and even observe some key areas continuously. Due to the very high orbit, the characteristics of GEO SAR are very different from those of the typical LEO SAR, which are listed as follows.

A. Earth rotation effect

The angular speed of the Earth rotation is about 7.29×10^{-5} rad/s, mainly influencing the direction of relative velocity in Earth-Centered, Earth-Fixed (ECEF) frame (in ECEF frame, the point (0,0,0) is defined as the center of mass of the Earth, and the axes are fixed with respect to the surface of the Earth). The comparisons of LEO SAR and GEO SAR are listed in Table 2.1. In LEO SAR, the orbital angular speed of the satellite is much larger than the Earth rotation angular speed, so the impact of Earth rotation is relatively small. However, in GEO SAR, the orbital angular speed is quite approximate to the Earth rotation angular speed. These bring great difficulties to the attitude steering.

B. Fixed longitude range

The angular speed of LEO SAR is much larger than that of the Earth, so LEO SAR is suitable for the global observation. However, the observation time for specific area is quite short and revisit period is several days in LEO SAR. The orbital angular speed of GEO SAR is similar to that of the Earth, and the relative velocity component between GEO SAR and the Earth in the equator direction changes periodically around zero, thus GEO SAR runs above the region with fixed longitudes. This means that GEO SAR has a longer observation time on the specific region, and the revisit

Table 2.1 Comparison of LEO SAR and GEO SAR

Parameter	LEO SAR	GEO SAR
Orbital angular speed (m)	1.24×10^{-3}	7.29×10^{-5}
The ratio of the orbital angular speed and earth rotation angular speed	>10	=1
Longitude range	Global	Local
Orbital speed (km/s)	7.25	2.74

time is less than one day. Although single GEO SAR cannot achieve global observation, it suits long time and high revisit observation for key areas.

C. Curved satellite track

In LEO SAR, the integration time is normally less than 10 s, and the relationship between the satellite and target can be described with a straight line model in the short integration time. However, in GEO SAR, the integration time is hundreds times of that of LEO SAR. In such long integration time, the satellite track is apparently curved. The curved satellite track leads to the great difference between the satellite speed and Doppler speed, which is moving speed of the Zero Doppler plane on the ground. This causes significantly different Doppler features from those of LEO SAR, including the sign variance of the Doppler rates, the slant range history following the "near-far-near" form, and so on.

D. Large scene

Limited by the orbital height and antenna size, the scene size of LEO SAR is normally several tens of kilometers. However, due to the promotion of orbital height, the scene size of GEO SAR is greatly increased. For an L-band GEO SAR, the scene size of the strip mode can be 600 km, and that of scan SAR mode can be over 4000 km. In such large scene, the spatial variance of the imaging parameters is considerably large, which is a great problem of GEO SAR imaging.

E. Squint observation

LEO SAR normally works in the broadside mode, and the parameter design and imaging is relatively simple. However, due to the influence of Earth's rotation and the requirement of maximizing the revisit time, GEO SAR usually works in a high squint mode. The high squint mode brings great challenge to SAR parameter design, and it also increases the spatial variance of the imaging parameters and the coupling of the range and azimuth directions.

2.1.2 Coverage and Revisiting

To analyze the coverage and revisit performance of GEO SAR, an evaluation standard is needed. The Global Earthquake Satellite System (GESS) project released by National Aeronautics and Space Administration (NASA) made comparisons of many aspects between GEO SAR and LEO SAR, including the strip width, the maximum revisit time and the orbital repeat cycle. On this basis, this chapter uses coverage time per day, maximum revisit time and orbital repeat cycle to analyze the characteristics of coverage and revisit.

Coverage time per day refers to the length of time that the beam covers the specified area within one day. Revisit time is the time interval between the observations on the same area, that is, the coverage interval. For GEO SAR, the maximum revisit time is the maximum time interval between the radar beams to the

Table 2.2 Parameters of GEO SAR and ALOS PALSAR

Parameter	GEO SAR	ALOS PALSAR
Orbit height (km)	35,793	691.65
Eccentricity	0.07	~0
Orbital inclination (deg)	53	98.16
Orbit repeat period (day)	1	46
Carrier frequency (GHz)	1.25	1.257

same area in a cycle. This parameter reflects the rapid response capability of space-borne SAR. In addition, the orbital repeat cycle is the interval at which the satellite nadir track is completely coincident. If the along-track interference is required, the time is the shortest time to obtain the data of along-track interference.

In the following analysis, the GEO SAR parameters shown in Table 2.2 were used for the analysis of coverage and revisit characteristics. We analyzed the LEO SAR satellite ALOS PALSAR's characteristic of revisit and coverage, to compare with GEO SAR.

For intuitive comparison, we calculated the coverage time per day and maximum revisit time of GEO SAR and ALOS PALSAR, and draw them on the map.

Figure 2.1 shows GEO SAR's daily coverage for China, Fig. 2.2 shows GEO SAR's maximum revisit time for China. The different colors represent different coverage time per days and maximum revisit times in hours. In Fig. 2.1, GEO SAR's coverage time per day for China ranges from 4 to 14 h with a minimum coverage time per day of 4 h in the central part of China. The coverage time per day is gradually extended from the central area to eastern and western region. The daily coverage of the eastern coastal areas is 8–10 h, and the maximum value is 14 h in Tibet. In Fig. 2.2 GEO SAR has a maximum revisit time of 2–16 h for China. A maximum revisit time in most of China is 6–10 h, 16 h in the northeastern part of China, and the shortest time 2 h in some parts of Tibet.

Figure 2.3 shows ALOS PALSAR's coverage time per day for China, and the time units are minutes. Figure 2.4 shows ALOS PALSAR's maximum revisit time for China, and the time units are hours. In Fig. 2.3, the coverage time per day of ALOS PALSAR ranges from 8 to 16 min, and is much shorter than that of GEO SAR. In Fig. 2.4 ALOS PALSAR has a maximum revisit time of 20–24 h for most areas of China, and it is much longer than that of GEO SAR. Table 2.3 shows the comparison of the coverage time per day and maximum revisit time for GEO SAR and also.

2.1.3 Motion Characteristics

The motion of space-borne SAR can be represented with the satellite track in ECEF frame. In GEO SAR, the satellite track in the ECEF frame can be designed to various shapes (such as circles, ellipses and shapes of figure "8") by using different orbital parameters [1, 2]. Because of the complex motion, the Doppler

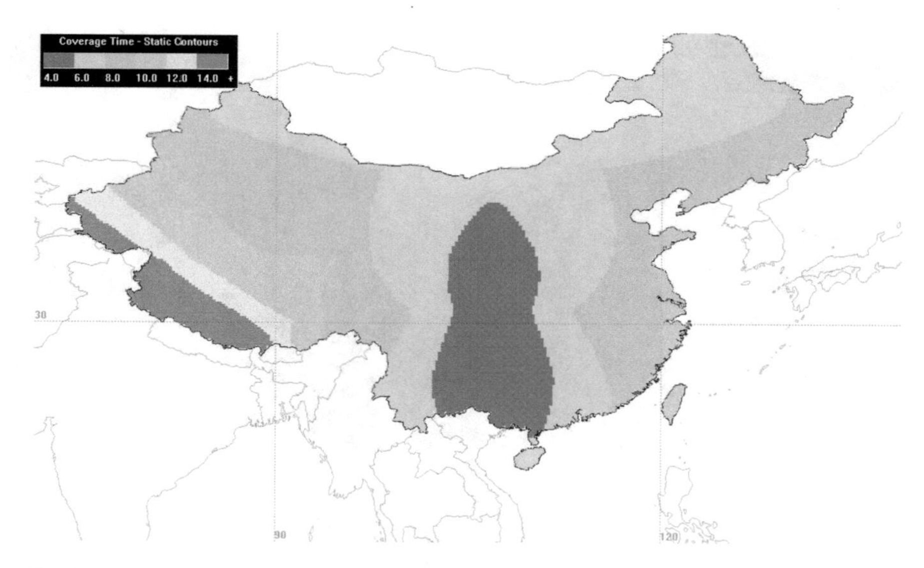

Fig. 2.1 Coverage time per day of GEO SAR in the land region of China (hours)

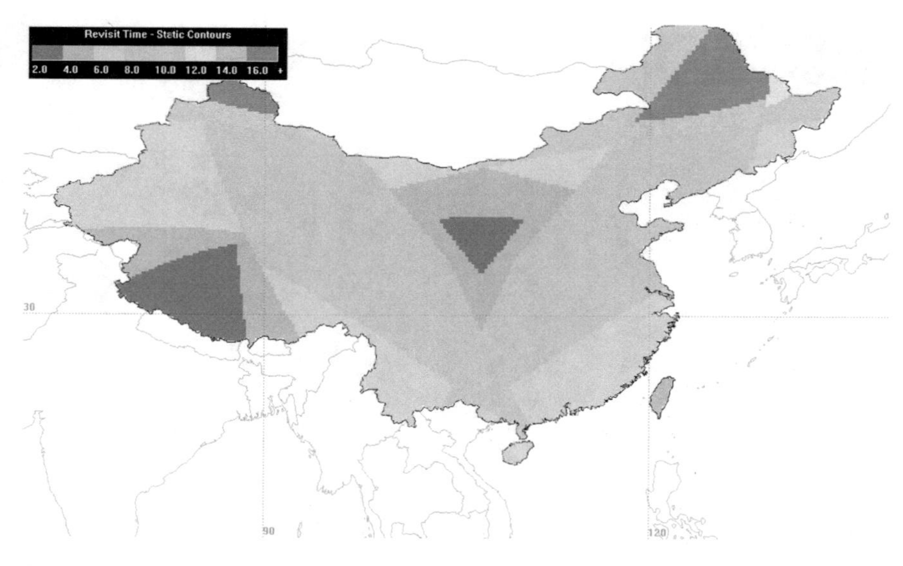

Fig. 2.2 Maximum revisit time of GEO SAR in the land region of China (hours)

characteristics of GEO SAR are also considerably complex. The sign of Doppler rate varies with orbital positions and elevation angles, the slant range history can be quadratic or cubic forms, and the Doppler parameters are 2-dimensional (2D) space-variant [3–5]. Furthermore, the work modes of SAR can be determined by the Doppler characteristics. Because of the complex Doppler characteristics, it is hard

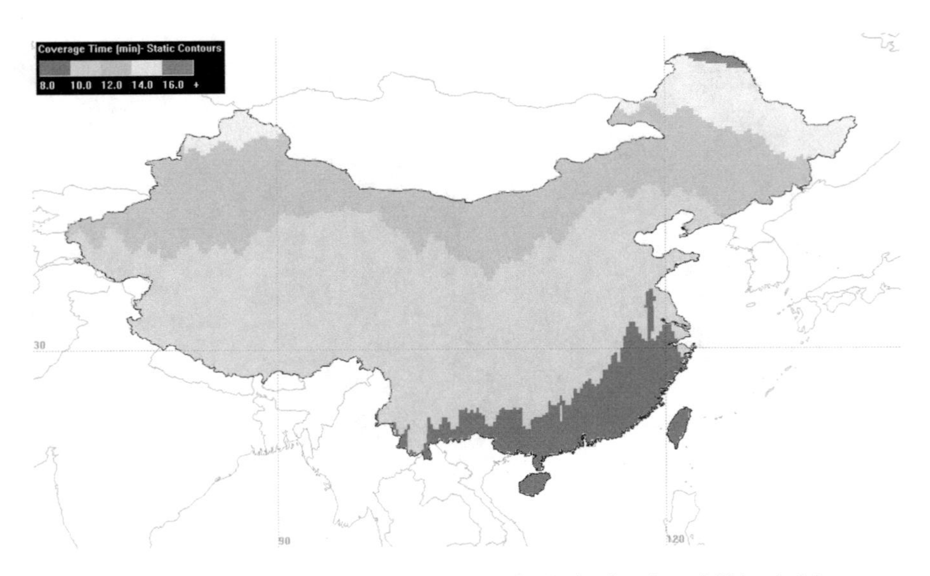

Fig. 2.3 Coverage time per day of ALOS PALSAR in the land region of China (min)

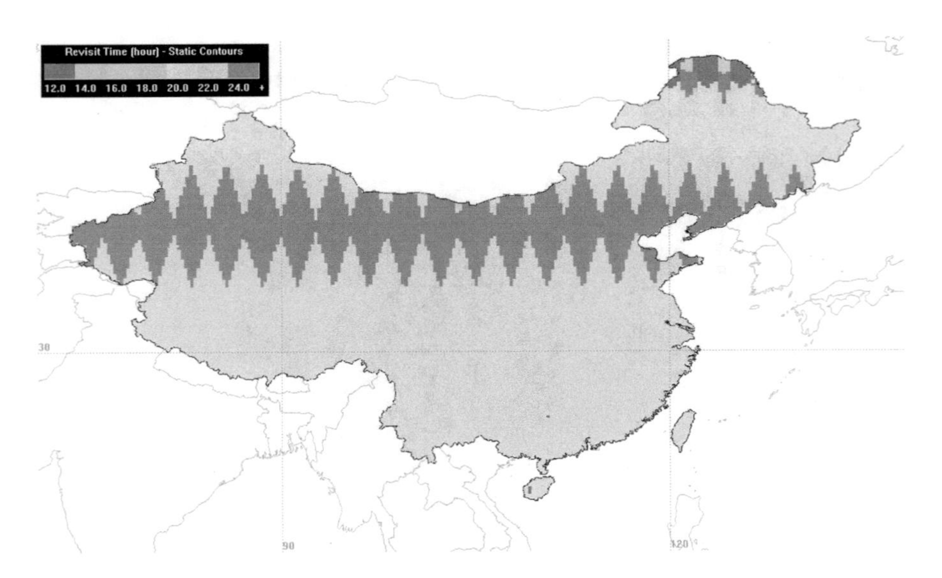

Fig. 2.4 Maximum revisit time of ALOS PALSAR in the land region of China (hours)

Table 2.3 GEO SAR versus ALOS PALSAR coverage and maximum revisit time comparison

	GEO SAR	ALOS PALSAR
Coverage time per day	4–14 h	8–16 min
Maximum revisit time	2–16 h	20–24 h

to know what mode GEO SAR works in, which affects the choice of the suitable applications and imaging algorithms. Therefore, a suitable theory is needed to provide a reasonable explanation for the complex Doppler characteristics and determine the work mode of GEO SAR.

2.1.3.1 Circular Motion Model

Currently, the satellite track of LEO SAR is described with the linear motion model (LMM), in which the satellite track is approximate to a straight line in the integration time [6]. However, the integration time of GEO SAR is much longer than that of LEO SAR, and can be several thousand seconds. In such long time, the curved satellite track cannot be approximate to a straight line. What's worse, the LMM only corresponds to a negative Doppler rate, whereas GEO SAR has sign-variant Doppler rates.

To describe the curved satellite track of GEO SAR, a circular motion model (CMM) based on the mathematical concept of curvature circle is used. The geometry of GEO SAR is shown in Fig. 2.5. The point O represents the Earth center, the point S represents the satellite position, and the point T represents the target position on the Earth surface. The satellite position vector $\mathbf{R_s}$ is from the Earth center to the satellite position, and the slant range vector $\mathbf{R_{ST}}$ is from the satellite position to the target position. Elevation angle β is the angle between the slant range vector and the satellite position vector. $\mathbf{V}$ is the satellite velocity vector and $\mathbf{A}$ is the satellite acceleration vector in ECEF frame.

The motion of space-borne SAR normally refers to the relative motion between the platform and the ground target, which can be described approximately with the satellite velocity and the satellite acceleration in the ECEF frame.

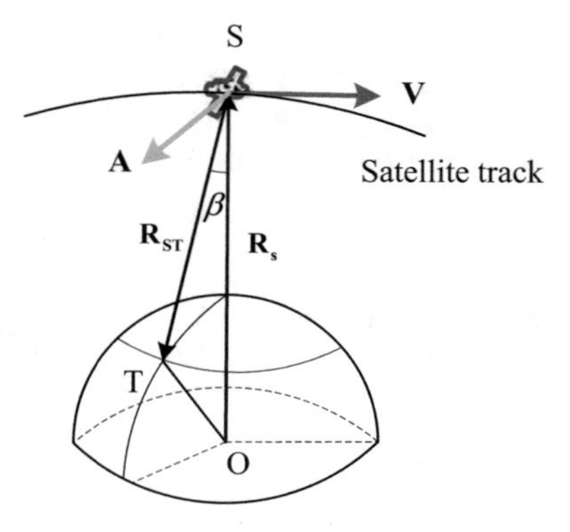

Fig. 2.5 Geometry of GEO SAR in ECEF frame

As shown in Fig. 2.6a, the satellite acceleration can be decomposed to the along-track acceleration $\mathbf{A_a}$ and the cross-track acceleration $\mathbf{A_c}$. Their values satisfy

$$\begin{cases} a_a = \mathbf{A}^{\mathbf{T}}\boldsymbol{\Gamma} \\ a_c = \mathbf{A}^{\mathbf{T}}\boldsymbol{\Phi} \end{cases} \tag{2.1}$$

where $\boldsymbol{\Gamma}$ is the unit vector along the satellite velocity direction, and $\boldsymbol{\Phi}$ is the unit vector along the cross-track acceleration direction. They can be written as

$$\begin{cases} \boldsymbol{\Gamma} = \mathbf{V}/v \\ \boldsymbol{\Phi} = \dfrac{(\mathbf{I} - \boldsymbol{\Gamma}\boldsymbol{\Gamma}^{\mathbf{T}})\mathbf{A}}{|(\mathbf{I} - \boldsymbol{\Gamma}\boldsymbol{\Gamma}^{\mathbf{T}})\mathbf{A}|} \end{cases} \tag{2.2}$$

where $\mathbf{I}$ is the unit matrix, v is the value of the satellite velocity vector. The along-track acceleration $\mathbf{A_a}$ changes the value of the satellite velocity vector, and the cross-track acceleration $\mathbf{A_c}$ changes its direction. In a short time, the effect of the along-track acceleration $\mathbf{A_a}$ can be neglected, and the space-borne SAR runs along the curvature circle determined by the centripetal acceleration $\mathbf{A_c}$ and the tangent velocity $\mathbf{V}$. The curvature radius r_c is given by

$$r_c = \frac{v^2}{a_c} \tag{2.3}$$

where a_c is the value of the satellite cross-track acceleration $\mathbf{A_c}$. Besides, the satellite acceleration can be written as

$$\mathbf{A} = \mathbf{A_s} + \boldsymbol{\omega}_\mathbf{e} \times [(\boldsymbol{\omega}_\mathbf{e} \times \mathbf{R_s}) - 2\mathbf{V_s}] \tag{2.4}$$

where $\boldsymbol{\omega}_\mathbf{e}$ is the angular velocity of the Earth rotation, $\mathbf{V_s}$ is the inertial satellite velocity, $\mathbf{A_s}$ is the inertial satellite acceleration and it is along the nadir direction. According to (2.4), the Earth rotation deflects the satellite acceleration and the curvature circle plane from the nadir direction. The angle between the curvature

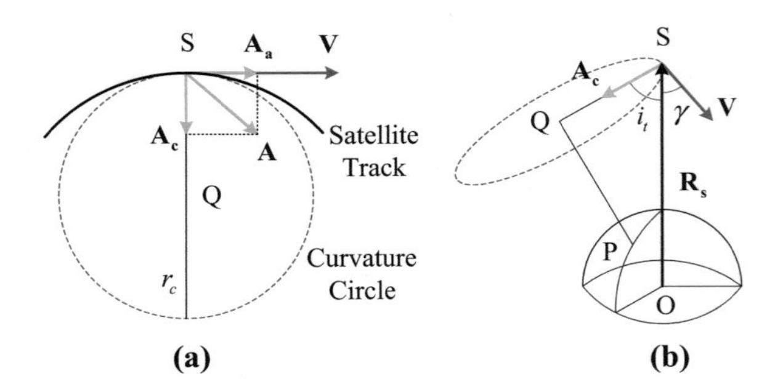

Fig. 2.6 Circular motion model **a** the curvature circle determined by the centripetal acceleration and the satellite velocity **b** the track inclination and pitch in ECEF frame

circle plane and the nadir direction can be defined as the track inclination angle α, which is given by

$$i_t = \operatorname{acos}\left(-\frac{\mathbf{A_c^T R_s}}{a_c r_s}\right) \tag{2.5}$$

where the orbital radius r_s is the value of the satellite position vector $\mathbf{R_s}$. Besides, the satellite velocity is not parallel to the local ground plane in case of an elliptical orbit. This causes pitches of the track plane, which can be described with a track pitch γ as shown in Fig. 2.6.

$$\gamma = \operatorname{acos}\left(-\frac{\mathbf{V^T R_s}}{v r_s}\right) \tag{2.6}$$

The track pitch equals to 90° when GEO SAR runs in a circular orbit, and varies around 90° in an elliptical orbit.

During the integration time T_a, the platform rotation angle $\Delta\theta$ on the curvature circle plane can be expressed approximately as

$$\Delta\theta \approx \frac{vT_a}{r_c} \tag{2.7}$$

Thus, the motion of the space-borne SAR can be described with a circular motion model in a short time. In the CMM, the satellite is considered to run along the curvature circle which is determined by the centripetal acceleration $\mathbf{A_c}$ and the tangent velocity $\mathbf{V}$. The satellite motion in the integration time can be described approximately with the curvature radius r_c, the track inclination i_t, the track pitch γ and the platform rotation angle $\Delta\theta$. Although the satellite track of GEO SAR is not an exact circle due to the effect of the along-track acceleration $\mathbf{A_a}$, the CMM is accurate enough for the analyses of the motion and Doppler characteristics.

2.1.3.2 Motion Comparison Between LEO SAR and GEO SAR

Based on the CMM, the motion characteristics of GEO SAR are analyzed and compared with those of ALOS PALSAR, including the curvature radius, the track inclination and the platform rotation angle. The orbital parameters of GEO SAR used are listed in Table 2.2. The orbital inclination of 50° can provide a fine latitudinal coverage [7]. A zero eccentricity can facilitate feature comparisons between GEO SAR and LEO SAR without loss of generality, since LEO SAR mostly runs on near-circular orbits. The large antenna size of 30 m can reduce the requirement for transmitter power, and the L-band carrier frequency can provide excellent performance of surface deformation measurement [8]. The parameters of ALOS PALSAR are also listed as a comparison [9].

According to (2.3) and (2.5), the curvature radius of ALOS PALSAR is closely approximate to the orbital radius, and the track inclination α is less than 8°. Therefore, ALOS PALSAR rotates around the Earth center with a curvature radius r_s approximately as shown in Fig. 2.7a. In GEO SAR, the satellite acceleration direction is affected significantly by the Earth rotation. The track inclination α varies with orbital positions from 0° to 83°, and the rotation radius varies from 5000 to 42,164 km. It indicates that the curvature circle plane is deflected far away from the nadir direction in GEO SAR, and the curvature circle center Q is not the Earth center at most orbital positions as shown in Fig. 2.7b.

In ALOS PALSAR, the full aperture integration time for the strip-map mode is less than 5 s, and the platform rotation angle is less than 0.2°. Therefore, the motion of LEO SAR follows the LMM approximately, namely the CMM is compatible to the LMM in case of LEO SAR. Compared with LEO SAR, GEO SAR has much longer integration time and much larger rotation angles as shown in Fig. 2.8. Especially in the red color region, the full aperture integration time is longer than 2000 s and its maximum value reaches 6600 s; the rotation angle is larger than 15° approximately and the maximum value reaches 40°. Therefore, the satellite track of GEO SAR cannot be approximate to a straight line. What is more important, the

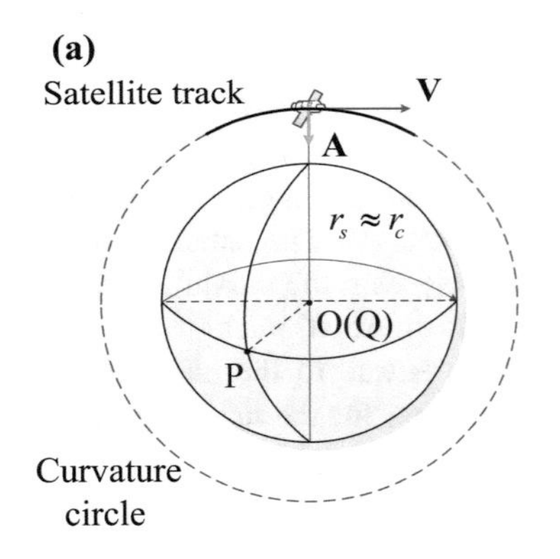

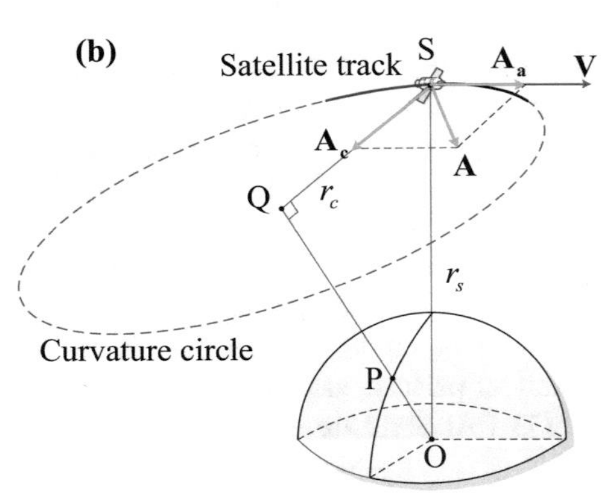

Fig. 2.7 Motion characteristics of LEO SAR and GEO SAR **a** LEO SAR **b** GEO SAR

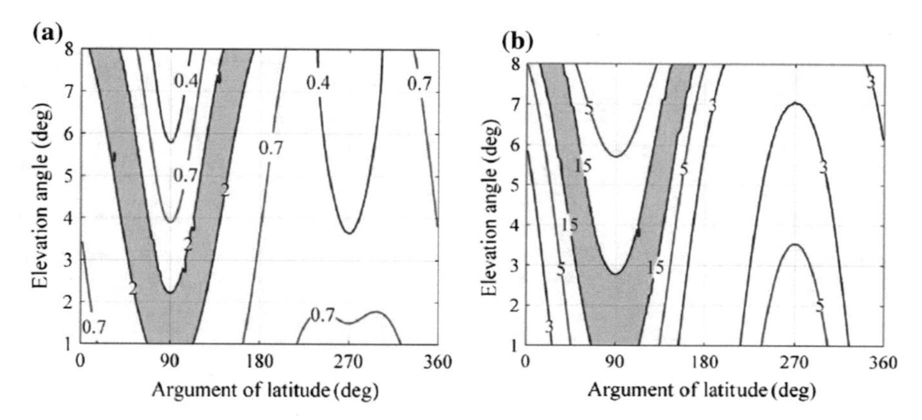

Fig. 2.8 Contour results of the full aperture integration time (10^3 s) and the rotation angles (deg) varying with the orbital angles and elevation angles in GEO SAR. The red color region corresponds to the integration time longer than 2000 s and the rotation angle larger than 15°, respectively. **a** the full aperture integration time (10^3 s) **b** the rotation angles (deg)

LMM only corresponds a negative Doppler rate and linear time-frequency relationship, and it cannot explain the sign-variant Doppler rate and nonlinear time-frequency relationship in GEO SAR.

2.1.4 Doppler Characteristics

SAR obtains the azimuth resolution by using the Doppler frequency caused by the relative motion between the platform and target, so the Doppler characteristics are essential to the SAR system design and imaging processing. An important parameter to describe the Doppler characteristics is the Doppler rate, which describes the changing rate of the Doppler frequencies.

The Doppler rate can be expressed by using the Doppler velocity. Based on [10], the Doppler rate of the space-borne SAR is given by

$$f_r = -\frac{2}{\lambda r}\left(v^2 - a_r r\right) = -\frac{2vv_d}{\lambda r} \tag{2.8}$$

where λ is the wave length, r is the value of slant range, a_r is the value of acceleration, v_d is the Doppler velocity, and can be expressed as

$$v_d = v - \frac{a_r r}{v} \tag{2.9}$$

The Doppler velocity is different from the beam velocity. The beam velocity v_g is the velocity of the beam or the zero-Doppler line on the ground [11]. It can be

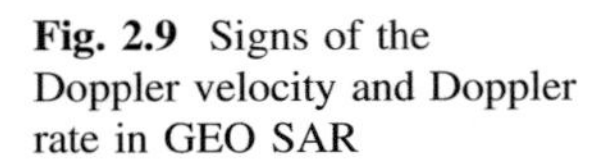

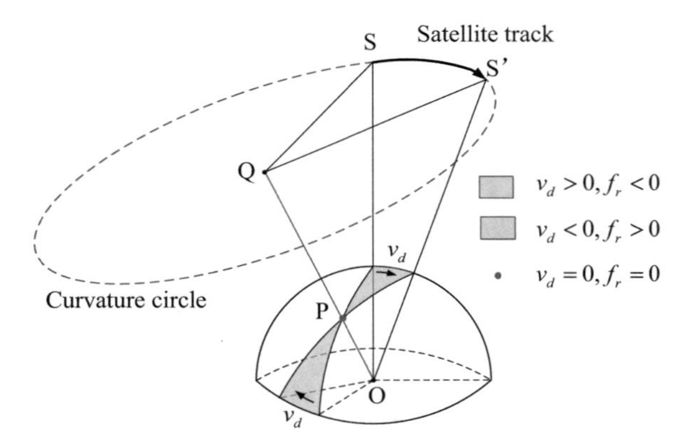

Fig. 2.9 Signs of the Doppler velocity and Doppler rate in GEO SAR

decomposed to the tangential component v_{gt} (on the zero-Doppler plane) and the normal component v_{gn} (normal to the zero-Doppler plane). The Doppler velocity is the rotation velocity of the zero-Doppler plane at the target position, and it has the same value and direction as the normal component v_{gn} of the beam velocity.

According to (2.8), the sign of the Doppler rate is opposite to those of the Doppler velocity v_d. Furthermore, the rotation center Q changes with orbital positions due to the variance of the curvature radius and the track inclination. The signs of the Doppler velocity and Doppler rate change with both orbital positions and elevation angles as shown in Fig. 2.9. In LEO SAR, the curvature circle center is the Earth center approximately, and the illumination region is always nearer the nadir than the ground center P. Thus, the Doppler rate is always negative in LEO SAR.

2.1.4.1 Linear Time-Frequency Relationship

When the Doppler rate is non-zero, GEO SAR has a linear time-frequency relationship. The Doppler characteristics of a point array at the orbital position with an argument of latitude of 90° are simulated. For a 30-m-diameter L-band antenna, the beam footprint is about 320 km (range) × 260 km (azimuth). Five targets are placed at the center and edges of the beam footprint with a range interval of 130 km and an azimuth interval of 160 km as shown in Fig. 2.10. T_1, T_2 and T_3 are distributed along the range direction, and T_2, T_4, and T_5, are distributed along the azimuth direction.

The simulation results are shown in Fig. 2.11. The five targets have positive Doppler rate and "near-far-near" form of the slant range history, which are opposite to those in LEO SAR. The Doppler frequencies increase linearly with the azimuth time, and the spatial variation of the Doppler characteristics arises mainly in the range direction. Thus, the Doppler characteristics when the Doppler rate is non-zero have no essential difference from those of LEO SAR.

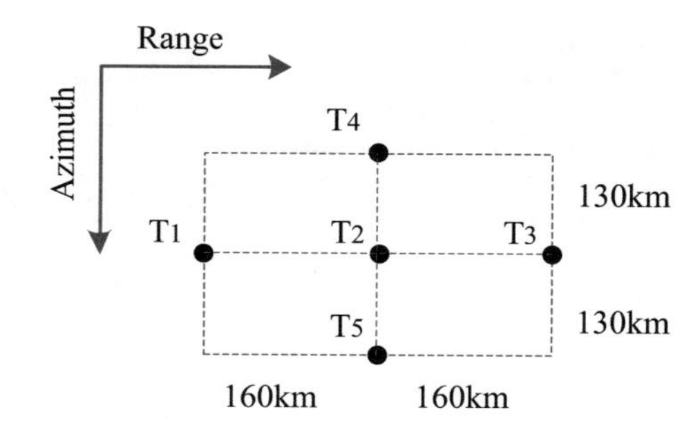

Fig. 2.10 Distributions of a point array

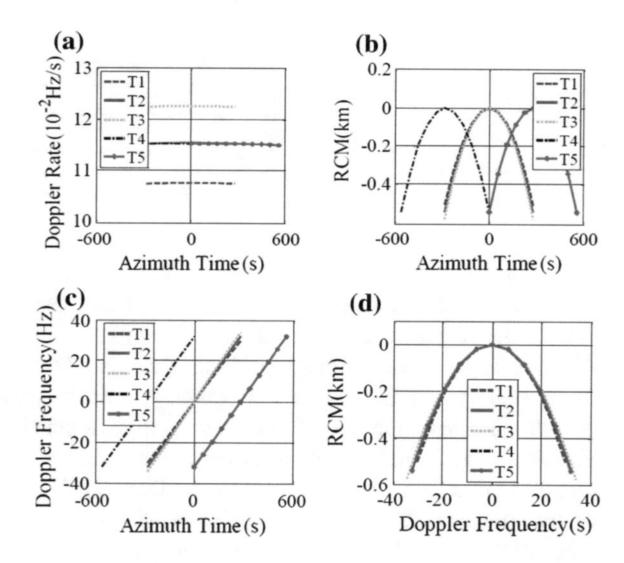

Fig. 2.11 Doppler characteristics at the orbital position with an argument of latitude of 90° **a** Doppler rate **b** range cell migration (RCM) **c** time-frequency relationship **d** properties in the range-Doppler (RD) domain

2.1.4.2 Nonlinear Time-Frequency Relationship

When the Doppler rate is zero or near zero, GEO SAR has a nonlinear time-frequency relationship. Figure 2.12 shows the Doppler characteristics of the point array when Doppler rate is zero or near zero. The spatial variance of the Doppler characteristics arises in both the azimuth and the range directions, especially in the azimuth direction. The Doppler rates of the azimuth targets T_4 and T_5 are negative and positive, respectively. The Doppler rates of targets T_1, T_2 and T_3 change signs in the integration time. The slant range histories of three azimuth targets firstly change from the quadratic form of "far-near-far" to the cubic form, and then change to the quadratic form of "near-far-near". The Doppler frequencies are not linear but nearly quadratic functions of the azimuth time, and the range cell migration (RCM) curves are folded to the right side of the range-Doppler (RD) plane. Thus, the Doppler characteristics when Doppler rate is zero or near zero are totally different from those in LEO SAR.

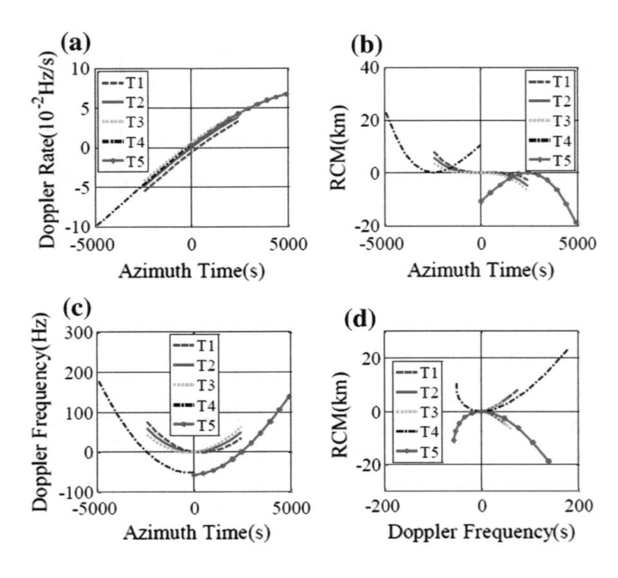

Fig. 2.12 Doppler characteristics when the Doppler rate is near zero **a** the Doppler rate **b** the RCM **c** the time-frequency relationship **d** the RCM properties in the RD domain

Although the slant range history of various forms can be accurately described with a high-order slant range history model (such as the series expansion model), it is only the fundamental requirement of the frequency domain imaging algorithms. The frequency domain imaging algorithms also demand that the RCM is a single value function of the Doppler frequency, namely one Doppler frequency corresponds to only one RCM. However, this demand is not satisfied when the time-frequency relationship is nonlinear as shown in Fig. 2.12d. Thus, the correction at one Doppler frequency in the Doppler frequency domain will take effect on two different RCMs. If one of the RCM can be corrected properly, the absolute value of the second RCM will be even larger. For this reason, the frequency-domain imaging algorithms cannot be directly used for the case of nonlinear time-frequency relationship. To perform the imaging in the Doppler frequency domain, the nonlinear time-frequency relationship should be firstly transformed into a linear one to avoid the folded RCM in the RD domain.

Another solution is to perform the RCMC and the azimuth focusing in the time domain, which can be done by the back projection algorithm (BPA) [12]. However, the time complexity of BPA is much higher than that of the frequency-domain algorithms. In GEO SAR, the operations required by the standard BPA can be several thousand times of those required by the frequency-domain algorithms [13]. Besides, BPA is also sensitive to the phase noise and the antenna center phase error, which mainly cause the deterioration of the peak side lobe ratio (PSLR) and the integrated side lobe ratio (ISLR). Large antenna center phase error also causes the defocusing [14].

2.1.5 *Work Modes of GEO SAR*

The SAR work modes are normally determined by the platform motion and the Doppler characteristics. With the analyses above, we can find that GEO SAR works in two different modes. When the Doppler rate is non-zero, the time-frequency relationship is nearly linear, GEO SAR works in the strip-map mode. When the Doppler rate is zero or near zero, the time-frequency relationship is nonlinear. Thus, the satellite track is approximate to but not an exact partial circle. GEO SAR can be considered as a pseudo-circular SAR (PCSAR), which is similar to a partial aperture CSAR and can provide a height resolution.

To completely show the characteristics, we provide a design example of the two modes based on the CMM. The arguments of latitude are set to be 0°, 90° and 51.06°, which correspond to the negative, positive and zero Doppler rates, respectively.

The design results are illustrated in Table 2.4. In the strip-map mode, the Doppler velocity and the beam velocity are non-zero. Considering the large swath width and non-zero beam velocity, the strip-map mode can be used for extensive region mappings with a resolution of 2–6 m approximately.

In the PCSAR mode, GEO SAR continuously illuminates a specific spot for about 80 min. In this case, PCSAR mode can provide an azimuth resolution better than 1 m and a height resolution better than 10 m as shown in Table 2.4. Besides, the PCSAR mode can also obtain images from multiple directions with a modest resolution, which may reveal the dynamic change of the target region. For example, ten images can be obtained with a resolution of 5 m and a time interval of 8 min for one region. Thus, the PCSAR mode can be used for continuous observation to obtained high 3-D resolutions or for multiple-direction observation on specific region.

Table 2.4 Design results of different work modes with CMM

Work mode	Strip-map		PCSAR
Argument of latitude (deg)	0	90	51.06
Full aperture integration time (s)	739	563	4886
Doppler rate f_r (Hz/s)	−0.21	0.12	0
Doppler velocity v_d (m/s)	351	−461	0
Doppler bandwidth B_d (Hz)	153.57	64.92	0
Target radius r_t (km)	5695	−2156	0
Curvature radius r_c (km)	42,164	5136	18,995
Platform rotation angle $\Delta\theta$ (deg)	2.61	6.90	27.17
Azimuth resolution ρ_a (m)	2.03	6.29	0.43
Height resolution ρ_h (m)	4534.58	426.75	8.57
Slant range history form	Far-near-far	Near-far-near	Various forms
Imaging algorithm	CSA, etc.	CSA, etc.	BPA

Furthermore, the work modes of GEO SAR actually are determined by the signs of the Doppler velocity and Doppler rate, and the length of integration time is not an essential criterion. Even if the integration time and the platform rotation angle are reduced when the Doppler rate is zero or near zero, GEO SAR still works in the PCSAR mode.

GEO CSAR mode has been studied, and these studies are all based on the circular or near-circular satellite track obtained by using specific orbital parameters [10, 15]. Different from these studies, this research indicates that any GEO SAR is approximate to a partial aperture CSAR at specific orbital positions and elevation angles, even if the satellite track is not an exact circle in an orbital period.

The analyses of this Section are based on a zero Doppler centroid, which is the basic geometry of SAR imaging and can be realized by a total zero Doppler steering method [16].

2.2 System Parameters Analysis and Design

SAR is a kind of complex microwave remote sensing radar, and the SAR system performance is described with a number of parameters. Among the radar parameters, the most closely related with the radar image quality are the image resolution, the radar transmission power and the image ambiguity-to-signal ratio. Image resolution is the minimum distance between two targets that can be distinguished in the SAR image. Radar transmit power determines the maximum distance of radar detection, which is also closely related with the signal to noise ratio of image or noise equivalent sigma zero (NESZ). The ambiguity-to-signal ratio of image is the ratio of the aliasing signal in the echo to the main signal power, and characterizes the degree of degradation of the image quality caused by the aliasing signal.

In conventional LEO SAR, the payload usually works in the broadside mode. In GEO SAR, high squint mode is widely used to improve the performance of revisit. In the case of high squint mode, the resolution characteristic and the power budget are quite different from those of the broadside mode. Therefore, the system parameter analysis and design method of LEO SAR are not suitable. The high squint mode is also widely used in traditional airborne SAR, but the geometric model of airborne SAR is far from the space-borne SAR and cannot be directly applied. Based on the geometric relationship of GEO SAR, this section discusses the analysis and design method of resolution, power and ambiguity.

2.2.1 Resolution

The resolution is one of the most important factors to characterize the radar system capability and imaging results, but there is no accurate resolution expression at present for GEO SAR. Because of the large orbit height and non-negligible Earth rotation, the calculation method of Doppler bandwidth and equivalent velocity for LEO SAR will result in a big error when directly used for GEO SAR [16, 17], thus it is difficult to find an appropriate formula to characterize the resolution for GEO SAR. Furthermore the range resolution direction and azimuth resolution direction are regarded to be mutually normal in ground plane in monostatic SAR, but the conclusion does not hold anymore in GEO SAR. Therefore, we must derive the resolution expression for GEO SAR.

The most comprehensive information regarding the radar resolution can be obtained from the generalized ambiguity function (GAF) [6, 18]. According to [6], the GAF can be expressed as

$$\chi = \frac{\iint E_A(t,nT)\cdot E_B^*(t,nT)\mathrm{d}t\mathrm{d}n}{\sqrt{\iint|E_A(t,nT)|^2\mathrm{d}t\mathrm{d}n}\sqrt{\iint|E_B(t,nT)|^2\mathrm{d}t\mathrm{d}n}} \tag{2.10}$$

where, the symbol * is the complex conjugate operator, A and B are adjacent points, $E_A(t,nT)$ and $E_B(t,nT)$ are the received signals from target A and B, t is range time, nT is azimuth time, T is the Pulse Repetition Time (PRT), n is the index of transmitted pulse corresponding to the azimuth slow time. For convenience, the symbol nT is replaced by the symbol n. Equation (2.10) indicates that the similarity of received signal from target A and B determines the spatial resolution of radar system. Furthermore the similarity is presented by the correlation coefficient. When target A and B are very close, the received signatures are very similar, we cannot distinguish the target A and B and vice versa.

In the following, ambiguity function on the slant plane is firstly derived according to the ambiguity and the relationship between the satellite and the ground. Then the expressions of ambiguity on ground plane and resolution on ground plane are obtained according to the projection relationship between slant plane and ground plane.

2.2.1.1 Resolution on the Slant Range Plane

The geometry of GEO SAR in ECI is shown in Fig. 2.13. The Earth center O is one of the focuses of elliptical orbit and is also the origin of Earth inertial coordinate system $OX_iY_iZ_i$. $\boldsymbol{\omega}_\mathbf{e}$ is the angular velocity of the Earth rotation, pointed to the North Pole. The satellite position vector is $\mathbf{R_S}$, velocity vector is $\mathbf{V}_S$, and ω_s is the satellite angular velocity, perpendicular to the orbital plane. The position vectors of adjacent point target A and B are $\mathbf{R_A}$ and $\mathbf{R_B}$, slant range vectors are $\mathbf{R_{SA}}$ and $\mathbf{R_{SB}}$. In the above vector, slant range vector, satellite position vector, velocity vector and angular velocity vector are functions of azimuth time, and slant range vector can be expressed as

$$\begin{aligned}\mathbf{R_{SA}}(n) &= \mathbf{R_s}(n) - \mathbf{R_A}\\ \mathbf{R_{SB}}(n) &= \mathbf{R_s}(n) - \mathbf{R_B}\end{aligned} \tag{2.11}$$

$R_{SA}(n)$ and $R_{SB}(n)$ represent vector $\mathbf{R_{SA}}$ and $\mathbf{R_{SB}}$ values.

The received signals from target A and B can also be written as

$$E_A(t,n) = W_a(n) \cdot s(t - \tau_A(n)) \exp\{j2\pi f_c(t - \tau_A(n))\} \tag{2.12}$$

$$E_B(t,n) = W_a(n) \cdot s(t - \tau_B(n)) \exp\{j2\pi f_c(t - \tau_B(n))\} \tag{2.13}$$

where $W_a(\cdot)$ is the antenna pattern function in azimuth direction, f_c is the carrier frequency, $\tau_A(n)$ and $\tau_B(n)$ are the propagation delay time of target A and B, and can be approximately written as

$$\tau_A(n) = \frac{2R_{SA}(n)}{c} \tag{2.14}$$

$$\tau_B(n) = \frac{2R_{SB}(n)}{c} \tag{2.15}$$

The GAF is directly related to the similarity of target received signals. However, (2.12) and (2.13) shows that the similarity of received signals is mainly determined by the propagation delay time and Doppler phase. Considering (2.10) and (2.13), the GAF can be written as

$$\chi = \frac{\int \left\{ \begin{array}{l} p(f) \exp[j2\pi f_c[\tau_B(n) - \tau_A(n)]] \\ \int \sqrt{W_A(n) W_B(n)} \exp[j2\pi f_c[\tau_B(n) - \tau_A(n)]] \mathrm{d}n \end{array} \right\} \mathrm{d}f}{\sqrt{\iint p(f) W_A(n) \mathrm{d}f \mathrm{d}n} \sqrt{\iint p(f) W_B(n) \mathrm{d}f \mathrm{d}n}} \tag{2.16}$$

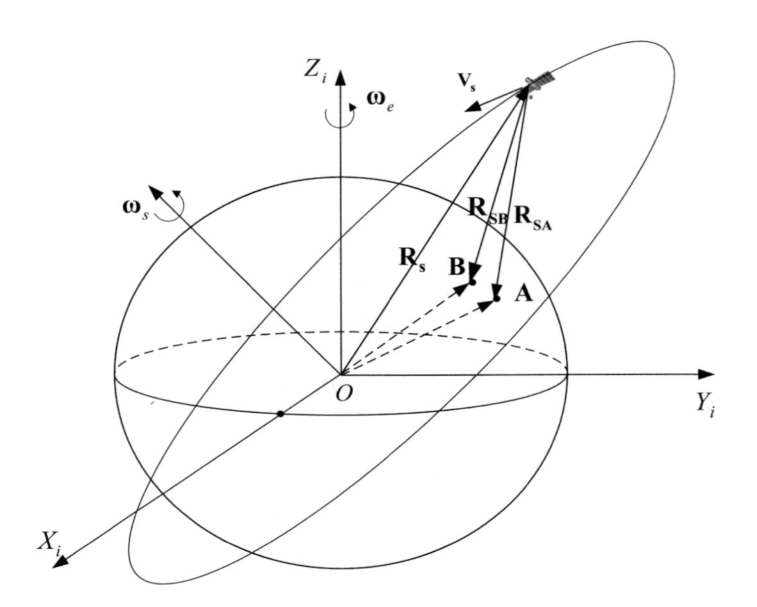

Fig. 2.13 Geometry of GEO SAR in ECI

where $p(f)$ is the power spectrum of the normalized baseband signal $s(t)$, $W_A(n)$ and $W_B(n)$ are the antenna pattern functions in azimuth direction at target A and B. After further simplification of (2.16), it can be expressed as

$$\chi = \exp\{j2\pi f_c \tau_d(0)\} \cdot P(\tau_d(0)) \cdot Q(f_d(0)) \tag{2.17}$$

where the function $P(\cdot)$ is the Inverse Fourier Transform (IFT) of normalized power spectrum $\bar{p}(f) = p(f) / \int_{-\infty}^{\infty} p(f) \mathrm{d}f$, the function $Q(\cdot)$ is the IFT of normalized antenna power pattern in azimuth direction, $\chi(\cdot)$ is the GAF, $\tau_d(0)$ is delay time difference between target A and B at aperture center moment (ACM), $f_d(0)$ is Doppler frequency difference between target A and B at ACM. According to (2.14) and (2.15), $\tau_d(0)$ can be expressed as

$$\tau_{d0} = \frac{2R_{SA}(n)}{c} - \frac{2R_{SB}(n)}{c} \tag{2.18}$$

Doppler frequency can be expressed as

$$\begin{aligned} f_{dA}(n) &= \frac{2}{\lambda R_{SA}(n)} \mathbf{V_s}(n)^{\mathbf{T}} \mathbf{R_{SA}}(n) \\ f_{dB}(n) &= \frac{2}{\lambda R_{SB}(n)} \mathbf{V_s}(n)^{\mathbf{T}} \mathbf{R_{SB}}(n) \end{aligned} \tag{2.19}$$

where the superscript T means the transpose operator, and $f_d(0)$ can be expressed as

$$f_d(0) = \frac{2}{\lambda} \mathbf{V_s}(n)^{\mathrm{T}} \left[\frac{\mathbf{R_{SA}}(n)}{R_{SA}(n)} - \frac{\mathbf{R_{SB}}(n)}{R_{SB}(n)} \right] \tag{2.20}$$

In order to obtain the spatial resolution, we should express the delay time differences and Doppler frequency difference as the function of the target position. In the resolution analysis, the target positions are very close and first order approximation has enough accuracy. Using first order Taylor expansion with respect to the target A in (2.18) and (2.20), we have

$$\tau_d(0) \approx \tau_d(0)|_{\mathbf{B=A}} + \nabla \tau_d(0) \cdot (\mathbf{R_B} - \mathbf{R_A}) \tag{2.21}$$

$$f_d(0) \approx f_d(0)|_{\mathbf{B=A}} + \nabla f_d(0) \cdot (\mathbf{R_B} - \mathbf{R_A}) \tag{2.22}$$

where ∇ means the gradient operator. The constant terms in (2.21) and (2.22) are equal to

$$\tau_d(0)|_{\mathbf{B=A}} = 0, \quad f_d(0)|_{\mathbf{B=A}} = 0 \tag{2.23}$$

The Eq. (2.23) indicates that the delay time difference and Doppler frequency difference both are equal to zero when target positions are identical, i.e., we cannot distinguish these two targets.

Due to the Earth rotation, the gradient operators are determined by the relative rotation between satellite and Earth. According to the geometry relationship shown in Fig. 2.1, the gradient vectors can be expressed as

$$\nabla \tau_d(0) = \frac{\partial \tau_{d0}}{\partial \mathbf{B}}\bigg|_{\mathbf{B}=\mathbf{A}} = \frac{2}{c} \cdot \frac{\mathbf{R_{SA}}(0)}{R_{SA}(0)} \tag{2.24}$$

$$\begin{aligned} \nabla f_d(0) &= \frac{\partial f_d(0)}{\partial \mathbf{B}}\bigg|_{\mathbf{B}=\mathbf{A}} \\ &= \frac{2}{\lambda} \cdot \left[\begin{array}{l} \dfrac{\mathbf{R_{SA}}(0)}{R_{SA}(0)} \begin{pmatrix} 0 & -\omega_e & 0 \\ \omega_e & 0 & 0 \\ 0 & 0 & 0 \end{pmatrix} \\ + \dfrac{[\mathbf{V_s}(0) - \mathbf{V_{ga}}(0)]\left(\mathbf{I} - \frac{1}{R_A^2(0)} \mathbf{R_{SA}}(0)^T \cdot \mathbf{R_{SA}}(0)\right)}{R_{SA}(0)} \end{array} \right] \\ &= \frac{2}{\lambda R_{SA}(0)} \cdot \left[\begin{array}{c} \boldsymbol{\omega}_\mathbf{e} \times \mathbf{R_{SA}}(0) + [\mathbf{V_s}(0) - \mathbf{V_{ga}}(0)] \\ \left(\mathbf{I} - \dfrac{1}{R_{SA}^2(0)} \mathbf{R_{SA}}(0)^T \cdot \mathbf{R_{SA}}(0)\right) \end{array} \right] \end{aligned} \tag{2.25}$$

where $\mathbf{V_{ga}}(0)$ is the velocity vector at ACM caused by Earth rotation, and can be expressed as

$$\mathbf{V_{ga}}(0) = \boldsymbol{\omega}_\mathbf{e} \times \mathbf{R_A} \tag{2.26}$$

where the symbol '$\times$' means the cross product. In addition, satellite velocity in ECEF frame can be expressed as

$$\mathbf{V}(n) = \mathbf{V_s}(n) - \boldsymbol{\omega}_\mathbf{e} \times \mathbf{R_s}(n) \tag{2.27}$$

Hence

$$\nabla \tau_d(0) = \frac{2}{c} \mathbf{X_s} \tag{2.28}$$

$$\nabla f_d(0) = \frac{\partial f_d(0)}{\partial \mathbf{B}}\bigg|_{\mathbf{B}=\mathbf{A}} = \frac{2}{\lambda} \cdot \omega \mathbf{Y_s} \tag{2.29}$$

where ω is the slant range rotating angular velocity.

$$\omega = \frac{\left|\left(\mathbf{I} - \mathbf{X}_s\mathbf{X}_s^{\mathbf{T}}\right)\mathbf{V}(0)\right|}{R_{SA}(0)} \tag{2.30}$$

$\mathbf{X}_s$ and $\mathbf{Y}_s$ can be expressed as

$$\begin{cases} \mathbf{X}_s = \dfrac{\mathbf{R}_{\mathbf{SA}}(0)}{R_{SA}(0)} \\ \mathbf{Y}_s = \dfrac{\left(\mathbf{I} - \mathbf{X}_s\mathbf{X}_s^{\mathbf{T}}\right)\mathbf{V}(0)}{\left|\left(\mathbf{I} - \mathbf{X}_s\mathbf{X}_s^{\mathbf{T}}\right)\mathbf{V}(0)\right|} \end{cases} \tag{2.31}$$

The plane consisting of $\mathbf{X}_s$ and $\mathbf{Y}_s$ is the slant range plane which is the imaging plane in monostatic SAR. In low Earth orbit (LEO) SAR or airborne SAR cases, the slant range plane is directly determined by the satellite velocity vector and the direction of line of sight, while in GEO SAR case, we know that the slant range plane is affected a lot by the Earth rotation. In (2.25), it shows that the Earth rotation ω_e plays a very important role in the Doppler gradient operator, it is different from the LEO SAR and airborne SAR, thus we must consider the effects of Earth rotation on the resolution analysis in GEO SAR.

- When the Earth rotation is equal to zero, i.e., when $\omega_e = 0$, the Doppler gradient operator is reduced to $\frac{2}{\lambda} \cdot \left(\mathbf{I} - \mathbf{X}_s\mathbf{X}_s^{\mathbf{T}}\right)\mathbf{V}_s(0)/R_A(0)$. The same is in the cases of the conventional LEO SAR and airborne SAR.
- When the satellite orbit increases, namely when the satellite position vector $\mathbf{R}_s$ increases and the satellite velocity vector $\mathbf{V}_s$ decreases, the satellite angular velocity with respect to target becomes smaller and smaller. Especially under the case of GEO SAR, the satellite angular velocity is nearly equal to the Earth angular velocity, the resolution along some special direction will increase or decrease 50% or even more.
- Considering (2.25) and (2.28), no matter whether the Earth rotation exists or not, the inner product of $\nabla\tau_d(0)$ and $\nabla f_d(0)$ is always equal to zero, i.e., that $\langle\nabla\tau_d(0),\ \nabla f_d(0)\rangle = 0$, it means that the side-lobe of GAF is always orthogonal in the slant range plane.

Based on (2.18) through (2.25), the GAF shown in (2.17) can be further rewritten as

$$\chi \approx \exp\left\{j2\pi \cdot \frac{2\cdot}{\lambda}\mathbf{X}_s^T\Delta\mathbf{T}\right\} \cdot P\left(\frac{2}{c} \cdot \mathbf{X}_s^T\Delta\mathbf{T}\right) \cdot Q\left(\frac{2}{\lambda} \cdot \omega\mathbf{Y}_s^T\Delta\mathbf{T}\right) \tag{2.32}$$

where λ is the transmitted signal wavelength, the phase term of ambiguity function has no contribution to the system resolution, but it does play a significant role in the interferometric SAR performance analysis, which is out of the scope of this chapter. In the monostatic SAR, the transmitted signal is chirp signal, the bandwidth of transmitted signal is denoted as B, under the case of rectangular envelope, the function $P(\ )$ can be written as $\sin(\pi Bx)/(\pi Bx)$, thus the −3 dB width is approximately equal to $0.886/B$. Likewise, when the antenna power pattern in azimuth direction is approximately rectangular during observation time, the function $Q(*)$ can be written as $\sin(\pi T_c x)/(\pi T_c x)$, the −3 dB width is approximately equal to $0.886/T_a$, where T_a is coherent accumulative time which should consider the 'pseudo-spotlight' effect, namely that curved synthetic aperture will result in longer accumulation time. Based on (2.32), the resolution along different directions can be expressed as follows.

When P, Q satisfies the Sinc function, ambiguity function can be expressed as

$$\chi \approx \exp\left\{ j2\pi \cdot \frac{2\cdot}{\lambda} \mathbf{X}_{\mathbf{s}}^{T} \Delta\mathbf{T} \right\} \cdot \mathrm{sinc}\left(\frac{2B}{c} \mathbf{X}_{\mathbf{s}}^{T} \Delta\mathbf{T} \right) \cdot \mathrm{sinc}\left(\frac{2\omega T_a}{\lambda} \mathbf{Y}_{\mathbf{s}}^{T} \Delta\mathbf{T} \right) \tag{2.33}$$

−3 dB resolution can be written as

$$\begin{cases} \rho_r = 0.886 \dfrac{c}{2B} \\ \rho_a = 0.886 \dfrac{\lambda}{2\omega T_a} \end{cases} \tag{2.34}$$

According to (2.34), the azimuth resolution and range resolution formulas in GEO SAR are similar to that in LEO SAR, which also verifies the resolution analysis in GEO, but the Earth rotation shown in (2.30) must be considered in the calculation of rotation angle, it also verifies the correctness of the resolution analysis. Meanwhile, it is clear from (2.32) to (2.34) that the azimuth resolution is equal to the resolution along iso-range line direction, and the range resolution is equal to the resolution along iso-doppler line direction in slant range plane. Note that the conclusion only comes into existence in the slant range plane. In most cases, the resolution is considered in the ground plane. In GEO SAR case, the side-lobe direction of point spread function (PSF) in ground plane is not orthogonal, which is a big difference compared to LEO SAR.

2.2.1.2 Resolution on the Ground Plane

Figure 2.14 shows the projection of slant range plane to ground plane. The degree of squint looking can be represented with a squint angle θ or an azimuth angle ϕ. As shown in Fig. 2.14, O is the Earth center, S is satellite position, and T is the target position. Slant range vector $\mathbf{R_{SA}}$ in Fig. 2.13 was adapted to $\mathbf{R}$. The azimuth angle ϕ is the angle between the X axis and the projection direction of the slant range vector on the XY plane, varying from 0° to 360°. The squint angle θ is the angle between the slant range vector $\mathbf{R}$ and the zero-Doppler plane, and it satisfies

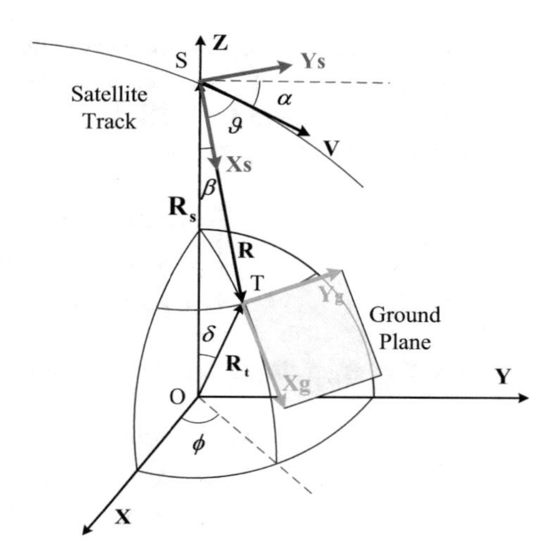

Fig. 2.14 Projection of slant range plane to ground plane

$$\sin\theta = \cos\vartheta = \cos\alpha\sin\beta\sin\phi + \sin\alpha\cos\beta \tag{2.35}$$

Vectors $\mathbf{X_s}$ and $\mathbf{Y_s}$ composes a slant range coordinate system. The azimuth vector $\mathbf{Y_s}$ is perpendicular to $\mathbf{X_s}$ on the slant range plane composed by the vectors $\mathbf{V}$ and $\mathbf{R}$.

Ground coordinate system $\mathbf{X_g Y_g}$ is on the ground plane that is tangential to the Earth surface at the target position T. The ground range vector $\mathbf{X_g}$ is along the projection direction of the slant range vector $\mathbf{R}$ on the ground plane, and the ground azimuth vector $\mathbf{Y_g}$ is perpendicular to $\mathbf{R}$ as well as $\mathbf{X_g}$. They can be expressed as

$$\begin{cases} \mathbf{Y_g} = \dfrac{\mathbf{R_t}\times\mathbf{R}}{|\mathbf{R_t}\times\mathbf{R}|} \\ \mathbf{X_g} = \dfrac{\mathbf{Y_g}\times\mathbf{R_t}}{|\mathbf{Y_g}\times\mathbf{R_t}|} \end{cases} \tag{2.36}$$

where $\mathbf{R_t}$ is the target position vector.

Based on the geometry of GEO SAR shown in Fig. 2.14 the adjacent positions vector can be written as

$$\begin{cases} \mathbf{T} = R_e\sin\delta\cos\phi\cdot\mathbf{X} + R_e\sin\delta\sin\phi\cdot\mathbf{Y} + R_e\cos\delta\cdot\mathbf{Z} \\ \mathbf{T}' = R_e\sin(\delta+\Delta\delta)\cos(\phi+\Delta\phi)\cdot\mathbf{X} \\ \qquad + R_e\sin(\delta+\Delta\delta)\sin(\phi+\Delta\phi)\cdot\mathbf{Y} + R_e\cos(\delta+\Delta\delta)\cdot\mathbf{Z} \end{cases} \tag{2.37}$$

where R_e is the Earth radius, δ is the Earth center angle, $\Delta\delta$ is the Earth center angle difference between two adjacent targets $\mathbf{T}'$ and $\mathbf{T}$, and $\Delta\phi$ are the azimuth angle difference.

In the resolution analyses, the target displacement $|\Delta \mathbf{T}|$ is on the same level as the ground resolution. The angle differences $\Delta\phi$ and $\Delta\delta$ are no larger than $|\Delta \mathbf{T}|/R_e$, which is approximate to 1.8×10^{-4} degrees for a resolution of 20 m. Therefore, the angle differences satisfy the approximation relationship as follows.

$$\begin{cases} \cos\Delta\phi \approx 1, & \sin\Delta\phi \approx \Delta\phi \\ \cos\Delta\delta \approx 1, & \sin\Delta\delta \approx \Delta\delta \\ \sin\Delta\delta\sin\Delta\phi \approx 0 \end{cases}. \tag{2.38}$$

Thus, the target displacement $\Delta\mathbf{T}$ can be written as

$$\begin{aligned} \Delta\mathbf{T} \approx\; & R_e(\Delta\delta\cos\delta\cos\phi - \Delta\phi\sin\delta\sin\phi)\mathbf{X} \\ & + R_e(\Delta\delta\cos\delta\sin\phi + \Delta\phi\sin\delta\cos\phi)\mathbf{Y} \\ & - R_e(\Delta\delta\sin\delta)\mathbf{Z} \end{aligned}. \tag{2.39}$$

The range vector $\mathbf{X_s}$ and the azimuth vector $\mathbf{Y_s}$ on the slant range plane are given by

$$\begin{cases} \mathbf{X_s} = \sin\beta\cos\phi\cdot\mathbf{X} + \sin\beta\sin\phi\cdot\mathbf{Y} - \cos\beta\cdot\mathbf{Z} \\ \mathbf{Y_s} = \dfrac{1}{\cos\theta}\begin{bmatrix} -\sin\beta\cos\phi\sin\theta\cdot\mathbf{X} \\ +(\cos\alpha - \sin\beta\sin\phi\sin\theta)\cdot\mathbf{Y} \\ +(-\sin\alpha + \cos\beta\sin\theta)\cdot\mathbf{Z} \end{bmatrix} \end{cases} \tag{2.40}$$

Combing (2.39) with (2.40), we can obtain the equation as follows

$$\begin{bmatrix} \mathbf{X_s^T}\Delta\mathbf{T} \\ \mathbf{Y_s^T}\Delta\mathbf{T} \end{bmatrix} \approx \begin{bmatrix} k_1 & 0 \\ k_2 & k_3\sin\delta \end{bmatrix} \begin{bmatrix} R_e\Delta\delta \\ R_e\Delta\phi \end{bmatrix} \tag{2.41}$$

where the coefficient k_1, k_2, and k_3 can be written as

$$\begin{cases} k_1 = \sin\eta \\ k_2 = \dfrac{\cos\eta(\cos\alpha\cos\beta\sin\phi - \sin\alpha\sin\beta)}{\cos\theta} \\ k_3 = \dfrac{\cos\alpha\cos\phi}{\cos\theta} \end{cases} \tag{2.42}$$

The incident angle η can be written as

$$\eta = \delta + \beta \tag{2.43}$$

where δ is the angle between the target position T and the $\mathbf{Z}$ axis. On the ground plane $\mathbf{X_g Y_g}$, the targets displacement $\Delta\mathbf{T}$ can be decomposed as $\mathbf{X_g^T}\Delta\mathbf{T}$ and $\mathbf{Y_g^T}\Delta\mathbf{T}$. They satisfies

$$\begin{bmatrix} \mathbf{X_g^T}\Delta\mathbf{T} \\ \mathbf{Y_g^T}\Delta\mathbf{T} \end{bmatrix} = \begin{bmatrix} 1 & 0 \\ 0 & \sin\delta \end{bmatrix} \begin{bmatrix} R_e\Delta\delta \\ R_e\Delta\phi \end{bmatrix} \tag{2.44}$$

Combining (2.41) with (2.44), we can get the projection relationship between the slant range plane and the ground plane as follows

$$\begin{bmatrix} \mathbf{X_s^T}\Delta\mathbf{T} \\ \mathbf{Y_s^T}\Delta\mathbf{T} \end{bmatrix} \approx \begin{bmatrix} k_1 & 0 \\ k_2 & k_3 \end{bmatrix} \begin{bmatrix} \mathbf{X_g^T}\Delta\mathbf{T} \\ \mathbf{Y_g^T}\Delta\mathbf{T} \end{bmatrix} \tag{2.45}$$

The ground GAF can be expressed as

$$\begin{aligned} \chi \approx & \exp\left\{ j2\pi \cdot \frac{2\cdot}{\lambda} \mathbf{X_s^{\mathit{T}}}\Delta\mathbf{T} \right\} \cdot \mathrm{sinc}\left(\frac{2B}{c} k_1 \mathbf{X_g^T}\Delta\mathbf{T} \right) \\ & \cdot \mathrm{sinc}\left[\frac{2\omega T_a}{\lambda} \left(k_2 \cdot \mathbf{X_g^T}\Delta\mathbf{T} + k_3 \cdot \mathbf{Y_g^T}\Delta\mathbf{T} \right) \right] \end{aligned} \tag{2.46}$$

Based on the projection relationship, the azimuth resolution ρ_{ga} and the range resolution ρ_{gr} on the ground can be expressed as

$$\begin{cases} \rho_{gr} = K_{gr}\rho_r \\ \rho_{ga} = K_{ga}\rho_a \end{cases} \tag{2.47}$$

where ρ_r and ρ_a are the −3 dB range resolution and azimuth resolution on the slant range plane. K_{gr} and K_{ga} are the range and azimuth projection coefficients, respectively. They can be written as

$$\begin{cases} K_{gr} = \dfrac{1}{k_1}\sqrt{1 + \dfrac{k_2^2}{k_3^2}} \\ K_{ga} = \dfrac{1}{|k_3|} \end{cases} \tag{2.48}$$

The angles between the sidelobe directions and the $\mathbf{Y_g}$ axis are given by

$$\begin{cases} \psi_r = \mathrm{atan}\left(-\dfrac{k_3}{k_2} \right) \\ \psi_a = 0 \end{cases} \tag{2.49}$$

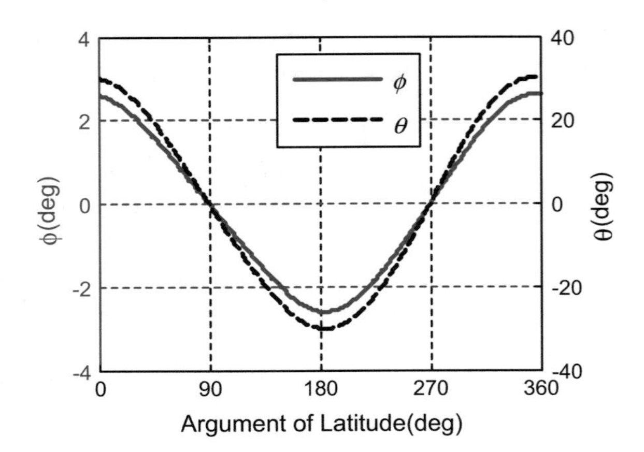

Fig. 2.15 Azimuth angle ϕ and squint angle θ corresponding to orthogonal sidelobes

Equation (2.49) indicates that the azimuth direction on the ground is always along the $\mathbf{Y_g}$ axis. Thus, the sidelobe angle ψ_r is also the angle between the range sidelobes and the azimuth sidelobes on the ground. Equation (2.49) also indicates that the sidelobes of GEO SAR may be non-orthogonal, which means the directions of range the range sidelobes and the azimuth sidelobes are not perpendicular. Equations (2.47) and (2.49) provide the values and directions of the range and azimuth ground resolutions, but they cannot describe the ground resolution completely.

According to (2.49), the sidelobes are orthogonal only when

$$\phi = \operatorname{asin}(\tan\alpha\tan\beta) \tag{2.50}$$

Besides, these orthogonal sidelobes correspond to a squint angle as follows

$$\theta = \operatorname{asin}\left(\frac{\sin\alpha}{\cos\beta}\right) \tag{2.51}$$

Figure 2.15 shows the azimuth angle and the squint angle which correspond to orthogonal sidelobes with an elevation angle of 4.5°. The used orbit parameters of GEO SAR are shown in Table 2.5. Because of the small elevation angle in GEO SAR, the azimuth angle is no larger than 2.6°. Meanwhile, the squint angle can reach 30°. This indicates that GEO SAR can obtain the orthogonal sidelobes by using a nearly zero azimuth angle, even if the squint angle is quite high. For other beam pointing directions, the sidelobes are normally non-orthogonal.

Table 2.5 Parameters of GEO SAR

Parameter	Value	Unit
Semi-major	42,164	km
Eccentricity	0.1	
Inclination	10	degree
Argument of perigee	90	degree
Band	L	

2.2.2 Power Budget

According to the radar equation, the received signal power can be expressed as [19, 20].

$$S = \frac{P_{av}\lambda^3 G^2 T_a B\sigma^0 A_{res}}{(4\pi)^3 R^4 L} \tag{2.52}$$

where P_{av} is the average transmitter power, G is the antenna gain, σ^0 is the normalized radar cross section, and L is the system and transmission loss factor, A_{res} is the target resolution cell area (RCA).

The RCA is written as

$$A_{res} = \rho_a \rho_{gr} = \frac{\rho_a \rho_r}{\sin\eta} \tag{2.53}$$

Currently, the calculation of transmitter power in GEO SAR is mainly based on (2.53), and does not consider the non-orthogonal sidelobes. For the non-orthogonal sidelobes, the resolution cell can be considered as a parallelogram with edge lengths of ρ_{gr} and ρ_{ga} as shown in Fig. 2.16b. The RCAs of the two type representations actually are the same. Thus, the RCA can be expressed as

$$A_{res} \approx \rho_{gr}\rho_{ga}\sin\psi_r = \frac{\rho_r\rho_a}{|k_1 k_3|} \tag{2.54}$$

According to (2.52) and (2.54), the required average transmitter power P_{av} can be expressed as

$$\begin{aligned} P_{av} &= \frac{(4\pi)^3 R^4 LkT_0 F_n}{\lambda^3 G^2 T_a \sigma_N A_{res}} \\ &= \frac{2(4\pi)^3 R^3 VLkT_0 F_n}{\lambda^3 G^2 \sigma_N \rho_{gr}} \cdot \frac{1}{K_{ga}\sin\psi_r} \end{aligned} \tag{2.55}$$

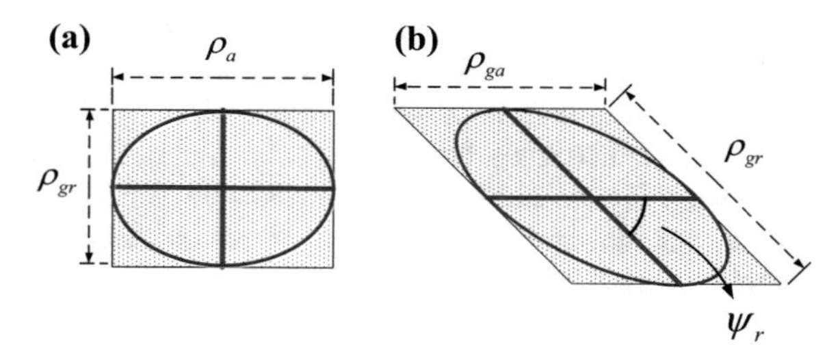

Fig. 2.16 Ground resolution cell **a** rectangle for orthogonal sidelobes **b** parallelogram for non-orthogonal sidelobes

Table 2.6 Power-related parameters of GEO SAR

Parameter	Value	Unit
System and transmission loss factor	5	dB
Receiver noise factor	5	dB
Antenna gain	47.7	dB
NESZ	−20	dB

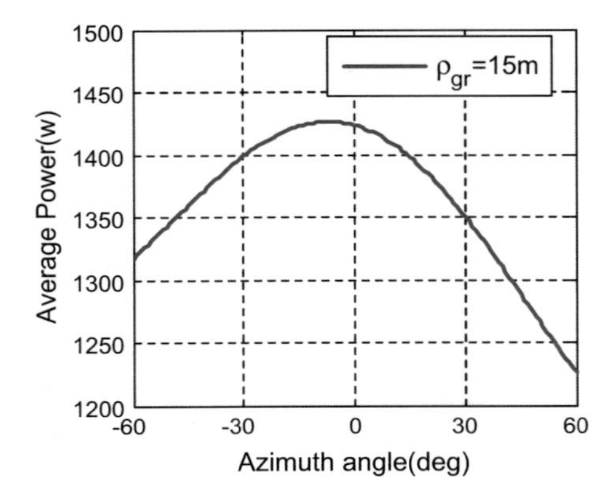

Fig. 2.17 Average power required for a resolution of 15 m and an elevation angle of 4.5°

where k is the Boltzmann's constant, F_n is the receiver noise factor, T_0 is the standard temperature, and σ_N is noise equivalent reflectivity.

With the parameters in Tables 2.5 and 2.6, the required average power for a resolution of 15 m on the ground plane and an elevation angle of 4.5° is shown in Fig. 2.17. Normally, the power requirements decrease when the squint angle increases due to the promotion of the RCA.

2.2.3 *Ambiguity*

2.2.3.1 Ambiguity Definition

The antenna of SAR system usually has directivity, that is, the antenna has different gain in different direction. The antenna pattern is usually used to describe this directivity. Based on the gain value, the antenna pattern is divided into the mainlobe and side lobe areas. Echo signal from antenna main lobe direction is expected by SAR receiver, and it is called main signal. Meanwhile, echo from side lobe direction also reaches the SAR receiver, superimposes on the main signal. Some echo from the sidelobe direction cannot be eliminated by data processing, and it is called the ambiguity signal. Processed by SAR system, ambiguity signal will interfere with the SAR image or cause fake target, which is called image ambiguity.

Ambiguity is divided into range ambiguity, azimuth ambiguity and cross ambiguity according to their source directions. Range ambiguity means the ambiguity signal has the same Doppler frequency with main signal, but the echo delay difference between the ambiguity signal and the main signal is an integer multiple of the pulse repetition time (PRT). Azimuth ambiguity means that the ambiguity signal has the same echo delay with main signal, but the Doppler frequency difference between the ambiguity signal and the main signal is integer multiples of pulse repetition frequency (PRF), which is the reciprocal of PRT. Cross ambiguity means the echo delay difference is the integer multiples of PRT, and Doppler frequency difference is the integer multiples of PRF.

Ambiguity-to-signal ratio (ASR), defined as the ratio between ambiguity signal and the main signal, is often used to describe SAR system's ambiguity. According to the difference of ambiguity source direction, ambiguity-to-signal ratio can be divided into the range ASR (RASR), azimuth ASR (AASR) and cross ASR (XASR), ASR is the sum of the three items. The definition of ASR is as follows.

$$ASR = \frac{\sum\limits_{\substack{m,n=-\infty \\ m\neq 0, n\neq 0}}^{\infty} \int_{-B_a/2}^{B_a/2} S\left(\tau + \frac{m}{PRF}, f + n \cdot PRF\right) \mathrm{d}f}{\int_{-B_a/2}^{B_a/2} S(\tau, f) \mathrm{d}f} \tag{2.56}$$

where S is the signal power, m denotes the range ambiguities number, and n denotes the azimuth ambiguities number.

When m or n is equal to zero and the other not, we obtain the azimuth ambiguities-to-signal ratio or the range ambiguities-to-signal ratio. If neither m nor n is equal to zero, we get the cross-term ambiguities. τ is the time delay for a range ambiguity number m, and f is the Doppler frequency for an azimuth ambiguity number n. τ_0 is the time delay for a zero range ambiguity number, and f_0 is Doppler frequency for a zero azimuth ambiguity number.

2.2.3.2 Calculation of Signal Power

Signal power can be expressed as

$$S(m,n) = G(m,n) \cdot \sigma(m) / R^4(m) \tag{2.57}$$

where G is the antenna gain, sigma is normalized scattering coefficient of typical target, R is the slant range, m is the range ambiguity number and n is the azimuth ambiguity number. The ambiguity numbers can be written as

$$\begin{cases} m = floor([\tau(m) - \tau_0]PRF) = floor\left(\frac{2}{c}[R(m) - R(0)]PRF\right) \\ n = floor\left[\frac{f(n) - f_0}{PRF}\right] \end{cases} \tag{2.58}$$

The main difficulty of signal power calculation lies on the calculations of the antenna gain and the Doppler frequency, which are discussed in the following part.

Antenna Gain

A Satellite local coordinates System (SCS) which is illustrated in detail in [16]. The unit vector of the beam direction in SCS can be denoted as

$$\mathbf{Q} = -\cos\theta_R \mathbf{u_r} + \sin\theta_R \cos\theta_A \mathbf{u_t} - \sin\theta_R \sin\theta_A \mathbf{u_p} \tag{2.59}$$

where θ_R and θ_A are described in Fig. 2.18.

The power gain of a rectangle array is calculated through

$$\begin{aligned} G(\theta_R, \theta_A) = \operatorname{sinc}^2&\left(\frac{L_a}{\lambda}(\sin\theta_R\cos\theta_A - \sin\theta_{R0}\cos\theta_{A0})\right) \\ &\cdot \operatorname{sinc}^2\left(\frac{L_r}{\lambda}(\sin\theta_R\sin\theta_A - \sin\theta_{R0}\sin\theta_{A0})\right) \end{aligned} \tag{2.60}$$

where θ_R and θ_A depict the normal direction of the antenna array. L_r and L_a are antenna size in range and azimuth direction.

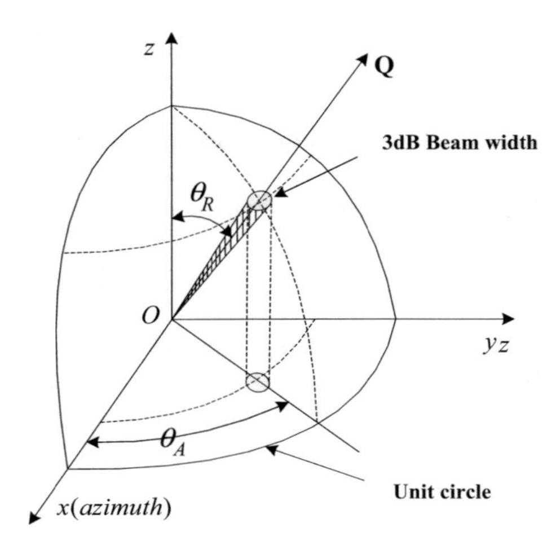

Fig. 2.18 Satellite local coordinates system

Doppler Frequency

For space-borne SAR, the Doppler frequency depends on the velocity vector relative to the target on the ground and the beam direction vector. The velocity vector relative to the target can be described in the Earth-centered rotation coordinates and then transformed into the Satellite local coordinates System (SCS).

The Doppler centroid frequency can be calculated through

$$\begin{aligned} f_{Dc} = &-\frac{2}{\lambda}\big[R_s(\omega_s - \omega_e \cos i)\sin\theta_R \cos\theta_A \\ &-\dot{R}_s \cos\theta_R - R_s\omega_e \sin i \cos\theta_{lat} \sin\theta_R \sin\theta_A\big] \end{aligned} \tag{2.61}$$

Calculation of ω_s, R_s and $\dot{R}_s$ could be found in [6]. ω_e is the angular velocity of Earth rotation. i is the orbit inclination. θ_{lat} is the argument of latitude, which could determine the position of the satellite.

2.2.3.3 Ambiguity Area and the ASR

Figure 2.19 shows an iso rang-Doppler sketch map. The black squares illustrate the beam shape of an antenna, in range and azimuth direction. The yellow squares are the primary ambiguities locations, and the black points represent other cross-term ambiguities, such as an ambiguity deviating from the useful echo signal by a PRF in azimuth frequency and a PRT in echo time. In bore-sight mode, the ambiguities locate in the two main lobes of antenna. A 30° squint mode is obtained after attitude control. However, the primary azimuth ambiguities would not lie in the main lobes of antenna, which might improve the performance of the primary azimuth ambiguities. By contrast, the cross-term ambiguities would lie in the main lobes of antenna, which would make cross-term ambiguities deteriorate sharply. The cross-term ambiguities could become more serious in space-borne SAR.

In high squint mode SAR, the XASR becomes comparable with RASR and cannot be neglected. When a high squint mode SAR system is designed or analyzed, XASR should be taken into account. The ambiguity performance of a typical X-band system is illustrated in Fig. 2.20.

Because of the non-orthogonality between the range direction and the azimuth direction, the ambiguities do not lie in the main lobe of antenna. Consequently, the XASRs deteriorate sharply, even worse than the RASR. In high squint mode, the XASR should be taken more attention in performance analysis. In simulations of this section, the XASRs are 15 dB worse than the RASR. In addition, the AASR do not keep the same in the whole swath as in the traditional stripmap mode. The reason is that the squint mode brings the inequality of Doppler centroid and the azimuth Doppler bandwidth within the whole swath of some beam orientation.

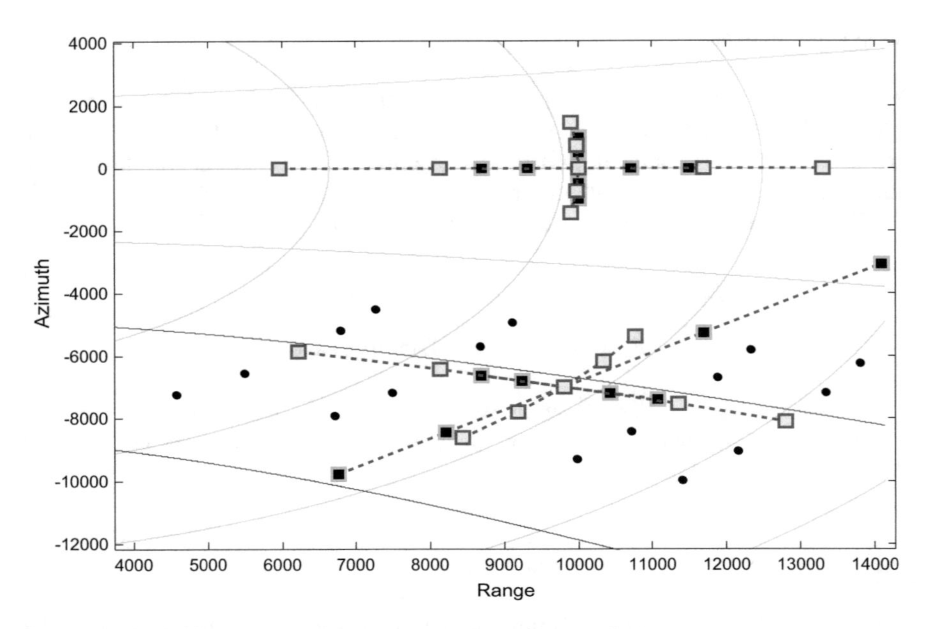

Fig. 2.19 Ground projections of beam shape and ambiguities locations

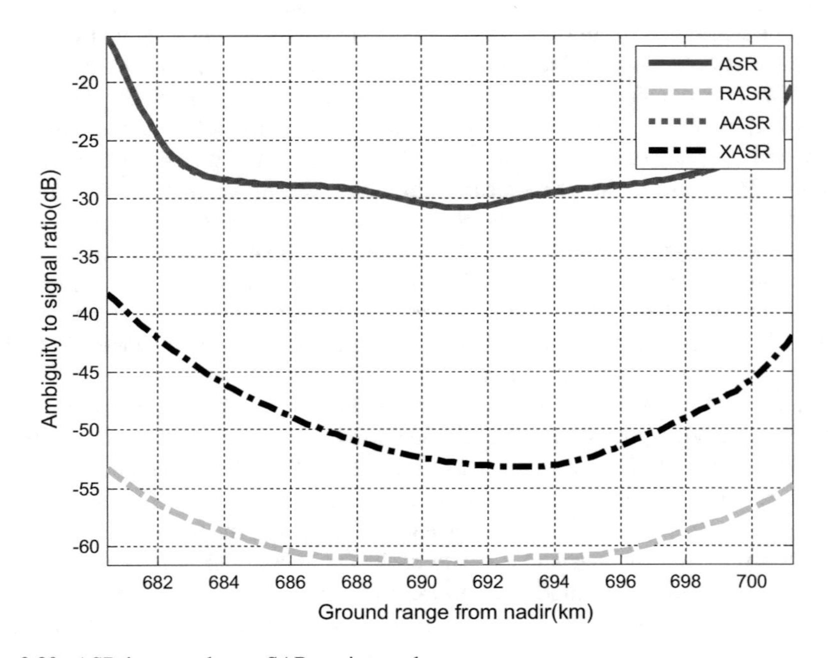

Fig. 2.20 ASR in space-borne SAR squint mode

2.3 Strategies of Attitude Steering

In spaceborne SAR, the non-zero Doppler centroid is induced by the Earth's rotation and the elliptical satellite orbit, and will cause a large range migration in echoes. Attitude steering is induced to steer the attitude of the satellite to change the direction of the beam and obtain zero Doppler centroid.

In 1986, Raney [21] first proposed a one-dimensional attitude steering method to compensate for the Doppler shift induced by Earth's rotation in the case of circular orbits. Subsequently, a method of two-dimensional (2D) attitude steering named Total Zero Doppler Steering (TZDS) is applied to LEO SAR [22], [23]. The method in [22] and [23] works well in the TerraSAR-X, and can reduce the Doppler centroid to a considerably low level.

However, the TZDS method does not work well in GEO SAR, and a large residual Doppler centroid exists, owing to the fact that the coupling of Earth's rotation and the elliptical orbit is not taken into account in the calculation of the pitch and yaw angles. Here an improved 2D attitude steering approach is derived considering the coupling effects between the Earth rotation and the elliptical orbit. Then, the employment of 2D phased array antenna scanning ability is presented to avoid the large angles steering and keep the platform stable.

In addition, GEO SAR has non-orthogonal and non-uniform resolution distribution, which is related to the beam direction. Thus, attitude steering can be used to change the beam direction and improve the resolution. At last, an optimal resolution steering method is proposed aiming to obtain the optimal resolution and decrease the attitude steering angle by pitch-yaw 2D steering.

2.3.1 2D Attitude Steering in Space-Borne SAR

The TZDS method discussed in this section is used to maintain zero Doppler centroid frequency. Conventional TZDS method is applied in LEO SAR (TerraSAR-X), but the effect of Earth rotation is ignored while the pitching angle is derived, so the absolute zero Doppler frequency is not satisfied. The effect of Earth rotation is weak in LEO SAR, thus the residual Doppler centroid is small and can be neglected. But the effect of Earth rotation is strong because the angular speed of the orbit and Earth rotation is similar and the residual Doppler centroid cannot be neglected. The improved TZDS method discussed in this section consider the Earth rotation and theoretically can attain zero Doppler centroid frequency.

The geometry of ECI coordinates is shown in Fig. 2.21. $\mathbf{R}_s$ and $\mathbf{R}_t$ are the position vectors of the satellite and the target. $\dot{\mathbf{R}}_s$ and $\dot{\mathbf{R}}_t$ are the first-order time derivative. The relative position vector between the satellite and the target is expressed as $\mathbf{R} = \mathbf{R}_s - \mathbf{R}_t$, and $R = |\mathbf{R}|$ is the slant range.

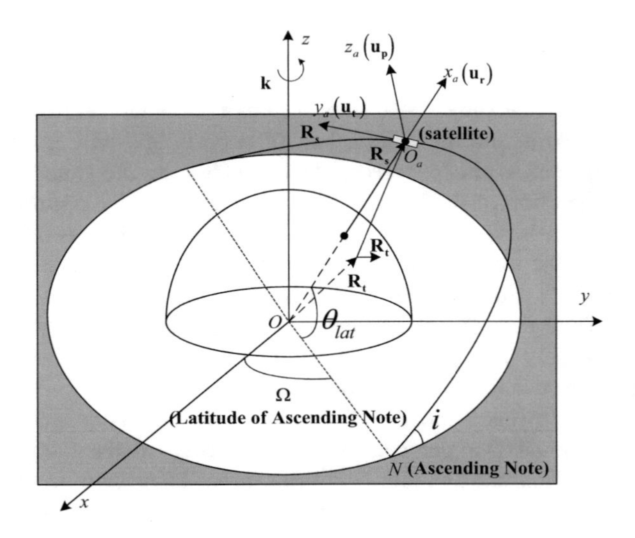

Fig. 2.21 The geometry of earth-centered inertial (ECI) coordinates and satellite local coordinates

The Doppler centroid f_{Dc} of a target fixed on the earth is given by [10]

$$f_{Dc} = (-2/\lambda R)\left[(\mathbf{R_s} - \mathbf{R_t}) \cdot \dot{\mathbf{R}}_\mathbf{s} - \mathbf{R_t} \cdot (\mathbf{R_s} \times \omega_e \mathbf{k})\right] \tag{2.62}$$

where **k** represents the vector of Earth's rotation pointing to the North Pole, ω_e is the rotational angular velocity of the Earth. According to properties of the vector cross product, we obtain

$$\mathbf{R_s} \cdot (\mathbf{R_s} \times \omega_e \mathbf{k}) = \mathbf{0} \tag{2.63}$$

As a result, (2.62) can be transformed into

$$f_{Dc} = (-2\mathbf{R}/\lambda R) \cdot \left[(\mathbf{R_s} \times \omega_e \mathbf{k}) + \dot{\mathbf{R}}_\mathbf{s}\right] \tag{2.64}$$

For the convenience of clarity, the vector P is denoted as

$$\mathbf{P} = (\mathbf{R_s} \times \omega_e \mathbf{k}) + \dot{\mathbf{R}}_\mathbf{s} \tag{2.65}$$

The P-vector, which is a function of the platform nadir position along the entire orbit, is fairly independent of platform attitudes and beam directions as well as the target status. It can be considered as the satellite-relative velocity vector in ECEF frame. The Doppler centroid frequency can be calculated through the projection of the vector P on the beam direction. Moreover P-vector which is uniquely determined by the satellite position plays an important role in analytical derivation of the 2D attitude steering angles, which will be illuminated greater detail in the next section. The parameters used in this section are listed in the Table 2.7.

Table 2.7 Orbit and antenna parameters of GEO SAR

Parameters	Specification	Unit
Semi major axis	42,164,170	m
Eccentricity	0.07	
Argument of perigee	270	Deg
Antenna fixed angle	4.65	Deg
Down looking angle	4.65	Deg
Antenna diameter	15	m

2.3.1.1 Doppler Property Analysis in SCS

The geometry of satellite local coordinate system (SCS) is shown in Fig. 2.21. The SCS can be defined as: the x-axis is along the satellite position vector away from the Earth center, the z-axis is along the angular momentum direction, and the y-axis obeys a right-hand rule. The unit vectors of the three axes are denoted as $\mathbf{u_r}$, $\mathbf{u_t}$ and $\mathbf{u_p}$ [6]. The beam direction of a broadside-looking space-borne SAR can be expressed as $\mathbf{Q} = \mathbf{R}/R$. The calculation formula of the Doppler centroid f_{Dc} can be rewritten as

$$f_{Dc} = -2\mathbf{P} \cdot \mathbf{Q}/\lambda \tag{2.66}$$

To find the analytic formula for f_{Dc}, we should obtain the denotation of **P** and **Q** in SCS. If the down-looking angle is β and the satellite is right-looking, **Q** can be expressed as

$$\mathbf{Q} = -\cos\beta\mathbf{u_r} - \sin\beta\mathbf{u_p} \tag{2.67}$$

Likewise, vector **P** can be expressed as

$$\mathbf{P} = (R_s\mathbf{u_r} \times \omega_e\mathbf{k}) + \left(\dot{R}_s\mathbf{u_r} + R_s\dot{\theta}_{lat}\mathbf{u_t}\right) \tag{2.68}$$

According to the transformation between the ECI and the SCS, we can obtain

$$\mathbf{k} = \sin i \sin\theta_{lat}\mathbf{u_r} + \sin i \cos\theta_{lat}\mathbf{u_t} + \cos i\mathbf{u_p} \tag{2.69}$$

where θ_{lat} is the argument of latitude, and i is the orbit inclination angle.

Then, vector **P** in SCS can be expressed as

$$\mathbf{P} = \dot{R}_s\mathbf{u_r} + R_s\left(\dot{\theta}_{lat} - \omega_e\cos i\right)\mathbf{u_t} + R_s\omega_e \sin \mathrm{i} \cos\theta_{lat}\mathbf{u_p} \tag{2.70}$$

Taking the attitude steering into account, the beam direction can be expressed as

$$\mathbf{Q}' = -\cos\beta\mathbf{u'_r} - \sin\beta\mathbf{u'_p} \tag{2.71}$$

Through the transformation matrix between the SCS and the SCS with attitude steering we can be obtain

$$\begin{aligned} \mathbf{u'_r} &= \cos\gamma \mathbf{u_r} - \sin\gamma\cos\varphi \mathbf{u_t} + \sin\gamma\sin\varphi \mathbf{u_p} \\ \mathbf{u'_p} &= \sin\varphi \mathbf{u_t} + \cos\varphi \mathbf{u_p} \end{aligned} \tag{2.72}$$

where φ is the yaw angle and γ is the pitch angle.

Consequently the Doppler centroid f_{Dc} considering satellite attitude steering or attitude errors can be modified as

$$f_{Dc} = -2\mathbf{P} \cdot \mathbf{Q}'/\lambda \tag{2.73}$$

2.3.1.2 Analytical Calculation of 2D Attitude Steering Angles

In general, $\mathbf{R} \cdot \mathbf{P}$ is not equal to $\mathbf{0}$ in the case of broadside-looking space-borne SAR, because of the influence caused by Earth's rotation and the elliptical orbit. That is to say, the relative velocity is not perpendicular to the beam direction. Fortunately, the zero-Doppler plane perpendicular to the relative velocity can be easily found. Through the control of yaw and pitch angles, beam direction can be located in the zero-Doppler plane, which minimizes f_{Dc} to zero.

In the space-borne SAR system, the antenna is fixed on the satellite platform. As a result, the beam direction changes as the platform rotates. After attitude steering, the beam direction will locate in a new plane which can be determined by vector $\mathbf{u'_r}$ and $\mathbf{u'_p}$. Through (2.66) to (2.73), the Doppler centroid can be compensated into zero if (2.74) are satisfied.

$$\mathbf{P} \cdot \mathbf{Q}' = \mathbf{0} \tag{2.74}$$

Equation (2.74) means the new plane coinciding exactly with the zero-Doppler plane. The solution of (2.74) gives the yaw and pitch steering angles.

$$\varphi = -\arctan\left(\frac{\omega_e \sin i \cos\theta_{lat}}{\dot{\theta}_{lat} - \omega_e \cos i}\right) \tag{2.75}$$

$$\theta = \arctan\left(k\frac{\dot{R}_s/R_s}{\sqrt{(\omega_e \sin i \cos\theta_{lat})^2 + \left(\dot{\theta}_{lat} - \omega_e \cos i\right)^2}}\right) \tag{2.76}$$

where

$$k = sign\left(\dot{\theta}_{lat} - \omega_e \cos \mathrm{i}\right) = \begin{cases} 1 & \dot{\theta}_{lat} - \omega_e \cos i > 0 \\ -1 & \dot{\theta}_{lat} - \omega_e \cos i < 0 \end{cases} \tag{2.77}$$

Likewise, the attitude steering angles can be calculated if performing the pitch steering firstly. There are different orders of the steering angles: one is yaw steering first and then pitch steering, and the other is pitch steering first and then yaw steering. Different attitude steering orders induce different transformation matrix, and the equations for yaw and pitch angles are different. As a result, the pitch and yaw steering differ from the yaw and pitch steering in magnitudes of the angles. The latter one is better because of the smaller pitch steering angle.

2.3.1.3 Performance Evaluation

In space-borne SAR system, the Doppler centroid is non-zero which varies along the orbit, as shown in Fig. 2.22. The variations are caused by Earth's rotation (ER) and the elliptical orbit (EO), both of which are very large in most orbit positions. The results in Fig. 2.22 show that neither of them can be neglected.

On the contrast, neither the traditional yaw method [21] (shown in Fig. 2.23) nor the TerraSAR-X (TSX) method [22, 23], (shown in Fig. 2.23) can reduce the Doppler centroid to zero completely. The Doppler residuals are magnified in the GEO SAR system parameters. In Fig. 2.23 the Doppler residuals can reach more than 800 Hz. The TSX method uses the same yaw steering angles as the traditional ones. Simulations here validate the discussion in Sect. 2.3.1: when attitude steering is performed, the effects of elliptical orbits or the coupling between earth and elliptical orbits cannot be neglected.

The approach presented could perfectly compensate the Doppler centroid to zero (shown in Fig. 2.23a). The yaw and pitch angles vary along the orbit position. There is no demand on attitude steering when the platform lies in the perigee or apogee, because the relative velocity is perpendicular to the beam direction. Figure 2.23a depicts pitch angles of the 2D attitude steering method, which can

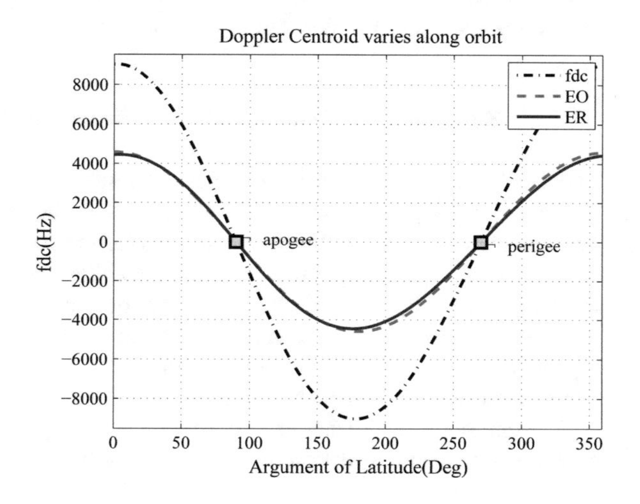

Fig. 2.22 Doppler property along the orbit

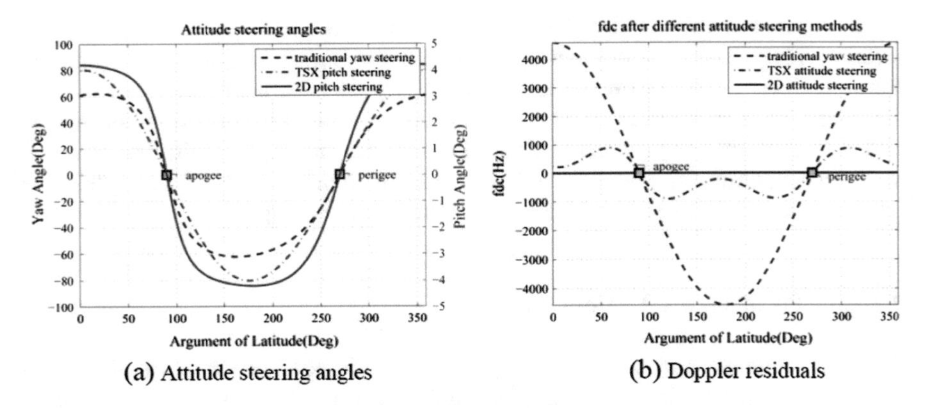

Fig. 2.23 Comparisons of three different attitude steering methods and their Doppler residuals. The dashed line is the traditional yaw steering angles (left) and the residuals (right) after it. TSX and 2D methods use the same yaw angles but different pitch angles. The dash-dot line represents the pitch angles of TSX method and the Doppler residuals after it. The solid line represents the 2D method and its perfect effect on Doppler centroid control

compensate the Doppler centroid frequency to zero. The yaw steering angles are same as the traditional ones.

The 2D attitude steering method is lack of operation in GEO SAR system, especially for large yaw angles of 60°, though it can theoretically compensate the Doppler centroid to zero. Several tens of degrees of yaw angles make the stabilization of GEO SAR unachievable within an acceptable time frame. In addition, the yaw angles vary along the orbit. Consequently attitude steering can hardly be achieved in real time through the platform rotation. In consequence, the next section will present a new zero-Doppler centroid method based on phase scan technique, which can achieve the equivalent effects as the 2D attitude steering method. In the new method, the phase scan angles can be calculated based on the 2D attitude steering angles, and the detailed derivations are shown in the next section.

2.3.1.4 Zero-Doppler Centroid Control Based on Phase Scan

The GEO SAR platform should be large enough to satisfy the needs of power and antenna size. However, the larger the platform is, the more difficult stabilization is. Attitude steering is difficult to perform with GEO SAR, especially under conditions of large yaw angle (shown in Fig. 2.23a). Another characteristic is the outstanding coverage capability, several degrees of the look angle for the whole Earth surface. Therefore phased array antenna is suitable for the GEO SAR. The 2D phase scan is a good substitution for the attitude steering in the zero-Doppler centroid control and can achieve the equivalent effects as the platform steering method.

The geometry of the antenna coordinate system [18] is shown in Fig. 2.24. The z-axis is the normal direction of the antenna arrays, the x-axis is along the antenna

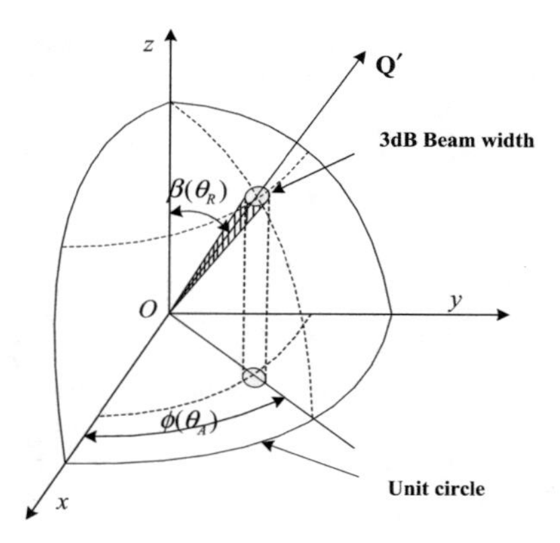

Fig. 2.24 The geometry of the antenna coordinates

azimuth direction and the y-axis is along the antenna elevation direction. They obey the right-hand rule. In this system, the angle between the normal direction (z-axis) and the nadir point is δ. In this coordinate system, the beam direction $\mathbf{Q}'$ can be uniquely derived from the azimuth angle ϕ and the elevation angle β, and furthermore can be represented by the satellite coordinate vectors as

$$\mathbf{Q_x} = \mathbf{u_t}, \mathbf{Q_y} = \sin\delta\mathbf{u_r} - \cos\delta\mathbf{u_p}, \mathbf{Q_z} = -\cos\delta\mathbf{u_r} - \sin\delta\mathbf{u_p}, \tag{2.78}$$

In (2.74), we have obtained the new beam direction $\mathbf{Q}'$ which located in the zero-Doppler plane after attitude steering based on (2.75) and (2.76). The same direction can be achieved by the 2D phase scan. The azimuth and elevation angles of phase scan can be calculated as follows.

$$\mathbf{Q}' = Q_x\mathbf{Q_x} + Q_y\mathbf{Q_y} + Q_z\mathbf{Q_z} \tag{2.79}$$

$$\begin{aligned} Q_x &= \cos\gamma\sin\theta\cos\varphi - \sin\gamma\sin\varphi \\ Q_y &= \sin\gamma\cos\delta\cos\varphi + \cos\gamma(\cos\delta\sin\theta\sin\varphi - \sin\delta\cos\theta) \\ Q_z &= \sin\gamma\sin\delta\cos\varphi + \cos\gamma(\cos\delta\cos\theta + \sin\delta\sin\theta\sin\varphi) \end{aligned} \tag{2.80}$$

Based on (2.79), the azimuth angle α and elevation angle β in the antenna coordinate system is

$$\phi = \arctan\left(Q_y/Q_x\right), \beta = \arcsin\left(\sqrt{Q_x^2 + Q_y^2}\right) \tag{2.81}$$

After we obtain the angles of ϕ and β through the 2D phase scan, the new beam direction $\mathbf{Q}'$ will lie in the zero-Doppler plane, and then the zero-Doppler centroid can be obtained. During the operation, β is the angle which is apart from the normal direction whose magnitude decides the performance of the method, such as the broadening of beam width or the loss of the antenna gain. The technique of phased array antenna makes the control of beam pointing easier and offers a feasible way to

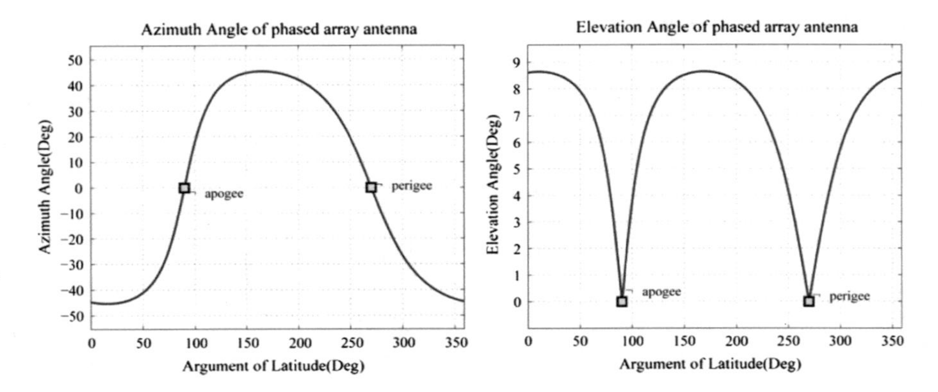

Fig. 2.25 Azimuth angle and elevation angle of phased array antenna along the orbit

GEO SAR zero-Doppler centroid control. The angle ϕ determines the scanning plane apart from the x-axis, which depicts the azimuth position of the beam direction and is irrelevant with the performance. As a result, there are no limits on the azimuth angle. The elevation angle is related to the performance of the antenna, such as beam width broadening and the gain loss. So the maximum elevation angle can be decided by the requirements of system performance.

The new method can realize the zero-Doppler centroid control, avoiding platform rotation. Phase scan angles, ϕ and β, can be obtained based on the 2D attitude steering angles (shown in Fig. 2.23a). The phase scan angles are shown in Fig. 2.25.

In simulations (shown in Fig. 2.25), elevation angle β is acceptable for the phase scan. Though azimuth angle ϕ is large, it describes the location of the beam pointing and has no influence on performance. In the GEO SAR system, the accurate beam pointing can be achieved through the amplitude and phase weighting in the phased array antenna. The 2D phase scan technique makes the zero-Doppler centroid realization feasible and reduces imaging difficulties. Therefore, besides GEO SAR, the proposed method is a good substitution for attitude steering method in the LEO or medium Earth orbit (MEO) SAR system.

2.3.2 *Optimized Resolution Attitude Control*

This section mainly discussed an optimal resolution steering (ORS) method for GEO SAR, which is used to optimize the resolution. In the ORS method, a beam pointing direction is derived that results in orthogonal two-dimensional sidelobes on the ground and optimal ground resolution. A roll-pitch strategy is derived to obtain small steering angles. The ORS method results in small Doppler variations in the range direction, which avoids aliasing of the Doppler spectra.

2.3.2.1 Ground Resolution Optimization

Ground resolution is an important parameter for SAR image. The two-dimensional sidelobes of the GAF on the ground are non-orthogonal due to the high eccentricity of GEO SAR. The ground resolution is not uniform; instead, it varies with direction. To describe the resolution characteristics completely, the ground resolution ellipse is introduced [24]. The ground resolution is represented by the area of an ellipse as follows:

$$S = \rho_a \rho_r \sqrt{\frac{1 - (\cos\phi \sin\beta \cos\varphi + \sin\alpha \cos\beta)^2}{[\cos\phi \sin(\beta + \delta) \sin\varphi]^2}} \tag{2.82}$$

where ρ_r and ρ_a are the range resolution and azimuth resolution on an inclined plane, respectively, and the angles are shown as Fig. 2.14. The resolutions are written as [25]

$$\begin{cases} \rho_r = 0.886 \cdot c/(2B) \\ \rho_a = 0.886 \cdot \lambda/(2\theta_{syn}) \end{cases} \tag{2.83}$$

where c is the speed of light, B is the signal bandwidth, λ is the wavelength, and θ_{syn} is the total synthetic rotation angle.

In (2.82), the diving angle α is determined by the orbital position. The elevation angle β and Earth center angle δ are determined by the illuminated area. The three angles have been illustrated in Fig. 2.14, and are also shown in Fig. 2.26. Therefore, the ground resolution area changes with the azimuth angle ϕ for a specific orbital position and a certain illuminated area. Take the derivation of (2.82) relative to the azimuth angle. When the derivation equals to zero, we can obtain the minimum ground resolution area, which can be expressed as

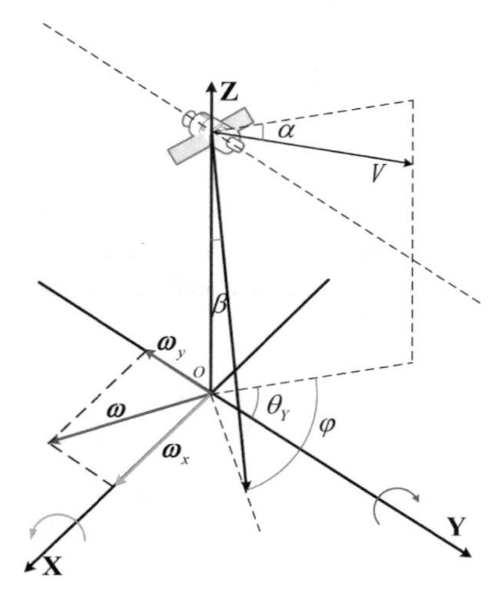

Fig. 2.26 Rotation strategy of the ORS method

$$S_{\min} = \pi \rho_a \rho_r / [4 \sin(\beta + \delta)] \tag{2.84}$$

The normalized ground resolution area is written as

$$S_N = \frac{S}{S_{\min}} = \sqrt{\frac{1 - (\cos\alpha \sin\beta \cos\phi + \sin\alpha \cos\beta)^2}{(\cos\alpha \sin\varphi)^2}} \tag{2.85}$$

The normalized ground resolution area is at least 1.0 and represents the degree of ground resolution deterioration.

2.3.2.2 Roll-Pitch Steering

To obtain optimal ground resolution with small steering angles, a new attitude control strategy is developed for GEO SAR. Because the ground resolution area is determined by the azimuth angle, we can derive an azimuth angle corresponding to the optimal ground resolution that satisfies

$$\mathrm{d}\left(S_N^2\right) / \mathrm{d}\varphi = 0 \tag{2.86}$$

Solving (2.86), we obtain the optimal azimuth angle

$$\varphi = \arccos(\tan\alpha \tan\beta) \tag{2.87}$$

According to (2.87), the optimal azimuth angle will change with the elevation angle along the swath. To eliminate this dependence, the median value of the elevation angle is used.

Using this approach, the beam can be steered to the optimal azimuth angle with small steering angles. Because the elevation angle in GEO SAR is less than 10°, we can rotate the beam directly from the nadir to the demanded position, as shown in Fig. 2.26.

The unit rotation vector is decomposed as follows:

$$\boldsymbol{\omega} = \boldsymbol{\omega}_{\mathbf{x}} + \boldsymbol{\omega}_{\mathbf{y}} \tag{2.88}$$

where $\boldsymbol{\omega}_x$ and $\boldsymbol{\omega}_y$ are the components in the X-axis and Y-axis, respectively. Therefore, the rotation is decomposed into a roll and pitch. If the platform is first rolled and then pitched to the demanded position, the transformation matrix is given by

$$T = \begin{bmatrix} \cos\vartheta_R & 0 & -\sin\vartheta_R \\ \sin\vartheta_P \sin\vartheta_R & \cos\vartheta_P & \sin\vartheta_P \cos\vartheta_R \\ \cos\vartheta_P \sin\vartheta_R & -\sin\vartheta_P & \cos\vartheta_P \cos\vartheta_R \end{bmatrix} \tag{2.89}$$

where ϑ_P is the pitch angle and ϑ_R is the roll angle in the ORS method. If the satellite is right-looking, the beam direction is expressed as

$$\mathbf{u} = [\sin\beta\sin(\varphi - \theta_Y), \sin\beta\cos(\varphi - \theta_Y), -\cos\beta]^T \tag{2.90}$$

where φ is the optimal azimuth angle and θ_Y is the yaw angle. With attitude steering, the beam direction is

$$\mathbf{u}' = [0, 0, -1]^T. \tag{2.91}$$

The rotation of the beam is described by

$$\mathbf{u}' = T\mathbf{u}. \tag{2.92}$$

Combining (2.89) through (2.92), we obtain the steering angles as follows:

$$\begin{cases} \vartheta_R = \arctan\{-\tan\beta \cdot \sin[\arccos(\tan\alpha\tan\beta) - \theta_Y]\} \\ \vartheta_P = \arcsin\{\sin\beta \cdot \cos[\arccos(\tan\alpha\tan\beta) - \theta_Y]\} \end{cases} \tag{2.93}$$

2.3.3 *Performance Comparison*

In this section, the performance of traditional yaw method, modified TZDS method and ORD method were compared, including the steering angle, the resolution area and the Doppler center frequency.

2.3.3.1 Theoretical Analysis

The differences in the beam directions and the ground resolutions of the ORS method, yaw steering method, and TZDS method are illustrated in Fig. 2.27. The beam directions are shown in the upper part of the figure, and the ground resolution are illustrated in the lower part.

In Fig. 2.27, point A is the satellite, point B represents the target, and the plane BCD is on the ground. Lines BD and CD represent the projections of the ECEF velocity vector and the slant range onto the ground plane, respectively. Line BC is the intersection of the slant plane and ground plane. The cross of solid lines represents the GAF in the slant plane, and the cross of dashed lines represents the GAF projected onto the ground plane.

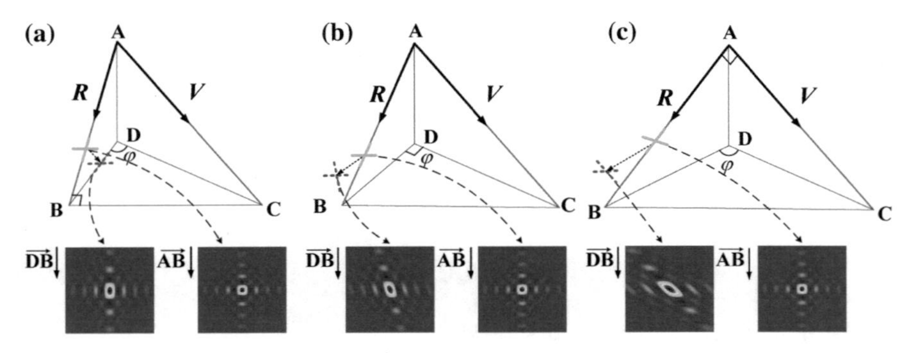

Fig. 2.27 Projection geometry with three attitude steering methods **a** the ORS method **b** the yaw steering method **c** the TZDS method

In the yaw steering method, the azimuth angle is 90°. In the TZDS method, the beam is perpendicular to the ECEF velocity vector in the slant plane in order to achieve a zero Doppler centroid. In the ORS method, the beam is perpendicular to the intersecting line of the ground and slant plane.

The two-dimensional sidelobes of the GAF in the slant plane are always orthogonal. However, they may or may not be orthogonal when projected onto the ground plane depending on the beam directions. Due to the different beam directions, the three methods obtain different ground resolution performance. In the ORS method, the two-dimensional sidelobes of the GAF in the ground plane are orthogonal, which results in optimal ground resolution. In the yaw steering method, the two-dimensional sidelobes in the ground plane are non-orthogonal, and the ground resolution is not optimal. In the TZDS method, because of the pitch angle in GEO SAR, the azimuth angle is considerably larger than that of the optimal method, which results in ground resolution deterioration.

Assuming no attitude steering and right side looking, the azimuth angle is written as

$$\varphi = 90^\circ + \theta_Y \tag{2.94}$$

where θ_Y is the yaw steering angle of TZDS method, and it is also the angle between the projections of the beam and ECEF velocity vector on the XOY plane. Because the elevation angle and diving angle are typically no more than 10° in GEO SAR, the normalized ground resolution area without attitude steering is expressed as

$$S_N = \frac{\sqrt{1 - (\cos\alpha \sin\beta \sin\theta_Y - \sin\alpha\cos\beta)^2}}{|\cos\alpha\cos\theta_Y|} \approx \frac{1}{|\cos\theta_Y|} \tag{2.95}$$

With yaw steering, the azimuth angle is 90°. The normalized ground resolution area is expressed as

$$S_N = \sqrt{1 - (\sin\alpha\cos\beta)^2} \Big/ |\cos\alpha| \approx 1 \tag{2.96}$$

The azimuth angle in the TZDS method satisfies

$$\varphi = \arccos(-\tan\alpha/\tan\beta) = \arccos(-\tan\theta_P/\tan\beta) \tag{2.97}$$

The normalized ground resolution area with the TZDS method is expressed as

$$S_N = \left| 1 \Big/ \left[\cos\alpha\sqrt{1 - (\tan\alpha/\tan\beta)^2}\right] \right| \tag{2.98}$$

The ORS method achieves optimal ground resolution, which is expressed as

$$S_N \approx 1. \tag{2.99}$$

With no attitude steering, the Doppler centroid can be written as

$$f_{dc} = \frac{2V_s}{\lambda}(-\cos\alpha\sin\beta\sin\theta_Y + \sin\alpha\cos\beta) \tag{2.100}$$

For yaw steering, the Doppler centroid can be written as

$$f_{dc} = \frac{2V_s}{\lambda}\sin\alpha\cos\beta \tag{2.101}$$

The residual Doppler centroid with the TZDS method is zero. The Doppler centroid with the ORS method is given by

$$f_{dc} = \frac{2V}{\lambda}\frac{\sin\alpha}{\cos\beta} \tag{2.102}$$

Because the diving angle α is non-zero in an elliptical orbit, the Doppler centroid using the ORS method and the yaw method is considerably large according to (2.102). The residual Doppler centroid will induce range migration, which must be considered in the imaging algorithm.

The ORS method and the yaw method can reduce the Doppler variation in the range direction. The Doppler variation in the range direction can be defined as the difference in the Doppler centroid at near and far ranges. The total Doppler bandwidth is the sum of the Doppler variation in the range direction and the target Doppler bandwidth. If the Doppler variation is sufficiently large in the range direction, the total Doppler bandwidth will exceed the pulse repetition frequency

(PRF). With the ORS method, the Doppler variation in the range direction is nearly zero, which avoids aliasing of the Doppler spectra.

2.3.3.2 Computer Simulation

The performance of the ORS method is evaluated in this section. The simulation parameters are shown in Table 2.8.

The steering angles with various attitude steering methods are shown in Fig. 2.28. The yaw angle in the yaw steering method and TZDS method varies from −68° to +68° in one orbital period. The steering angles in the ORS method are no greater than 7°, which are considerably smaller than those in the conventional methods.

Figure 2.29 presents the normalized ground resolution area with the various attitude steering methods in one orbital period. When the TZDS method is applied, the ground resolution is even worse than that without attitude steering. If the yaw steering method or ORS method is used, the ground resolution area will be nearly optimal throughout the orbital period.

Figure 2.30a presents the GAF on the slant plane, where the two-dimensional sidelobes are orthogonal. Figure 2.30b–e present the GAF in the ground plane with the various steering methods. When no attitude steering method or the TZDS method is used, the two-dimensional sidelobes nearly coincide, and there is significant ground resolution deterioration. When the yaw steering method or ORS method is implemented, the two-dimensional sidelobes in the ground plane are nearly orthogonal, and the ground resolution is nearly optimal.

The Doppler parameters with the various attitude steering methods are shown in Fig. 2.31. The TZDS method achieves a zero Doppler centroid; however, it does not reduce the Doppler variation in the range direction. In contrast, although the ORS method and yaw steering method produce a Doppler centroid of approximately 6000 Hz, the Doppler variation in the range direction is nearly zero.

Table 2.8 Simulation parameters

Parameter	Value	Unit
Semi-major axis a	42,164	km
Orbit inclination i	50	degrees
Eccentricity e	0.1	
Argument of perigee Ω	90	degrees
Wavelength λ	0.1	m
Elevation angle β	7	degrees
Slant range resolution ρ_r, ρ_a	1	m

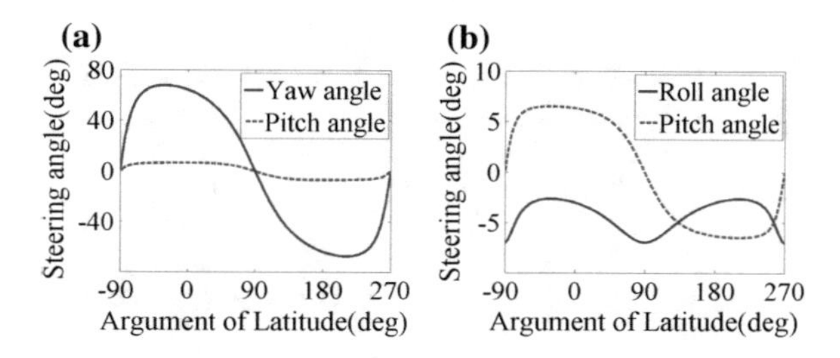

Fig. 2.28 Steering angles **a** yaw angle and pitch angle in the conventional steering methods **b** roll angle and pitch angle in the ORS method

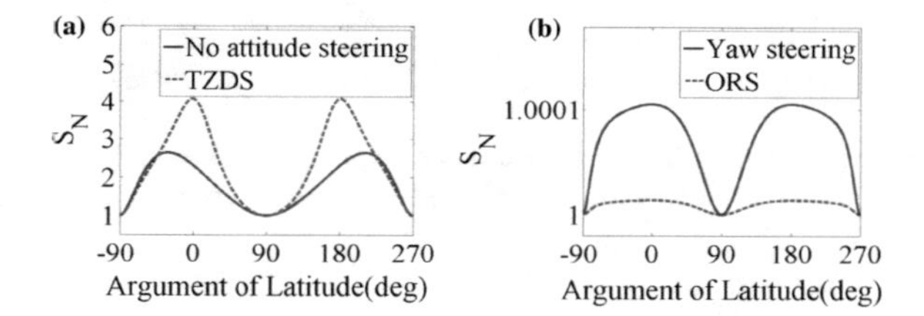

Fig. 2.29 Normalized ground resolution areas with various attitude steering methods **a** no attitude steering and the TZDS method **b** the yaw steering method and ORS method

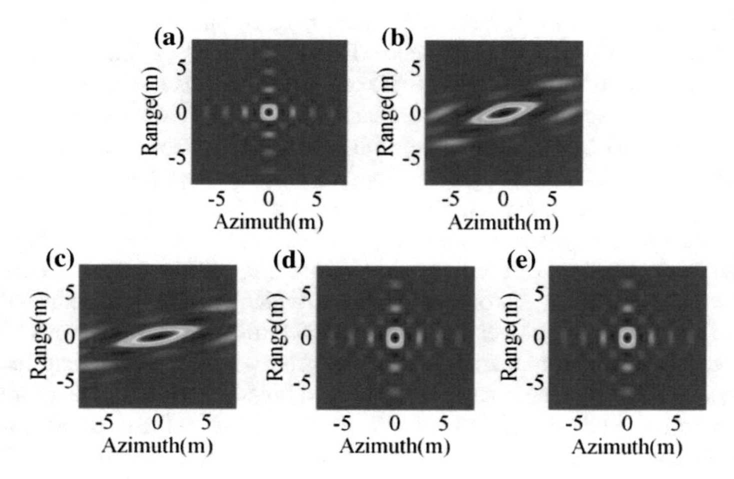

Fig. 2.30 GAF at the equator **a** in the slant plane **b** in the ground plane without attitude steering **c** in the ground plane with TZDS **d** in the ground plane with yaw steering **e** in the ground plane with ORS

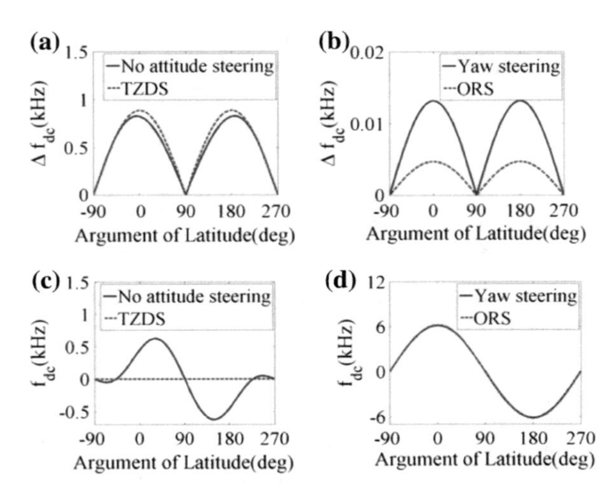

Fig. 2.31 Doppler parameters: (**a**) and (**b**) show the Doppler variation in the range direction with various attitude steering methods, and (**c**) and (**d**) show the Doppler centroid with the various attitude steering methods

2.4 Summary

Compared with the LEO SAR, GEO SAR is characterized by the fixed longitude range for observation, curved satellite track, high squint angle, serious effects of Earth rotation, and large imaging scenes. Due to the daily period and fixed longitude range, the revisit time of GEO SAR can be reduced to less than 16 h, and the coverage time for specific region can reach 14 h per day, which are much better than those of LEO SAR. Meanwhile, GEO SAR suffers from the curved satellite track, which produces non-negative Doppler rate and non-linear time-frequency relationship, and induces the complex work modes. These characteristics will affect the imaging processing and mode design of GEO SAR. Besides, the high squint mode and serious effects of Earth rotation induce non-orthogonal resolution feature and lifts cross ambiguity ratio of GEO SAR, which should be considered in the design of the resolution, power and ambiguities. Furthermore, the serious effects of Earth rotation induce a non-zero Doppler centroid and non-orthogonal resolution sidelobes, which can be compensated with the attitude steering. In this chapter, two attitude steering methods are introduced: the TZDS method uses a yaw-pitch 2-D steering to compensate for the effect of Earth rotation and obtain a zero Doppler centroid, while the ORS method utilizes a roll-pitch 2-D steering to optimize the resolution feature and reduce the required steering angle.

References

1. Tomiyasu K (1978) Synthetic aperture radar in geosynchronous orbit. In: Proc.Dig. Int. IEEE Antennas Propag. Symp., College Park, MD, May 1978, pp. 42–45
2. Ruiz-Rodon J, Broquetas A, Makhoul E, Monti Guarnieri A, Rocca F (2014) Nearly zero inclination geosynchronous SAR mission analysis with long integration time for earth observation. IEEE Trans Geosci Remote Sens 52(10):6379–6391
3. Gao Y, Hu C, Dong X, Long T (2012) Accurate system parameter calculation and coverage analysis in GEO SAR. In: Proc. EUSAR, Nuremberg, Germany, Apr 2012, pp 607–610
4. Hu C, Long T, Tian Y (2013) An improved nonlinear chirp scaling algorithm based on curve trajectory in geosynchronous SAR. Prog Electromagnet Res 135:481–513
5. Sun G, Xing M, Wang Y, Yang J, Bao Z (2014) A 2-D space-variant chirp scaling algorithm based on the RCM equalization and subband synthesis to process geosynchronous SAR data. IEEE Trans Geosci Remote Sens 52(8):4868–4880
6. Curlander JC, McDonough RN (1991) Synthetic aperture radar: systems and signal processing. Wiley, New York, MA
7. Madsen SN, Edelstein W, DiDomenico LD, LabBrecque J (2001) A geosynchronous synthetic aperture radar; for tectonic mapping, disaster management and measurements of vegetation and soil moisture. In Proc. IGARSS, Sydney, Australia, pp 447–449
8. Global earthquake satellite system: a 20-year plan to enable earthquake prediction. JPL Document, NASA and JPL, 2003 [Online]. Available: http://solidearth.jpl.nasa.gov/GESS/3123_GESS_Rep_2003.pdf
9. Rosenqvist A, Shimada M, Ito N, Watanabe M (2007) ALOS PALSAR: a pathfinder mission for global-scale monitoring of the environment. IEEE Trans Geosci Remote Sens 45(11): 3307–3316
10. Kou L, Wang X, Zhu M, Chong J, Xiang M (2010) Resolution analysis of circular SAR with partial circular aperture measurements. In Proc. EUSAR, Aachen, Germany, pp 1–4
11. Cumming IG, Wong FH (2005) Digital processing of synthetic aperture radar data. Artech House, Boston, MA
12. Li Z, Li C, Yu Z, Zhou J, Chen J (2011) Back projection algorithm for high resolution GEO-SAR image formation. In: Proc. IGARSS, Vancouver, BC, pp 336–339
13. Hu C, Long T, Liu Z, Zeng T, Tian Y (2014) An improved frequency domain focusing method in geosynchronous SAR. IEEE Trans Geosci Remote Sens 52(9):5514–5528
14. Kou L, Wang X, Xiang M, Chong J, Zhu M (2011) Effect of orbital errors on the geosynchronous circular synthetic aperture radar imaging and interferometric. J Zhejiang Univ Sci C 12(5):404–416
15. Kou L, Wang X, Xiang M, Zhu M (2012) Interferometric estimation of three-dimensional surface deformation using geosynchronous circular SAR. IEEE Trans Aerosp Electron Syst 48(2):1619–1635
16. Long T, Dong X, Hu C, Zeng T (2011) A new method of zero-Doppler centroid control in GEO SAR. IEEE Geosci Remote Sens Lett 8(3):513–516
17. Yu Z, Chen J, Li CS, Li Z, Zhang Y (2009) Concepts, properties and imaging technologies for GEO SAR. In: Proc. of SPIE, vol7494, pp 749407-1–749407-8
18. Skolnik M (1990) Radar handbook. McGraw Hill, New York
19. Prati C, Rocca F, Giancola D, Guarnieri AM (1998) Passive geosynchronous SAR system reusing backscattered digital audio broadcasting signals. IEEE Trans Geosci Remote Sens 36(6):1973–1976
20. Osipov IG, Neronskiy LB, Andrianov VI, Verba VS, Kozlov KV, Kurenkov VN, Pushkov DV (2006) Calculated performance of SAR for high orbit spacecraft using nuclear power supply. In: Proc. EUSAR, Dresden, Germany, pp 1–4
21. Raney RK (1986) Doppler properties of radar in circular orbits. Int J Remote Sens 7(9): 1153–1162

22. Fiedler H, Fritz T, Kahle R (2008) Verification of the total zero Doppler steering. In: Proc. IEEE Radar Conf, pp 340–342
23. Mittermayer J, Younis M, Metzig R, Wollstadt S, Márquez J, Meta A (2010) TerraSAR-X system performance characterization and verification. IEEE Trans Geosci Remote Sens 48(2):660–676
24. Long J, Yao D, Sun Y, Tian W (2013) The method of resolution analysis based on distinguishable ellipse in squinted SAR. Acta Electronic Sinica 41(12):2493–2498
25. Hu C, Long T, Zeng T, Liu F, Liu Z (2011) The accurate focusing and resolution and analysis method in geosynchronous SAR. IEEE Trans Geosci Remote Sens 49(10):3548–3562

Chapter 3
Algorithms for GEO SAR Imaging Processing

Abstract GEO SAR imaging processing faces serious difficulties because of the negative influence caused by high orbital altitude, curved trajectory, long synthetic aperture time (SAT) and large scene size, including complex slant range history, failure of "stop-and-go" assumption, two-dimensional (2D) spatially variant slant range model coefficients, etc. To solve these problems and achieve GEO SAR imaging, an accurate echo signal model, including the slant range model, the 2D spectrum and the spatially variant model of slant range coefficients based on the curved trajectory and Taylor series which take the error of "stop-and-go" assumption into account are first introduced in detail. Then, the time domain algorithm which can be used in any conditions is discussed detailedly, including its procedure flows and computational load. To improve the processing efficiency, the frequency domain algorithm is addressed. In this chapter, the difficulties of GEO SAR imaging in frequency domain are first analyzed in detail; then, the azimuth compensation which consists of a time domain compensation and a frequency domain compensation is derived to reduce the azimuth variance of the focus parameters and unfold the folded azimuth spectrum at certain specific orbital positions; after the azimuth compensation, a 2D nonlinear chirp scaling algorithm (NCSA) is introduced to finally obtain GEO SAR images; besides, the accuracy of the frequency algorithm, including its geometric distortion and available azimuth swath is discussed. At last, we summarize this chapter and give the conclusion.

T. Long et al., *Geosynchronous SAR: System and Signal Processing*,
https://doi.org/10.1007/978-981-10-7254-3_3

3.1 Introduction

Although GEO SAR has promising prospects of wide swath and short revisit time, realizing this still faces many problems, especially for imaging processing. For GEO SAR running on the inclined geosynchronous satellite orbit (IGSO), the propagation delay of signal can reach up to sub-seconds, the trajectory of GEO SAR is curved and the synthetic aperture time (SAT) of GEO SAR reaches up to hundreds of seconds. Therefore, the "stop-and-go" assumption is infeasible, the traditional "far-near-far" hyperbolic slant range model and the corresponding classical algorithms such as Range Doppler algorithm (RDA) [1–3], chirp scaling algorithm (CSA) [4–6], Omega-K algorithm (ωKA) [7] and Spectral Analysis (SPECAN) [8] are no longer valid. Furthermore, the swath width and the trajectory curvature of GEO SAR can be extremely huge and severe; thus, the slant range history of different targets is significantly different, and the slant range coefficients have severe 2D spatial variance, which increase the difficulty of imaging processing and must be taken into consideration. Besides, high squint mode which can further improve the revisiting time of GEO SAR greatly will be widely utilized in GEO SAR. This work mode will result in more curved trajectory, longer SAT and more severe space-variance. In summary, GEO SAR imaging processing faces problems of the significant spatial and temporal variance, and specific algorithms should be developed for GEO SAR imaging.

Time domain algorithms, such as back projection algorithm (BPA) [9–11], can be applied to any condition. Unfortunately, large swath width and long SAT of GEO SAR result in tremendous amount of echo data and further make the time domain algorithms inefficient and unsuitable for GEO SAR systems in engineering. Thus, to obtain excellent imaging efficiency, frequency domain algorithms will be widely used in GEO SAR imaging processing. Currently, most of the frequency domain algorithms for GEO SAR are the refinements of the classical airborne or low-earth-orbit (LEO) SAR algorithms. Usually, these algorithms utilize modified slant range model which can express the slant range history of GEO SAR precisely, such as refined hyperbolic slant range model [18], refined slant range model (RSRM) [25], slant range model based on Doppler parameters [12] and slant range model based on Taylor expansion [21]. Then, based on the slant range models, the refined frequency domain algorithms for GEO SAR imaging such as refined RDA [12–16], *ωKA* [17, 18], SPECAN [19, 20] and especially CSA [21–24] as well as nonlinear chirp scaling algorithm (NCSA) [25, 30, 31] are deducted.

Among all the modified slant range models, the refined hyperbolic slant range model is the most understandable one, and it is a simple refinement of the traditional hyperbolic slant range model which takes the influence of the curved trajectory on slant range model into consideration. However, this model cannot express the details of the complex slant range history of GEO SAR and the special "near-far-near" slant range history of GEO SAR at apogee. RSRM is a slant range model which can address these two difficulties. It adds linear, third as well as fourth order terms to the traditional hyperbolic slant range model to express the details of the slant range history; furthermore, this model considers the special "near-far-near" slant range history by appropriately change the sign of the second order term. Nevertheless, the order number of the RSRM is constant. Correspondingly, the higher order term (like fourth order term) of the RSRM may be redundant when the SAT as well as satellite trajectory are relatively short as well as straight and result in unnecessary computational load; furthermore, the RSRM may cannot express the details of the slant range history when the SAT as well as satellite trajectory are relatively long and curved due to the limited higher order terms (up to fourth order term). Thus, the slant range models based on Doppler parameters and Taylor expansion which can overcome all the aforementioned shortcomings have been widely used in GEO SAR signal modeling and imaging processing. These two kinds of slant range models have the same principle of series expansion. Through appropriately selecting the order number of series expansion, the best compromise between computational load and accuracy can be reached. The differences of them are the expressions of their slant range model coefficients. The former uses the Doppler parameters such as Doppler centroid and Doppler rate, and the latter uses the satellite and earth motion parameters, such as velocity and acceleration.

Based on these proposed slant range models, many modified frequency domain algorithms are utilized to obtain well focused GEO SAR images. However, these modifications of the algorithms are based on the refinement of the slant range model, and the essences of these algorithms do not change. Thus, the processing steps and typical characteristics, such as the capability of dealing with the spatially variant focus parameters, of the traditional frequency domain algorithms are inherited by these modified algorithms. Different from traditional airborne/LEO SAR which have 2D spatial invariant or only range variant as well as focus parameters, the focus parameters of GEO SAR have severe nonlinear 2D spatial variance which will be introduced detailedly later. Consequently, the modified RDA, ωKA, SPECAN and CSA which can only handle the spatially invariant or linear variant focus parameters cannot completely meet the requirements of GEO SAR imaging with long SAT, curved trajectory and large scene size. Because the NCSA which is the refinement of CSA can effectively cope with the nonlinear spatial variant of focus parameters, the modified 2D NCSA based on the improved slant range models are deeply introduced in this chapter.

In this chapter, the slant range model based on Taylor expansion and the corresponding refined 2D NCSA will be introduced. Section 3.2 introduces the echo signal model which consists of the accurate slant range model, 2D spectrum and the 2D spatially variant model of slant range coefficients based on the "non-stop-and-go" assumption. Then, Sect. 3.3 discussed the process flow and the computational budget of the BPA. Section 3.4 proposes and derives the azimuth compensation which can simplify the imaging processing and the 2D NCSA which achieve GEO SAR images with large scene size based on the analysis of the difficulties existing in frequency-domain GEO SAR imaging. Finally, the summary is given in Sect. 3.5.

3.2 Echo Signal Model

3.2.1 Error Analysis of the "Stop-and-Go" Assumption

A classical SAR echo signal model is created according to the "stop-and-go" assumption, as shown in Fig. 3.1a, namely, the transmitting point and receiving point can be considered the same position. However, with respect to the GEO SAR, the orbit height is 36,000 km, and the slant range is near the order of magnitude of 40,000 km. Then, the two-way propagation time delay approaches to the order of magnitude of seconds (such as 0.25−0.3 s); furthermore, the velocity of the satellite can be up to 3000 km/s. Thus, the range model or the signal model based on the "stop-and-go" assumption cannot fully reveal the behavioral characteristics of GEO SAR due to the great variation of the slant range when the radar and targets are moving. Therefore, the transmitting point is different from the receiving point in GEO SAR, as indicated in Fig. 3.1b. This difference will cause unneglected error and result in significant negative effect on GEO SAR imaging. Thus, the error caused by the "stop-and-go" assumption will be addressed in detail.

The SAR acquisition geometry is shown in Fig. 3.2. $O{-}xyz$ is the geocentric inertial coordinate system, and the Earth's center is set as origin O, The satellite position vector (SPV) and the velocity vector can be denoted as $\vec{\mathbf{r}}_{sn}$ and $\vec{\mathbf{v}}_{sn}$ respectively. During the process of signal propagation, the velocity vector can be approximately considered constant. Likewise, at the nth pulse transmission moment, the target position vector (TPV) and the velocity vector can be denoted as $\vec{\mathbf{r}}_{gn}$ and $\vec{\mathbf{v}}_{gn}$, respectively. During the data acquisition, the SAR platform transmits the signal at the nth PRT, and both the platform and the target are moving against the signal propagation. The transmitted signal impinges the target after propagation time τ_1 and is then reflected from the target. At this moment, the SPV and the TPV can be denoted as $\vec{\mathbf{r}}_{sn,\tau_1}$ and $\vec{\mathbf{r}}_{gn,\tau_1}$, respectively. The reflected signal reaches the satellite after propagation time τ_2. At the moment of $nT + \tau_1 + \tau_2$, the SPV and the

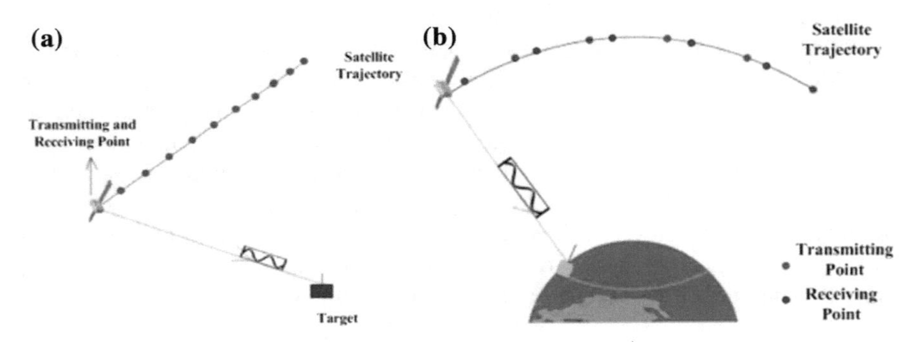

Fig. 3.1 Receiving point and transmitting point with: **a** "stop-and-go" assumption and straight trajectory, **b** non "stop-and-go" assumption and curved trajectory

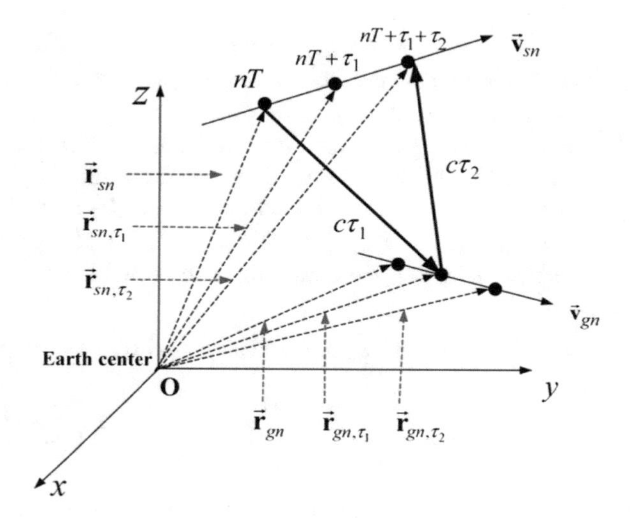

Fig. 3.2 The illustration of propagation range at nth PRT moment in GEO SAR

TPV can be denoted as $\overrightarrow{\mathbf{r}}_{sn,\tau_1}$ and $\overrightarrow{\mathbf{r}}_{gn,\tau_1}$, respectively. It should be noted that all vectors mentioned above are row vectors.

According to Fig. 3.2, the total propagation range can be expressed as $R_{sn} = c \cdot \tau_1 + c \cdot \tau_2$, and

$$\begin{aligned} R_1 = c\tau_1 = \left\| \overrightarrow{r}_{sn} - \overrightarrow{r}_{gn,\tau_1} \right\| \\ R_2 = c\tau_2 = \left\| \overrightarrow{r}_{gn,\tau_1} - \overrightarrow{r}_{sn,\tau_2} \right\| \end{aligned} \tag{3.1}$$

where c stands for the velocity of light.

The accurate roundtrip range can be written as

$$R_{sn}(t) = R_1(t) + R_2(t) = \left\| \overrightarrow{r}_{sn} - \overrightarrow{r}_{gn,\tau_1} \right\| + \left\| \overrightarrow{r}_{gn,\tau_1} - \overrightarrow{r}_{sn,\tau_2} \right\| \tag{3.2}$$

where the symbol '$\|\cdot\|$' is the norm operator.

It is known that $R_n(t)$ is related not only to slow time nT but also to fast time t, whereas it is only in correlation with the slow time in LEO SAR. Then, the relationships between $R_1(t)$, $R_2(t)$, and the fast time are

$$R_1(t) = \left\| \overrightarrow{\mathbf{r}}_{sn} - \overrightarrow{\mathbf{r}}_{gn} \right\| + \left(-\overrightarrow{\mathbf{v}}_{gn} \overrightarrow{\mathbf{u}}_{gs,n}^{T} \right) t + \frac{1}{2} \frac{\overrightarrow{\mathbf{v}}_{gn} \left(\mathbf{I} - \overrightarrow{\mathbf{u}}_{gs,n}^{T} \overrightarrow{\mathbf{u}}_{gs,n} \right) \overrightarrow{\mathbf{v}}_{gn}^{T}}{\left\| \overrightarrow{\mathbf{r}}_{sn} - \overrightarrow{\mathbf{r}}_{gn} \right\|} t^2 \tag{3.3}$$

$$R_2(t) = \left\| \overrightarrow{\mathbf{r}}_{gn,\tau_1} - \overrightarrow{\mathbf{r}}_{sn,\tau_2} \right\| + \left(-\overrightarrow{\mathbf{v}}_{sn} \overrightarrow{\mathbf{u}}_{gs,\tau_1,n}^{T} \right) t + \frac{1}{2} \frac{\overrightarrow{\mathbf{v}}_{sn} \left(\mathbf{I} - \overrightarrow{\mathbf{u}}_{gs,\tau_1,n}^{T} \overrightarrow{\mathbf{u}}_{gs,\tau_1,n} \right) \overrightarrow{\mathbf{v}}_{sn}^{T}}{\left\| \overrightarrow{\mathbf{r}}_{gn,\tau_1} - \overrightarrow{\mathbf{r}}_{sn,\tau_1} \right\|} t^2 \tag{3.4}$$

where t is corresponding to τ_1 and τ_2, $\vec{\mathbf{u}}_{gs,n} = \vec{\mathbf{r}}_{sn} - \vec{\mathbf{r}}_{gn} / \left\| \vec{\mathbf{r}}_{sn} - \vec{\mathbf{r}}_{gn} \right\|$ is the direction vector of the range between the target and satellite platform at the nth pulse moment, and $\vec{\mathbf{u}}_{gs,\tau_1,n} = \vec{\mathbf{r}}_{sn,\tau_1} - \vec{\mathbf{r}}_{gn,\tau_1} / \left\| \vec{\mathbf{r}}_{sn,\tau_1} - \vec{\mathbf{r}}_{gn,\tau_1} \right\|$ is the direction vector of the range between the target and satellite platform at the $nT + \tau_1$th pulse moment. In addition, it should be noted that the Taylor expansion of $R_2(t)$ is at moment τ_1. Thus, it is assumed that τ_1 is $R_2(t)$ at zero moment.

As $\left\| \vec{\mathbf{r}}_{gn,\tau_1} - \vec{\mathbf{r}}_{sn,\tau_1} \right\| = \left\| \left(\vec{\mathbf{r}}_{gn} - \vec{\mathbf{r}}_{sn} \right) + \left(\left(\vec{\mathbf{v}}_{gn} - \vec{\mathbf{v}}_{sn} \right) \tau_1 \right) \right\|$, after further expansion of $R_2(t)$, the total range can be written as

$$\begin{aligned} R_{sn}(t) = & 2\left\| \vec{\mathbf{r}}_{sn} - \vec{\mathbf{r}}_{gn} \right\| + \left(\vec{\mathbf{v}}_{sn} - \vec{\mathbf{v}}_{gn} \right) \vec{\mathbf{u}}^T_{gs,n} t + \frac{\vec{\mathbf{v}}_{sn} \left(\vec{\mathbf{v}}_{sn} - \vec{\mathbf{v}}_{gn} \right)^T}{\left\| \vec{\mathbf{r}}_{sn} - \vec{\mathbf{r}}_{gn} \right\|} \frac{t^2}{4} \\ & + \frac{\vec{\mathbf{v}}_{gn} \left(\mathbf{I} - \vec{\mathbf{u}}^T_{gs,n} \vec{\mathbf{u}}_{gs,n} \right) \vec{\mathbf{v}}^T_{gn}}{8\left\| \vec{\mathbf{r}}_{sn} - \vec{\mathbf{r}}_{gn} \right\|} \frac{t^2}{4} + \frac{1}{8} \left[\frac{\vec{\mathbf{v}}_{gn} \left(\mathbf{I} - \vec{\mathbf{u}}^T_{gs,n} \vec{\mathbf{u}}_{gs,n} \right) \vec{\mathbf{v}}^T_{sn}}{\left\| \vec{\mathbf{r}}_{sn} - \vec{\mathbf{r}}_{gn} \right\|} \right. \\ & \left. + \frac{\left(\vec{\mathbf{v}}_{sn} - \vec{\mathbf{v}}_{gn} \right) \left(\mathbf{I} - \vec{\mathbf{u}}^T_{gs,n} \vec{\mathbf{u}}_{gs,n} \right) \left(\vec{\mathbf{v}}_{sn} - \vec{\mathbf{v}}_{gn} \right)^T}{\left\| \vec{\mathbf{r}}_{sn} - \vec{\mathbf{r}}_{gn} \right\|} \right] t^2 \end{aligned} \tag{3.5}$$

In order to process conveniently in the imaging algorithm, the one-way propagation range is rewritten as (3.5) and express with $R_n(t)$ as follows:

$$\begin{aligned} R_n(t) = & \left\| \vec{\mathbf{r}}_{sn} - \vec{\mathbf{r}}_{gn} \right\| + \frac{\left(\vec{\mathbf{v}}_{sn} - \vec{\mathbf{v}}_{gn} \right) \vec{\mathbf{u}}^T_{gs,n} t}{2} + \frac{\vec{\mathbf{v}}_{sn} \left(\vec{\mathbf{v}}_{sn} - \vec{\mathbf{v}}_{gn} \right)^T}{\left\| \vec{\mathbf{r}}_{sn} - \vec{\mathbf{r}}_{gn} \right\|} \frac{t^2}{8} \\ & + \frac{\vec{\mathbf{v}}_{gn} \left(\mathbf{I} - \vec{\mathbf{u}}^T_{gs,n} \vec{\mathbf{u}}_{gs,n} \right) \vec{\mathbf{v}}^T_{gn}}{16\left\| \vec{\mathbf{r}}_{sn} - \vec{\mathbf{r}}_{gn} \right\|} t^2 + \frac{1}{16} \left[\frac{\vec{\mathbf{v}}_{gn} \left(\mathbf{I} - \vec{\mathbf{u}}^T_{gs,n} \vec{\mathbf{u}}_{gs,n} \right) \vec{\mathbf{v}}^T_{sn}}{\left\| \vec{\mathbf{r}}_{sn} - \vec{\mathbf{r}}_{gn} \right\|} \right. \\ & \left. + \frac{\left(\vec{\mathbf{v}}_{sn} - \vec{\mathbf{v}}_{gn} \right) \left(\mathbf{I} - \vec{\mathbf{u}}^T_{gs,n} \vec{\mathbf{u}}_{gs,n} \right) \left(\vec{\mathbf{v}}_{sn} - \vec{\mathbf{v}}_{gn} \right)^T}{\left\| \vec{\mathbf{r}}_{sn} - \vec{\mathbf{r}}_{gn} \right\|} \right] t^2 \end{aligned} \tag{3.6}$$

Equation (3.6) shows the relationship between the range variation and fast time t in a single pulse. Because fast time t is on the order of milliseconds, the range error is mainly determined by the first term in (3.6) as follows:

$$\Delta R_n(t) = \frac{\left(\vec{\mathbf{v}}_{sn} - \vec{\mathbf{v}}_{gn} \right) \vec{\mathbf{u}}^T_{gs,n} t}{2} \tag{3.7}$$

Equation (3.7) is a fixed value at the fixed PRT. However, with regard to different PRTs, the migration is varied. Equation (3.7) can be further expressed as

$$\Delta R_n(t) = \frac{\left(\vec{\mathbf{v}}_{sn} - \vec{\mathbf{v}}_{gn} \right) \vec{\mathbf{u}}^T_{gs,n}}{2} \cdot \frac{2\left\| \vec{\mathbf{r}}_{sn} - \vec{\mathbf{r}}_{gn} \right\|}{c} = \frac{\left(\vec{\mathbf{v}}_{sn} - \vec{\mathbf{v}}_{gn} \right) \left(\vec{\mathbf{r}}_{sn} - \vec{\mathbf{r}}_{gn} \right)^T}{c} \tag{3.8}$$

Equation (3.8) is the function of the slow time related to pulse number n. Supposing that the pulse serial number is from $-N$ to N, and the aperture center moment (ACM) corresponds to the index $n = 0$, Δr_n varies with the pulse number. The Taylor series expansion in (3.8) on the slow time at the ACM is

$$\Delta R_n = \Delta R + \Delta k_1(nT) + \Delta k_2(nT)^2 + \Delta k_3(nT)^3 + \cdots \tag{3.9}$$

where

$$\Delta R = \frac{\left(\vec{\mathbf{v}}_{s0} - \vec{\mathbf{v}}_{g0}\right)\left(\vec{\mathbf{r}}_{s0} - \vec{\mathbf{r}}_{g0}\right)^T}{c} \tag{3.10}$$

$$\Delta k_1 = \frac{\left(\vec{\mathbf{a}}_{s0} - \vec{\mathbf{a}}_{g0}\right)\left(\vec{\mathbf{r}}_{s0} - \vec{\mathbf{r}}_{g0}\right)^T + \left(\vec{\mathbf{v}}_{s0} - \vec{\mathbf{v}}_{g0}\right)\left(\vec{\mathbf{v}}_{s0} - \vec{\mathbf{v}}_{g0}\right)^T}{c} \tag{3.11}$$

$$\Delta k_2 = \frac{\left(\vec{\mathbf{b}}_{s0} - \vec{\mathbf{b}}_{g0}\right)\left(\vec{\mathbf{r}}_{s0} - \vec{\mathbf{r}}_{g0}\right)^T + 3\left(\vec{\mathbf{a}}_{s0} - \vec{\mathbf{a}}_{g0}\right)\left(\vec{\mathbf{v}}_{s0} - \vec{\mathbf{v}}_{g0}\right)^T}{2c} \tag{3.12}$$

$$\Delta k_3 = \frac{\left(\vec{\mathbf{d}}_{s0} - \vec{\mathbf{d}}_{g0}\right)\left(\vec{\mathbf{r}}_{s0} - \vec{\mathbf{r}}_{g0}\right)^T + 4\left(\vec{\mathbf{b}}_{s0} - \vec{\mathbf{b}}_{g0}\right)\left(\vec{\mathbf{v}}_{s0} - \vec{\mathbf{v}}_{g0}\right)^T}{6c} + \frac{\left\|\vec{\mathbf{a}}_{s0} - \vec{\mathbf{a}}_{g0}\right\|^2}{2c} \tag{3.13}$$

In (3.10)–(3.13), $\vec{\mathbf{r}}_{s0}$ and $\vec{\mathbf{r}}_{g0}$ represent the SPV and the TPV at the ACM, respectively; $\vec{\mathbf{v}}_{s0}$ and $\vec{\mathbf{v}}_{g0}$ denotes the velocity vectors of the satellite and the target at the ACM, respectively; $\vec{\mathbf{a}}_{s0}$ and $\vec{\mathbf{a}}_{g0}$ are the satellite acceleration vector and the target acceleration vector at the ACM, respectively; $\vec{\mathbf{b}}_{s0}$ and $\vec{\mathbf{b}}_{g0}$ are the time derivatives of $\vec{\mathbf{a}}_{s0}$ and $\vec{\mathbf{a}}_{g0}$ at the ACM, respectively; $\vec{\mathbf{d}}_{s0}$ and $\vec{\mathbf{d}}_{g0}$ are the time derivatives of $\vec{\mathbf{b}}_{s0}$ and $\vec{\mathbf{b}}_{g0}$ at the ACM, respectively.

As shown in the range difference simulated by the second-order approximation and third-order approximation of the "stop-and-go" assumption in Fig. 3.3, the error caused by the "stop-and-go" assumption and the second-order approximations is very small, and in particular, the error caused by the third-order approximation is nearly zero. Thus, the error caused by the "stop-and-go" assumption can be fully derived from (3.9).

According to (3.10)–(3.13), the differences of the Doppler centroid, the Doppler FM rate, and the derivative of the Doppler FM rate caused by the range difference can be expressed as

$$\Delta f_{dc} = \frac{2}{c\lambda}\left[\left(\vec{\mathbf{a}}_{s0} - \vec{\mathbf{a}}_{g0}\right)\left(\vec{\mathbf{r}}_{s0} - \vec{\mathbf{r}}_{g0}\right)^T + \left(\vec{\mathbf{v}}_{s0} - \vec{\mathbf{v}}_{g0}\right)\left(\vec{\mathbf{v}}_{s0} - \vec{\mathbf{v}}_{g0}\right)^T\right] \tag{3.14}$$

$$\Delta f_{dr} = \frac{2}{c\lambda}\left[\left(\vec{\mathbf{b}}_{s0} - \vec{\mathbf{b}}_{g0}\right)\left(\vec{\mathbf{r}}_{s0} - \vec{\mathbf{r}}_{g0}\right)^T + 3\left(\vec{\mathbf{a}}_{s0} - \vec{\mathbf{a}}_{g0}\right)\left(\vec{\mathbf{v}}_{s0} - \vec{\mathbf{v}}_{g0}\right)^T\right] \tag{3.15}$$

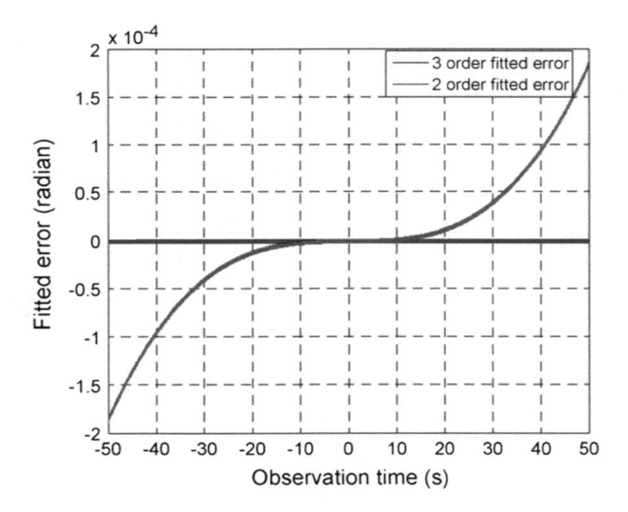

Fig. 3.3 Error of fitting caused by the second-order approximation and third-order approximation of the "stop-and-go" assumption

$$\Delta f_{drr} = \frac{2}{c\lambda}\left[\begin{array}{r}\left(\overrightarrow{\mathbf{d}}_{s0} - \overrightarrow{\mathbf{d}}_{g0}\right)\left(\overrightarrow{\mathbf{r}}_{s0} - \overrightarrow{\mathbf{r}}_{g0}\right)^T + 3\left\|\overrightarrow{\mathbf{a}}_{s0} - \overrightarrow{\mathbf{a}}_{g0}\right\|^2 \\ +4\left(\overrightarrow{\mathbf{b}}_{s0} - \overrightarrow{\mathbf{b}}_{g0}\right)\left(\overrightarrow{\mathbf{v}}_{s0} - \overrightarrow{\mathbf{v}}_{g0}\right)^T\end{array}\right] \tag{3.16}$$

where λ is the radar wavelength. Thus, the "stop-and-go" error can cause the position shift, the increase in the peak-to-side lobe ratio (PSLR), the asymmetry of the left side lobe and the right side lobe.

3.2.2 AccurateSlant Range Model in GEO SAR

The general geometric structure of GEO SAR is shown in Fig. 3.4. Taking the error caused by the "stop-and-go" assumption into account, the accurate range in GEO SAR can be written as

$$\begin{aligned} R_n =& \left\|\overrightarrow{\mathbf{r}}_{sn} - \overrightarrow{\mathbf{r}}_{gn}\right\| + \Delta R_n = (R+\Delta R) + (k_1+\Delta k_1)(nT) + (k_2+\Delta k_2)(nT)^2 \\ &+ (k_3+\Delta k_3)(nT)^3 + k_4(nT)^4 + \cdots \end{aligned} \tag{3.17}$$

which is called the curved trajectory model based on Taylor series, where ΔR, Δk_1,Δk_2 and Δk_3 depend on (3.10)–(3.13), and

$$R = \left\|\overrightarrow{\mathbf{r}}_{s0} - \overrightarrow{\mathbf{r}}_{g0}\right\| \tag{3.18}$$

$$k_1 = \overrightarrow{\mathbf{v}}_{s0} - \overrightarrow{\mathbf{v}}_{g0} \cdot \frac{\left(\overrightarrow{\mathbf{r}}_{s0} - \overrightarrow{\mathbf{r}}_{g0}\right)^T}{\left\|\overrightarrow{\mathbf{r}}_{s0} - \overrightarrow{\mathbf{r}}_{g0}\right\|} \tag{3.19}$$

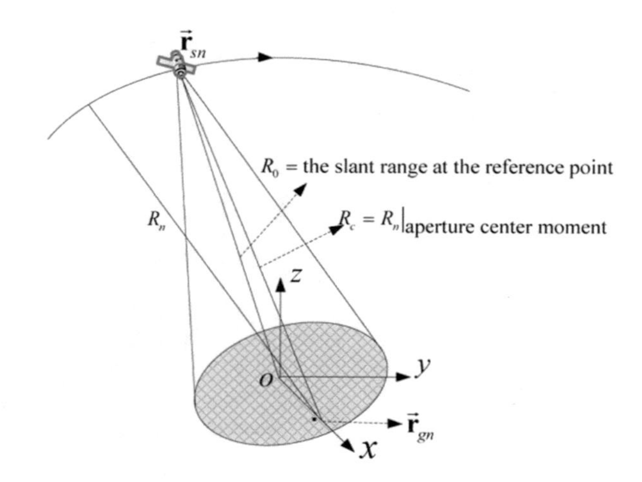

Fig. 3.4 General geometric structure of GEO SAR

$$k_2 = \frac{(\vec{\mathbf{a}}_{s0} - \vec{\mathbf{a}}_{g0})(\vec{\mathbf{r}}_{s0} - \vec{\mathbf{r}}_{g0})^T + \|\vec{\mathbf{v}}_{s0} - \vec{\mathbf{v}}_{g0}\|^2}{2\|\vec{\mathbf{r}}_{s0} - \vec{\mathbf{r}}_{g0}\|} - \frac{\left[(\vec{\mathbf{v}}_{s0} - \vec{\mathbf{v}}_{g0})(\vec{\mathbf{r}}_{s0} - \vec{\mathbf{r}}_{g0})^T\right]^2}{2\|\vec{\mathbf{r}}_{s0} - \vec{\mathbf{r}}_{g0}\|^3} \tag{3.20}$$

$$\begin{aligned} k_3 = & \frac{(\vec{\mathbf{b}}_{s0} - \vec{\mathbf{b}}_{g0})(\vec{\mathbf{r}}_{s0} - \vec{\mathbf{r}}_{g0})^T + 3(\vec{\mathbf{a}}_{s0} - \vec{\mathbf{a}}_{g0})(\vec{\mathbf{v}}_{s0} - \vec{\mathbf{v}}_{g0})^T}{6 \cdot \|\vec{\mathbf{r}}_{s0} - \vec{\mathbf{r}}_{g0}\|} \\ & + \frac{\left[(\vec{\mathbf{v}}_{s0} - \vec{\mathbf{v}}_{g0})(\vec{\mathbf{r}}_{s0} - \vec{\mathbf{r}}_{g0})^T\right]^3}{2 \cdot \|\vec{\mathbf{r}}_{s0} - \vec{\mathbf{r}}_{g0}\|^5} - \frac{(\vec{\mathbf{v}}_{s0} - \vec{\mathbf{v}}_{g0})(\vec{\mathbf{r}}_{s0} - \vec{\mathbf{r}}_{g0})^T \|\vec{\mathbf{v}}_{s0} - \vec{\mathbf{v}}_{g0}\|^2}{2 \cdot \|\vec{\mathbf{r}}_{s0} - \vec{\mathbf{r}}_{g0}\|} \\ & - \frac{(\vec{\mathbf{v}}_{s0} - \vec{\mathbf{v}}_{g0})(\vec{\mathbf{r}}_{s0} - \vec{\mathbf{r}}_{g0})^T (\vec{\mathbf{a}}_{s0} - \vec{\mathbf{a}}_{g0})(\vec{\mathbf{r}}_{s0} - \vec{\mathbf{r}}_{g0})^T}{2 \cdot \|\vec{\mathbf{r}}_{s0} - \vec{\mathbf{r}}_{g0}\|^3} \end{aligned} \tag{3.21}$$

$$\begin{aligned} k_4 = & \frac{(\vec{\mathbf{d}}_{s0} - \vec{\mathbf{d}}_{g0})(\vec{\mathbf{r}}_{s0} - \vec{\mathbf{r}}_{g0})^T + 4(\vec{\mathbf{b}}_{s0} - \vec{\mathbf{b}}_{g0})(\vec{\mathbf{v}}_{s0} - \vec{\mathbf{v}}_{g0})^T}{24 \cdot \|\vec{\mathbf{r}}_{s0} - \vec{\mathbf{r}}_{g0}\|} \\ & + \frac{\|\vec{\mathbf{v}}_{s0} - \vec{\mathbf{v}}_{g0}\|^2}{8\|\vec{\mathbf{r}}_{s0} - \vec{\mathbf{r}}_{g0}\|} - \frac{k_2^2 + 2k_1k_3}{2\|\vec{\mathbf{r}}_{s0} - \vec{\mathbf{r}}_{g0}\|} \end{aligned} \tag{3.22}$$

Furthermore, for the sake of simplicity, the slant range model can be rewritten as

$$R(t_a) = (R_0 + \Delta R) + (k_1 + \Delta k_1)t_a + (k_2 + \Delta k_2)t_a^2 + (k_3 + \Delta k_3)t_a^3 + k_4 t_a^4 + \cdots \tag{3.23}$$

where t_a is the azimuth time.

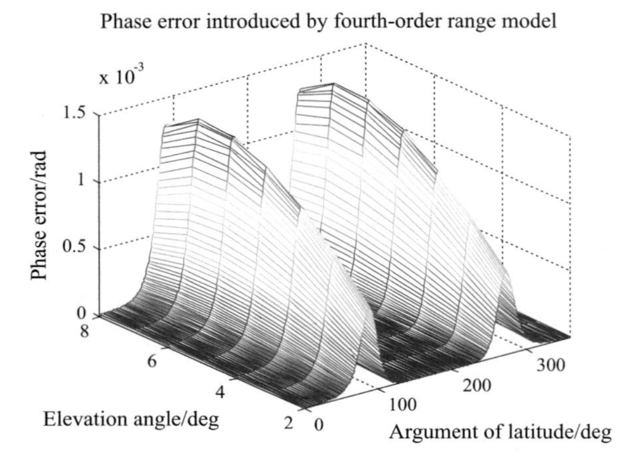

Fig. 3.5 Phase error introduced by the fourth-order slant range model with ground squint angle of 60°

To ensure that the fourth-order slant range model shown in (3.17) can meet the accuracy requirement of GEO SAR imaging, computer simulation of the phase error introduced by this model with ground squint angle of 60° was conducted. The orbital parameters are shown in Fig. 3.1, the required azimuth resolutions is 20 m, the wavelength is 0.24 m and the simulation result is shown in Fig. 3.5.

As shown in Fig. 3.5, the maximum phase error is far below $\pi/4$ rad which is the threshold value of the phase error, which obviously shows that the fourth-order slant range model is sufficiently precise to meet the needs of high squint GEO SAR imaging.

3.2.3 Two-Dimensional Spectrum of Echo Signal

In accordance with the curved trajectory model, classical imaging algorithms cannot be directly used for image formation in GEO SAR. Due to the expansion of the curved trajectory model, series reversion can be used to deduce the two-dimensional (2D) spectrum expression.

In the light of the SAR principle, the echo signal model can be expressed as

$$\begin{aligned} s(t_r, t_a) =& \sigma a_r(t_r - 2R(t_a)/c) \cdot a_n(t_a) \\ & \exp\left[j\pi K_r(t_r - 2R(t_a)/c)^2\right] \cdot \exp(-j4\pi R(t_a)/\lambda) \end{aligned} \tag{3.24}$$

where σ is the scattering coefficient; t_r is the range time, $a_r(\cdot)$ and $a_a(\cdot)$ stand for the range envelope and the azimuth envelope in time domain, respectively; $R(t_a)$ is the accurate range depending on (3.23) and K_r is the range FM rate.

Note that $R_0 + \Delta R$ and $(k_1 + \Delta k_1) - (k_3 + \Delta k_3)$ are abbreviated as R and $k_1 - k_3$ for the sake of simplicity, and this abbreviation will not affect following derivation.

The range-frequency expression of the echo signal can be obtained from the range fast Fourier transform (FFT) of (3.24) and can be expressed as

$$\begin{aligned} s(f_r, t_a) &= \sigma A_r(f_r) a_a(t_a) \exp\left(-\frac{j\pi f_r^2}{K_r}\right) \exp\left[-\frac{j4\pi R(t_a)(f_r + f_c)}{c}\right] \\ &= s_1(f_r, t_a) \exp\left[-\frac{j4\pi (f_r + f_c)}{c} k_1 t_a\right] \end{aligned} \tag{3.25}$$

where $A_r(\cdot)$ is the envelope function in the range-frequency domain, f_r is the range frequency, f_c is the carrier frequency, and $f_c = c/\lambda$, In (3.25), the first exponential term is the range modulation, and the second exponential term embodies the couple of the range signal and the azimuth signal. Accordingly, it is necessary to transform (3.25) into the 2D frequency domain. However, prior to the azimuth FFT, the following transformation has to be implemented:

$$s(f_r, t_a) = s_1(f_r, t_a) \exp\left[-\frac{j4\pi (f_r + f_c)}{c} k_1 t_a\right] \tag{3.26}$$

The linear phase in $s_1(f_r, t_a)$ is removed, compared with $s(f_r, t_a)$, to obtain the 2D spectrum of $s_1(f_r, t_a)$ using the series reversion. Finally, the 2D spectrum expression of $s(f_r, t_a)$ can be obtained through the following:

$$s_1(f_r, t_a) \Leftrightarrow s_1(f_r, f_a) \tag{3.27}$$

$$s_1(f_r, t_a) \exp\left[-\frac{j4\pi (f_r + f_c)}{c} k_1 t_a\right] \Leftrightarrow s_1\left(f_r, f_a + \frac{2k_1(f_r + f_c)}{c}\right) \tag{3.28}$$

$$s(f_r, t_a) \Leftrightarrow s(f_r, f_a) = s_1\left(f_r, f_a + \frac{2k_1(f_r + f_c)}{c}\right) \tag{3.29}$$

where f_a is the azimuth frequency, and $\Leftrightarrow$ denotes the Fourier transformation.

The next step is the azimuth FFT of $s_1(f_r, t_a)$. The corresponding integrated phase is

$$\Theta(f_r, t_a) = -\pi \frac{f_r^2}{K_r} - \frac{4\pi R'(t_a)}{c}(f_r + f_c) - 2\pi f_a(t_a) \tag{3.30}$$

where

$$R'(t_a) = R(t_a) - k_1 t_a = R_0 + k_2 t_a^2 + k_3 t_a^3 + k_4 t_a^4 \tag{3.31}$$

Through the derivation of (3.30), we can get

$$2k_2 t_a + 3k_3 t_a^2 + 4k_4 t_a^3 = \frac{c f_a}{2(f_r + f_c)} \tag{3.32}$$

Let us define

$$t_a = A_1\left[-\frac{cf_a}{2(f_r+f_c)}\right] + A_2\left[-\frac{cf_a}{2(f_r+f_c)}\right]^2 + A_3\left[-\frac{cf_a}{2(f_r+f_c)}\right]^3 \tag{3.33}$$

The following results are obtained by substituting (3.33) into (3.32):

$$\begin{aligned} A_1 &= 1/2k_2 \\ A_2 &= -3k_3/8k_2^3 \\ A_3 &= 9\left(k_3^2 - 4k_2k_4\right)/16k_2^5 \end{aligned} \quad . \tag{3.34}$$

Thus

$$\begin{aligned} s_1(f_r,f_a) = \sigma A_r(f_r)A_a(f_a)\exp\left(-j\frac{\pi f_r^2}{K_r}\right)\exp\left(-j\frac{4\pi(f_r+f_c)R}{c}\right) \\ \exp\left\{j2\pi\left[\begin{array}{c} +\frac{\pi}{4k_2}\left(\frac{c}{2(f_r+f_c)}\right)f_a^2 + \frac{k_3}{8k_2^3}\left(\frac{c}{2(f_r+f_c)}\right)^2 f_a^3 \\ +\frac{\left(9k_3^2-4k_2k_4\right)c^3}{64k_2^5}\left(\frac{c}{2(f_r+f_c)}\right)^3 f_a^4 \end{array}\right]\right\} \end{aligned} \tag{3.35}$$

Then, the 2D spectrum can be obtained from (3.29) and can be written as

$$\begin{aligned} s(f_r,f_a) = \sigma A_r(f_r)A_a(f_a)\exp\left(-j\frac{\pi f_r^2}{K_r}\right)\exp\left(-j\frac{4\pi(f_r+f_c)R}{c}\right) \\ \exp\left\{j2\pi\left[\begin{array}{r} +\frac{\pi}{4k_2}\left(\frac{c}{2(f_r+f_c)}\right)\left[f_a+\frac{2k_1(f_r+f_c)}{c}\right]^2 \\ +\frac{k_3}{8k_2^3}\left(\frac{c}{2(f_r+f_c)}\right)^2\left[f_a+\frac{2k_1(f_r+f_c)}{c}\right]^3 \\ +\frac{\left(9k_3^2-4k_2k_4\right)c^3}{64k_2^5}\left(\frac{c}{2(f_r+f_c)}\right)^3\left[f_a+\frac{2k_1(f_r+f_c)}{c}\right]^4 \end{array}\right]\right\} \end{aligned} \tag{3.36}$$

According to (3.36), there is a serious range-azimuth coupling in 2D spectrum, thus it is necessary to decouple and compensate spatial variation if we want to focus a large scene. After derivation and simplification to (3.36), we can obtain 2D spectrum as

$$\begin{aligned}
S_1(f_r, f_a) = & A_r\left(\frac{f_r}{K_r}\right)A_a\left[f_a + \tfrac{2k_1}{c}\cdot(f_r + f_c)\right] \\
& \underbrace{\exp\left[j2\pi\phi_{az}(f_a, R)\right]}_{\text{azimuth modulation term}}\underbrace{\exp(j2\pi\phi_{RP}(R))}_{\text{residual phase term}} \\
& \underbrace{\exp(j2\pi b(f_a)\cdot f_r)}_{\substack{\text{a part of migration phase} \\ \text{at the reference point}}}\underbrace{\exp\left(-j\frac{4\pi R}{cM(f_a)}\cdot f_r\right)}_{\text{range cell migration term}} \\
& \underbrace{\exp\left(-j\pi\frac{f_r^2}{K_s(f_a, R)}\right)}_{\text{range modulation term}}\underbrace{\exp\left[j2\pi\phi_3(f_a, R)\cdot f_r^3\right]}_{\text{two-dimensional coupling phase term}}
\end{aligned} \tag{3.37}$$

In the following, we will give the detailed expression of each phase term in (3.37).

(1) $\phi_{az}(f_a, R)$ is the azimuth modulation function and can be expressed as

$$\begin{aligned}
\phi_{az}(f_a, R) = & \left(\frac{k_1}{2k_2} + \frac{3k_3k_1^2}{8k_2^3} + \frac{k_1^3(9k_3^2 - 4k_2k_4)}{16k_2^5}\right)f_a \\
& + \left(\frac{\lambda}{8k_2} + \frac{3k_3k_1\lambda}{16k_2^3} + \frac{3\lambda k_1^2(9k_3^2 - 4k_2k_4)}{64k_2^5}\right)f_a^2 \\
& + \left(\frac{k_3\lambda^2}{32k_2^3} + \frac{k_1\lambda^2(9k_3^2 - 4k_2k_4)}{64k_2^5}\right)f_a^3 + \left(\frac{\lambda^3(9k_3^2\lambda^3 - 4k_4k_2)}{k_2^5}\right)f_a^4
\end{aligned} \tag{3.38}$$

$\phi_{az}(f_a, R)$ is the azimuth compression function and is only related to azimuth-Doppler frequency and target location. Thus, it can be compensated in range-Doppler domain.

(2) $\phi_{RP}(R)$ is the residual phase after Taylor expansion and can be expressed as

$$\phi_{RP}(R) = -\frac{2R}{\lambda} + \frac{k_3k_1^3}{4\lambda k_2^3} + \frac{k_1^2}{2\lambda k_2} + \frac{k_1^4\cdot(9k_3^2 - 4k_2k_4)}{32\lambda k_2^5} \tag{3.39}$$

Equation (3.39) has nothing to do with azimuth Doppler frequency and range frequency, but has relation with target location along range direction, thus it can be compensated in range-Doppler domain.

(3) $b(f_a)$ is a part of the migration phase at the reference point. It is a unique term introduced by the curved trajectory and must be compensated. Because it is spatially invariant, it can be compensated in 2D frequency domain.

$$\begin{aligned} b(f_a) = & \left(\frac{k_{10}^2}{2k_{20}c} + \frac{k_{10}^3 k_{30}}{4k_{20}^3 c} + \frac{k_{10}\left(9k_{30}^2 - 4k_{20}k_{40}\right)}{32ck_{20}^5}\right) \\ & + \left(-\frac{\lambda}{8k_{20}f_c} - \frac{3k_{30}k_{10}\lambda}{16k_{20}^3 f_c} + \frac{3k_{10}^2\lambda\left(4k_{40}k_{20} - 9k_{30}^2\right)}{64k_{20}^5 f_c}\right)f_a^2 \\ & + \left(-\frac{k_{30}\lambda^2}{16k_{20}^3 f_c} + \frac{k_{10}\lambda^2\left(4k_{40}k_{20} - 9k_{30}^2\right)}{32k_{20}^5 f_c}\right)f_a^3 \\ & + \left(-\frac{3\lambda^3\left(9k_{30}^2 - 4k_{20}k_{40}\right)}{512k_{20}^5 f_c}\right)f_a^4 + \left(\dot{B}_1 R_0 - 2\frac{R_0}{c}\right) \end{aligned} \tag{3.40}$$

where

$$\begin{aligned} \dot{B}_1 = & -\frac{2k_{10}k_{20}k_{11} - k_{10}^2 k_{21}}{2ck_{20}^2} - \frac{A_{30}k_{11} - k_{10}A_{31}}{32c} \\ & - \frac{\left(3k_{10}^2 k_{30}k_{11} + k_{10}^3 k_{31}\right)k_{20} - 3k_{10}^3 k_{30}k_{21}}{4ck_{20}^4} \\ & + \left\{ \begin{array}{l} -\frac{\lambda k_{21}}{8f_c k_{20}^2} + \frac{3\lambda\left(2k_{10}A_{30}k_{11} - k_{10}^2 A_{31}\right)}{64f_c} \\ + \frac{3\lambda[(k_{10}k_{31} + k_{30}k_{11})k_{20} - 3k_{10}k_{30}k_{21}]}{16f_c k_{20}^4} \end{array} \right\} \cdot f_a^2 \\ & + \left[\frac{\lambda^2(k_{20}k_{31} - 3k_{30}k_{21})}{16f_c k_{20}^4} + \frac{\lambda^2(A_{30}k_{11} - k_1 A_{31})}{32f_c}\right] \cdot f_a^3 + \frac{3\lambda^2 A_{31}}{512f_c} \cdot f_a^4 \end{aligned} \tag{3.41}$$

$$A_{30} = \frac{9\,k_{30}^2 - 4\,k_{20}\,k_{40}}{k_{20}^5} \tag{3.42}$$

$$A_{31} = \frac{[18\,k_{30}k_{31} - 4\,k_{20}k_{41} - 4\,k_{21}k_{40}]k_{20} - 5\left(9\,k_{30}^2 - 4\,k_{20}k_{40}\right)k_{21}}{k_{20}^6} \tag{3.43}$$

(4) The fourth term in (3.37) is RCM term and is range variant. $M(f_a)$ is referred as a migration factor and can be expressed as

$$M(f_a) = \frac{2}{\dot{B}_1 c} \tag{3.44}$$

where the expression of $\dot{B}_1$ is shown in (3.41).

(5) The fifth term in (3.37) is a frequency modulation function in range direction, where $K_s(f_a, R)$ is the new Frequency-Modulated (FM) factor because of the range-azimuth coupling, and $K_s(f_a, R)$ can be expressed as

$$K_s(f_a, R) = K_s(f_a, R_0) + \Delta k_s(f_a)[\tau(f_a, R) - \tau(f_a, R_0)], \tag{3.45}$$

where $K_s(f_a, R_0)$ is the reference FM factor and $\Delta k_s(f_a)$ is the derivatives of $K_s(f_a, R)$ in the range direction, and their expressions can be expressed as

$$\begin{aligned} \frac{1}{K_s(f_a, R_0)} = & \left(\frac{\lambda}{4k_{20}f_c^2} + \frac{3k_{30}k_{10}\lambda}{8k_{20}^3 f_c^2} - \frac{3\lambda k_{10}^2 (9k_{30}^2 - 4k_{20}k_{40})}{32k_{20}^5 f_c^2} \right) f_a^2 \\ & + \left(\frac{3}{16} \frac{k_{30}\lambda^2}{k_{20}^3 f_c^2} + \frac{3\lambda^2 k_{10} (9k_{30}^2 - 4k_{20}k_{40})}{32k_{20}^5 f_c^2} \right) f_a^3 \\ & + \left(\frac{3\lambda^3 (9k_{30}^2 - 4k_{40}k_{20})}{128k_{20}^5 f_c^2} \right) f_a^4 - \frac{1}{2k_r} \end{aligned} \tag{3.46}$$

$$\Delta k_s(f_a) = -\frac{K_s^2(f_a, R_0) c M(f_a)}{2} \cdot \left\{ \begin{aligned} & -\frac{\lambda k_{21}}{4f_c^2 k_{20}^2} f_a^2 + \frac{3\lambda[(k_{10}k_{31} + k_{30}k_{11})k_{20} - 3k_{10}k_{30}k_{21}]}{8f_c^2 k_{20}^4} \cdot f_a^2 \\ & + \frac{3\lambda(2k_{10}A_{30}k_{11} + k_{10}^2 A_{31})}{32f_c^2} f_a^2 + \frac{3\lambda^2 (k_{20}k_{31} - 3k_{30}k_{21})}{16f_c^2 k_{20}^4} \cdot f_a^3 \\ & + \frac{3\lambda^2 (A_{30}k_{11} + k_{10}A_{31})}{32f_c^2} \cdot f_a^3 + \frac{3\lambda^3 A_{31}}{128f_c^2} f_a^4 \end{aligned} \right\} \tag{3.47}$$

$$\tau(f_a, R) = 2R/cM(f_a) \tag{3.48}$$

$$\tau(f_a, R_0) = 2R_0/cM(f_a) \tag{3.49}$$

According to (3.45), the new FM factor $K_s(f_a, R)$ can be approximated as a linear term of target position, and it also depends on the high order terms of azimuth frequency.

(6) The sixth term in (3.37) is generated because of decoupling, and it can be expressed as

$$\begin{aligned}\phi_3(f_a,R) =& \left(-\frac{\lambda}{8k_2f_c^3} - \frac{3k_3k_1\lambda}{16k_2^3f_c^3} - \frac{3k_1^2\lambda\left(9k_3^2-4k_2k_4\right)}{64k_2^5f_c^3}\right)f_a^2 \\ &-\left(\frac{k_3\lambda^2}{8k_2^3f_c^3} + \frac{k_1\lambda^2\left(9k_3^2-4k_4k_2\right)}{16k_2^5f_c^3}\right)f_a^3 - \left(\frac{5\lambda^3\left(9k_3^2-4k_2k_4\right)}{256k_2^5f_c^3}\right)f_a^4\end{aligned} \tag{3.50}$$

The range variance of $\phi_3(f_a,R)$ is small enough so as to be ignored, thus we can use $\phi_3(f_a,R_0)$ instead of $\phi_3(f_a,R)$ in general.

3.2.4 *Spatially Variant Slant Range Model Coefficients*

The coefficients of the slant range model in (3.31) are based on scene central point. Commonly, the coefficients of slant range model are considered as 2D spatially invariant in tradition. However, in GEO SAR imaging processing, the coefficients of slant range model are considered as 2D space-variant because of the curved trajectory and large scene size. In order to express the real slant range of any target position in a large scene with curved trajectory, Taylor expansions on (3.31) along range direction and azimuth direction are further utilized.

First, we have the range variant model as:

$$\begin{aligned}k_1(R) &= k_{10} + k_{r11}\cdot(R-R_0)\\ k_2(R) &= k_{20} + k_{r21}\cdot(R-R_0) + k_{r22}\cdot(R-R_0)^2\\ k_3(R) &= k_{30} + k_{r31}\cdot(R-R_0)\\ k_4(R) &= k_{40} + k_{r41}\cdot(R-R_0)\end{aligned} \tag{3.51}$$

where R_0 is the distance from the satellite to reference point at ACM; k_{rij} (i = 1, 2, 3, 4) are the jth-order derivatives of k_i in the range direction, and their expressions are shown as:

$$\begin{aligned}R_c &= \left\|\overrightarrow{\mathbf{r}}_{s0} - \overrightarrow{\mathbf{r}}_{g0}\right\|\\ k_{11r} &= \frac{v_{s0x}}{r_{s0x}} - \frac{\left(\overrightarrow{\mathbf{v}}_{s0} - \overrightarrow{\mathbf{v}}_{g0}\right)\left(\overrightarrow{\mathbf{r}}_{s0} - \overrightarrow{\mathbf{r}}_{g0}\right)^T}{R_0^2}\end{aligned} \tag{3.52}$$

$$\begin{aligned}k_{21r} =& \frac{a_{s0x}R_0^2 - r_{s0x}\left[\left(\overrightarrow{\mathbf{a}}_{s0} - \overrightarrow{\mathbf{a}}_{g0}\right)\left(\overrightarrow{\mathbf{r}}_{s0} - \overrightarrow{\mathbf{r}}_{g0}\right)^T + \left\|\overrightarrow{\mathbf{v}}_{s0} - \overrightarrow{\mathbf{v}}_{g0}\right\|^2\right]}{2\cdot R_0^2\cdot r_{s0x}}\\ &- \frac{v_{s0x}\left(\overrightarrow{\mathbf{v}}_{s0} - \overrightarrow{\mathbf{v}}_{g0}\right)\left(\overrightarrow{\mathbf{r}}_{s0} - \overrightarrow{\mathbf{r}}_{g0}\right)^T}{R_0^2\cdot r_{s0x}} + \frac{3\left(\overrightarrow{\mathbf{v}}_{s0} - \overrightarrow{\mathbf{v}}_{g0}\right)\left(\overrightarrow{\mathbf{r}}_{s0} - \overrightarrow{\mathbf{r}}_{g0}\right)^T}{2\cdot R_0^4}\end{aligned} \tag{3.53}$$

$$k_{22r} = -\frac{a_{s0x}}{4\left\|\vec{\mathbf{r}}_{s0} - \vec{\mathbf{r}}_{g0}\right\|} - \frac{3r_{s0x}\left[\left(\vec{\mathbf{v}}_{s0} - \vec{\mathbf{v}}_{g0}\right)\left(\vec{\mathbf{r}}_{s0} - \vec{\mathbf{r}}_{g0}\right)^T\right]^2}{4\left\|\vec{\mathbf{r}}_{s0} - \vec{\mathbf{r}}_{g0}\right\|^5} + \frac{\left\{\begin{array}{l} 3r_{s0x}\left[\left(\vec{\mathbf{a}}_{s0} - \vec{\mathbf{a}}_{g0}\right)\left(\vec{\mathbf{r}}_{s0} - \vec{\mathbf{r}}_{g0}\right)^T + \left\|\vec{\mathbf{v}}_{s0} - \vec{\mathbf{v}}_{g0}\right\|^2\right]^2 \\ +2v_{s0x}\left(\vec{\mathbf{v}}_{s0} - \vec{\mathbf{v}}_{g0}\right)\left(\vec{\mathbf{r}}_{s0} - \vec{\mathbf{r}}_{g0}\right)^T \end{array}\right\}}{4\left\|\vec{\mathbf{r}}_{s0} - \vec{\mathbf{r}}_{g0}\right\|^3} \tag{3.54}$$

$$\begin{aligned} k_{31r} = & \frac{b_{s0x}R_0^2 - r_{s0x}\left[\left(\vec{\mathbf{b}}_{s0} - \vec{\mathbf{b}}_{g0}\right)\left(\vec{\mathbf{r}}_{s0} - \vec{\mathbf{r}}_{g0}\right)^T + 3\left(\vec{\mathbf{a}}_{s0} - \vec{\mathbf{a}}_{g0}\right)\left(\vec{\mathbf{v}}_{s0} - \vec{\mathbf{v}}_{g0}\right)^T\right]}{6R_0^2 r_{s0x}} \\ & + \frac{3v_{s0x}\left[\left(\vec{\mathbf{v}}_{s0} - \vec{\mathbf{v}}_{g0}\right)\left(\vec{\mathbf{r}}_{s0} - \vec{\mathbf{r}}_{g0}\right)^T\right]^2 - 5r_{s0x}\left[\left(\vec{\mathbf{v}}_{s0} - \vec{\mathbf{v}}_{g0}\right)\left(\vec{\mathbf{r}}_{s0} - \vec{\mathbf{r}}_{g0}\right)^T\right]^3}{2R_0^6 r_{s0x}} \\ & - \frac{v_{s0x}\left(\vec{\mathbf{a}}_{s0} - \vec{\mathbf{a}}_{g0}\right)\left(\vec{\mathbf{r}}_{s0} - \vec{\mathbf{r}}_{g0}\right)^T + a_{s0x}\left(\vec{\mathbf{v}}_{s0} - \vec{\mathbf{v}}_{g0}\right)\left(\vec{\mathbf{r}}_{s0} - \vec{\mathbf{r}}_{g0}\right)^T}{2R_0^2 r_{s0x}} \\ & + \frac{3\left(\vec{\mathbf{v}}_{s0} - \vec{\mathbf{v}}_{g0}\right)\left(\vec{\mathbf{r}}_{s0} - \vec{\mathbf{r}}_{g0}\right)^T\left(\vec{\mathbf{a}}_{s0} - \vec{\mathbf{a}}_{g0}\right)\left(\vec{\mathbf{r}}_{s0} - \vec{\mathbf{r}}_{g0}\right)^T}{2 \cdot R_0^4} \\ & + \frac{v_{s0x} \cdot \left\|\vec{\mathbf{v}}_{s0} - \vec{\mathbf{v}}_{g0}\right\|^2}{2 \cdot R_0^2 \cdot r_{s0x}} - \frac{3\left(\vec{\mathbf{v}}_{s0} - \vec{\mathbf{v}}_{g0}\right)\left(\vec{\mathbf{r}}_{s0} - \vec{\mathbf{r}}_{g0}\right)^T\left\|\vec{\mathbf{v}}_{s0} - \vec{\mathbf{v}}_{g0}\right\|^2}{2R_0^4} \end{aligned} \tag{3.55}$$

$$k_{41r} = \frac{\left\|\left(\vec{\mathbf{a}}_{s0} - \vec{\mathbf{a}}_{g0}\right)\right\|^2}{8R_0^2} + \frac{k_{20}k_{r21} + k_{30}k_{r11} + k_{10}k_{r31}}{r_{s0x}} + \frac{k_{20}^2 + 2k_{10}k_{30}}{2 \cdot R_0^2} \tag{3.56}$$

where r_{s0x} is the value of $\vec{r}_{s0}$ in the x direction; v_{s0x} is the value of $\vec{v}_{s0}$ in the x direction; a_{s0x} is the value of $\vec{a}_{s0}$ in the x direction; b_{s0x} is the value of $\vec{b}_{s0}$ in the x direction; d_{s0x} is the value of $\vec{d}_{s0}$ in the x direction.

The same as the range variant model, the azimuth variant model can be obtained using the Taylor series expansion. Thus, the 2D spatially variant model can be shown as:

$$\begin{aligned} k_1(R) &= k_{10}(R) \\ k_2\left(R, t_p\right) &= k_{20}(R) + k_{a21}(R)t_p + k_{a22}(R)t_p^2 \\ k_3\left(R, t_p\right) &= k_{30}(R) + k_{a31}(R)t_p + k_{a32}(R)t_p^2 \\ k_4\left(R, t_p\right) &= k_{40}(R) + k_{a41}(R)t_p \end{aligned} \tag{3.57}$$

where k_{aij}(i = 1,2,3,4) are the jth-order derivatives of k_i in the azimuth direction, and t_p is the beam center crossing time for an arbitrary target. Note that k_1 is considered to be azimuth-invariant because the azimuth residual phase will not affect the image quality. In reality, the orders of the spatially variant model are not constant, and they are determined by the orbital parameters, orbital positions, wavelength, resolution, etc.

3.3 Time Domains Algorithm

3.3.1 Traditional BP Algorithm

A. Details of BP algorithm

Commonly, most of the SAR imaging algorithms are achieved by matched filtering. In reality, SAR imaging processing can also be achieved by the convolution of the echo signal and the time shifted filters, and the back-projection (BP) algorithm is based on this principle.

Assuming that the coordinates of the imaging scene of interest are (x_i, y_j), the transmitting signal is $s_t(t)$, the location of the radar platform is $(0, u)$, and the echo signal of the coordinates (x_i, y_j) can be expressed as

$$s_t\left(t - 2\sqrt{x_i^2 + (y_j - u)^2} \Big/ c\right) \tag{3.58}$$

where t is the fast time.

Thus, the received echo signal can be expressed as:

$$s(t, u) = \sum_n \sigma_n s_t\left(t - 2\sqrt{x_n^2 + (y_n - u)^2} \Big/ c\right) \tag{3.59}$$

where (x_n, y_n) are the coordinates of different targets. Furthermore, it should be noted that the pixels and the targets may not coincide, and the pixel interval cannot be larger than the resolution.

After the range compression, the echo signal can be written as:

$$s_M(t, u) = s(t, u) * p^*(-t) = \int_\tau s(\tau, u) p^*(\tau - t) d\tau \tag{3.60}$$

Then, the pixel value $f(x_i, y_j)$ in the SAR image can be obtained by accumulating all the $s_M(t, u)$ with different u:

$$f(x_i, y_j) = \int_u s_M\left[2\sqrt{x_i^2 + (y_j - u)^2} \Big/ c, u\right] du = \int_u s_M\left[t_{ij}(u), u\right] du \tag{3.61}$$

where $t_{ij}(u) = 2\sqrt{x_i^2 + (y_j - u)^2} \Big/ c$ is the round-trip time delay of the pixel (x_i, y_j) when the platform is at $(0, u)$, and $s_M\left[t_{ij}(u), u\right]$ is the interpolation of the matched filtered signal $s_M(t, u)$. According to (3.61), the SAR image can be obtained by changing the coordinates of the pixels.

The flowchart of the traditional BP algorithm is shown as Fig. 3.6.

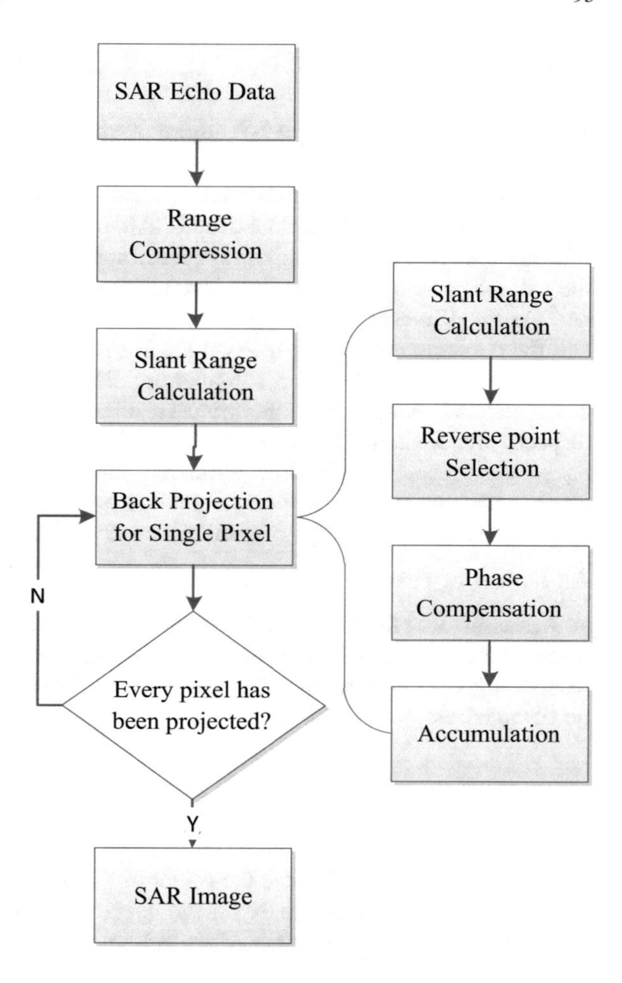

Fig. 3.6 Flowchart of the traditional BP algorithm

B. Analysis of Computational Budget

Assuming that the sampling numbers of echo data in range direction and azimuth direction are N_r and N_a, respectively, the pixel numbers of ground grid in range direction and azimuth direction are N_x and N_y, respectively, which are usually equal to N_r and N_a due to the oversampling, and the multiple of rise sampling is in range direction N_{up}. Correspondingly, the computational budget of the traditional BP is shown in Table 3.2.

According to Table 3.2, the total computational budget can be obtained as

$$5N_aN_r\log_2 N_r + 6N_aN_r + 5N_aN_{up}N_r\log_2(N_{up}N_r) + 22N_aN_xN_y \tag{3.62}$$

3.3.2 Fast BP Algorithm

There are many kinds of fast BP algorithms currently, and the fast BP algorithm introduced in this section is a most common one. The principle of it is dividing the whole aperture into many sub-apertures and synthesizing the sub-aperture based SAR images into a full-aperture based SAR image.

After the aperture division, the synthetic aperture length decreases. As we know, the azimuth resolution of SAR is inversely proportional to the synthetic aperture, and the pixel interval is smaller than the resolution. Thus, the azimuth resolution and the corresponding pixel interval are both enlarged. Consequently, the number of the pixels utilized in the sub-aperture BP processing will reduce; thus, the computational budget and the process efficiency are lessened and increased, respectively. It should be noted that the SAR images based on the sub-apertures have worse resolution and image quality, and rise sample rate processing is necessary when we synthesize the sub-aperture based SAR images into a full-aperture SAR image which has high resolution and good image quality.

A. Details of Fast BP algorithm

Assuming that the length of the whole synthetic aperture is L, the whole aperture is divided into N_{sub} sub-apertures, and the sub-aperture length is $l = L/N_{sub}$ correspondingly. Therefore, the pixel value of the sub-aperture based SAR image can be obtained as:

$$I_n(p) = \int_{-l/2}^{l/2} F\left(s_n + \varepsilon, \frac{2}{c}|\vec{p} - \vec{q}(s_n + \varepsilon)|\right) d\varepsilon \tag{3.63}$$

where $s_n = (n - 1/2)l$ denotes the center of the n th sub-aperture, $\vec{q}(s)$ is the radar platform position and $\vec{p}$ stands for the target position.

Furthermore, the pixel value of the full-aperture SAR image can be written as:

$$I(p) = \sum_{n=1}^{N_{sub}} I_n(p) = \int_0^L F\left(s, \frac{2}{c}|\vec{p} - \vec{q}(s)|\right) ds \tag{3.64}$$

Note that the sub-aperture based SAR image has coarse azimuth resolution which cannot be acceptable. Thus, polar coordinate system whose polar axis is the line between the sub-aperture center and the scene center is adopted to achieve better resolution and higher efficiency. Furthermore, the pixels should be re-distributed in the polar coordinate system according to the Nyquist Sampling Theorem as written:

$$\Delta r \leq \frac{c}{2(f_{\max} - f_{\min})} \tag{3.65}$$

$$\Delta\beta \leq \frac{c}{2f_{\max} l} \tag{3.66}$$

where (3.65) and (3.66) are the Nyquist Sampling Theorem in the range and azimuth directions, respectively. Based on (3.65) and (3.66), the sub-aperture based SAR image in polar coordinate system can be obtained.

After we obtain the sub-aperture based SAR image in the polar coordinate system, up-sampling processing will be conducted to obtain the pixel values on the final imaging plane. Larger multiple of up-sampling has higher accuracy but lower efficiency; thus, the multiple of up-sampling is usually between 2 and 8.

When all the sub-aperture based SAR images on the final imaging plane are obtained, the full-aperture based SAR image can be obtained as shown in (3.64).

The flowchart of the fast BP algorithm is shown as Fig. 3.7.

B. Analysis of Computational Budget

Assuming that the sampling number of echo data in azimuth direction is N_a, the full aperture is divided into N_a/M sub-aperture, and each sub-aperture contains M pulses. Furthermore, in the sub-aperture imaging, the sampling numbers in the range as well as angle directions are N and M, and the multiples of up-sampling in the range and azimuth direction are N_{up} and M_{up}, respectively. Thus, the computational budget can be obtained as shown in Table 3.3.

Thus, total calculation of the fast BP algorithm is shown as:

$$\begin{aligned} &5N_aN_r\log_2 N_r + 6N_aN_r + 5N_aN_{up}N_r\log_2\left(N_{up}N_r\right) + 22N_aNM \\ &\quad + 5N_aN\log_2(NM) + 5M_{up}N_aN\log_2\left(M_{up}NM\right) + 19\frac{N_a}{M}N^2 \end{aligned} \tag{3.67}$$

Furthermore, as shown in (3.67), when $M = \sqrt{N_a}$, the total calculation of the fast BP algorithm reaches the minimum.

3.3.3 Computer Simulation

Computer simulations based on SAR raw data are conducted to verify the traditional BP algorithm and the fast BP algorithm. The orbital parameters as well as simulation parameters are shown in Fig. 3.1 and Table 3.4, respectively. The SAR raw data have a scene size of 150 km × 150 km. Correspondingly, the sampling number in range and azimuth directions are respectively 8×10^4 and 26×10^4, and the multiples of up-sampling in the range and azimuth direction are both 8. Furthermore, the pulse number of the sub-aperture is 200 which is close to $\sqrt{2.6 \times 10^4}$, and the number of sub-aperture is 130. Then, the imaging results of the traditional BP algorithm and the fast BP algorithm are shown in Figs. 3.8 and 3.9, respectively.

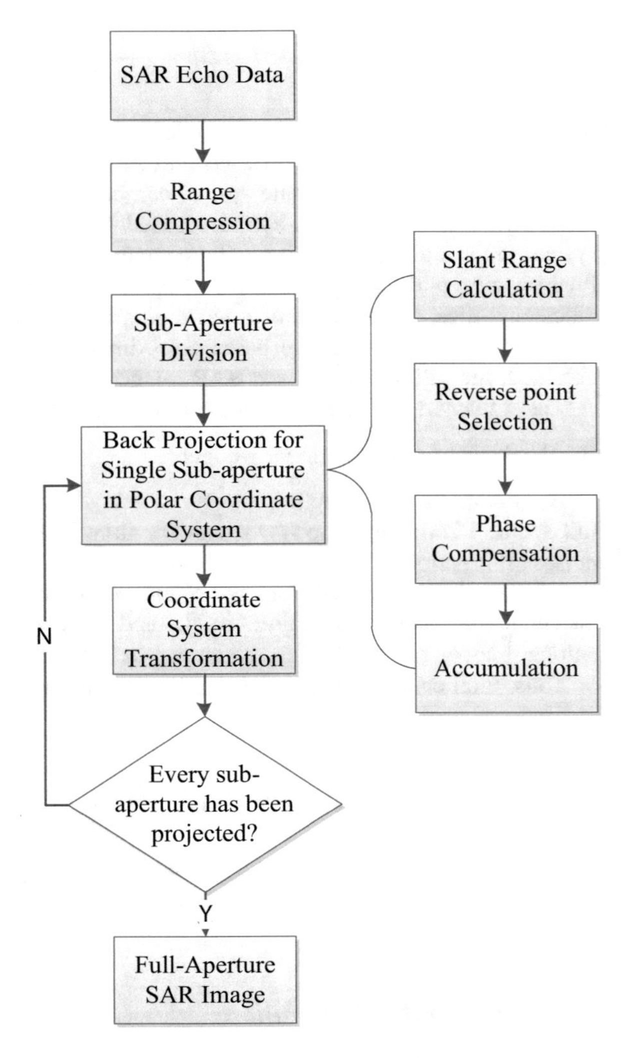

Fig. 3.7 Flowchart of the fast BP algorithm

Fig. 3.8 Imaging result of the traditional BP algorithm

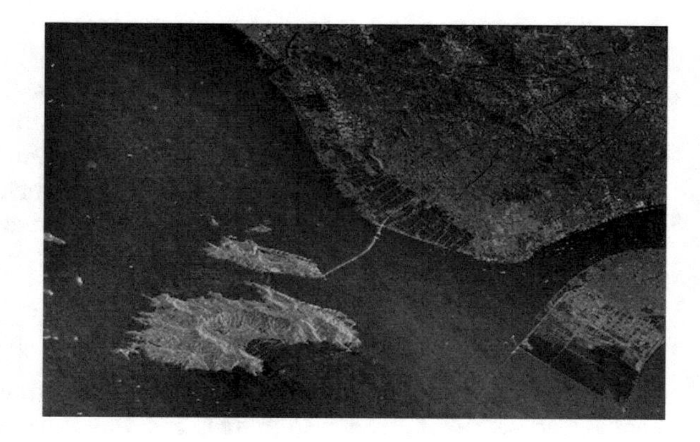

Fig. 3.9 Imaging result of the fast BP algorithm

As shown in Figs. 3.8 and 3.9, the imaging results of the traditional BP and fast BP algorithms are almost the same. However, the calculations of them are nearly 1.19×10^{15} and 3.45×10^{13}, and the former is 34.55 times as much as the latter. In other words, under the same calculation condition, the operation time of the fast BP is 1/34 of the operation time of traditional BP.

3.4 Frequency Domain Algorithm

3.4.1 Analysis of Difficulties

The frequency domain algorithms have better efficiency performance compared with the time domain algorithms, especially under the circumstance that the echo data of GEO SAR can range up to hundreds of gigabyte (GB). Thus, the frequency domain algorithms will be widely used in GEO SAR imaging processing.

However, different from the time domain algorithms, traditional frequency domain algorithms cannot directly be utilized in GEO SAR imaging because of several additional problems, such as the curved trajectory, 2D spatial variance of the slant range coefficients as well as RCM, etc. Moreover, high squint mode which will be widely used in GEO SAR system to improve the ability to revisit specific areas will result in tremendous range walk, longer SAT as well as more curved trajectory and further increase the difficulty of GEO SAR imaging. Thus, to improve the traditional frequency domain algorithms and make them suitable for GEO SAR imaging, the aforementioned difficulties existing in GEO SAR imaging processing should be analyzed in detail first.

Considering the broadside mode and the high squint mode, the difficulties of GEO SAR imaging processing can be summarized as range walk (for high squint mode), 2D spatially variant slant range model coefficients, small azimuth time-bandwidth product (TBP) and azimuth variant RCM. In this part, these four problems will be discussed in detail.

(1) Range walk

The most direct effect of high squint mode is range walk which increases the difficulty of range cell migration correction (RCMC). In broadside mode, the range walk is zero or nearly zero. However, in high squint mode the range walk can involve thousands of range cells and be far greater than half of the range cell.

In reality, range walk is linear RCM caused by non-zero values of k_1. Therefore, the range walk expression can be written as follows:

$$R_w = k_1 \cdot \left(t_a - t_p\right) \approx k_{10} \cdot \left(t_a - t_p\right) \tag{3.68}$$

Under medium resolution and relatively small trajectory curvatures, the 2D spatial variance of k_1 can be ignored, and the concept of approximate equality holds.

A computer simulation was conducted to illustrate the range walk of high squint mode. The orbital and simulation parameters are shown in Fig. 3.1 and Table 3.5, respectively. The satellite is at the equator, the required SAT is 150 s, and the simulation result is shown in Fig. 3.10, which shows that the range walk of high squint mode at the equator exceeds 2000 range cells.

(2) 2D spatially variant slant range model coefficients

A curved trajectory and a large scene size will cause 2D spatial variance of the slant range coefficients, which will result in a 2D residual phase, decrease the image quality and even cause defocusing. Compared with broadside mode, the 2D spatial variance of the slant range model coefficients in high squint mode will be more severe because the squint angle will result in a longer SAT and a more curved trajectory. Thus, the 2D spatial variance model of the slant range model coefficients for high squint mode has higher accuracy compared with that for broadside mode. Thus, the analysis in this section is based on high squint mode. Furthermore, the analysis results and corresponding model are both suitable for broadside mode.

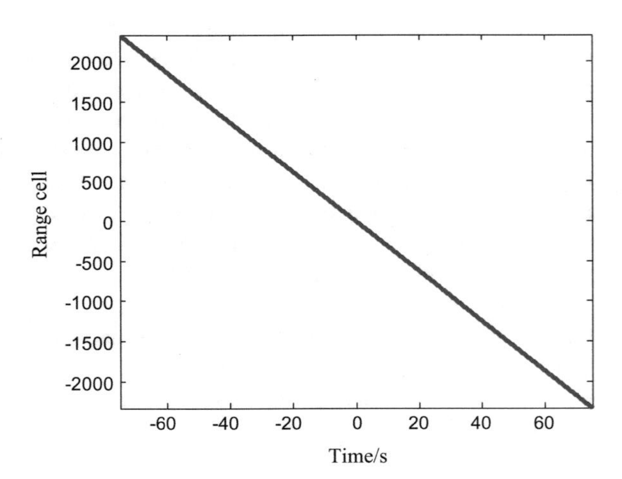

Fig. 3.10 Range walk of high squint mode

Assuming that the range walk has been removed via the linear compensation proposed in Sect. 3.4.2, a computer simulation was conducted to illustrate the residual phase caused by the 2D spatially variant slant range model coefficients. The orbital parameters and simulation parameters are shown in Fig. 3.1 and Table 3.5, respectively. The satellite is at the equator, the required SAT is 150 s and the simulation results are shown in Fig. 3.11. As shown in Fig. 3.11, the 2D nonlinear residual phase are far more than $\pi/4$ rad which is the threshold of nonlinear residual phase.

According to Fig. 3.11, a 2D spatially variant model of slant range model coefficients sufficiently accurate for high squint GEO SAR imaging can be obtained as follows:

$$\begin{aligned} k_1(R) &= k_{10}(R) \\ k_2(R, t_p) &= k_{20}(R) + k_{a21}(R)t_p + k_{a22}(R)t_p^2 \\ k_3(R, t_p) &= k_{30}(R) + k_{a31}(R)t_p + k_{a32}(R)t_p^2 \\ k_4(R, t_p) &= k_{40}(R) + k_{a41}(R)t_p \end{aligned} \tag{3.69}$$

where

$$\begin{aligned} k_1(R) &= k_{10} + k_{r11} \cdot (R - R_0) \\ k_2(R) &= k_{20} + k_{r21} \cdot (R - R_0) + k_{r22} \cdot (R - R_0)^2 \\ k_3(R) &= k_{30} + k_{r31} \cdot (R - R_0) + k_{r32} \cdot (R - R_0)^2 \\ k_4(R) &= k_{40} + k_{r41} \cdot (R - R_0) \end{aligned} \tag{3.70}$$

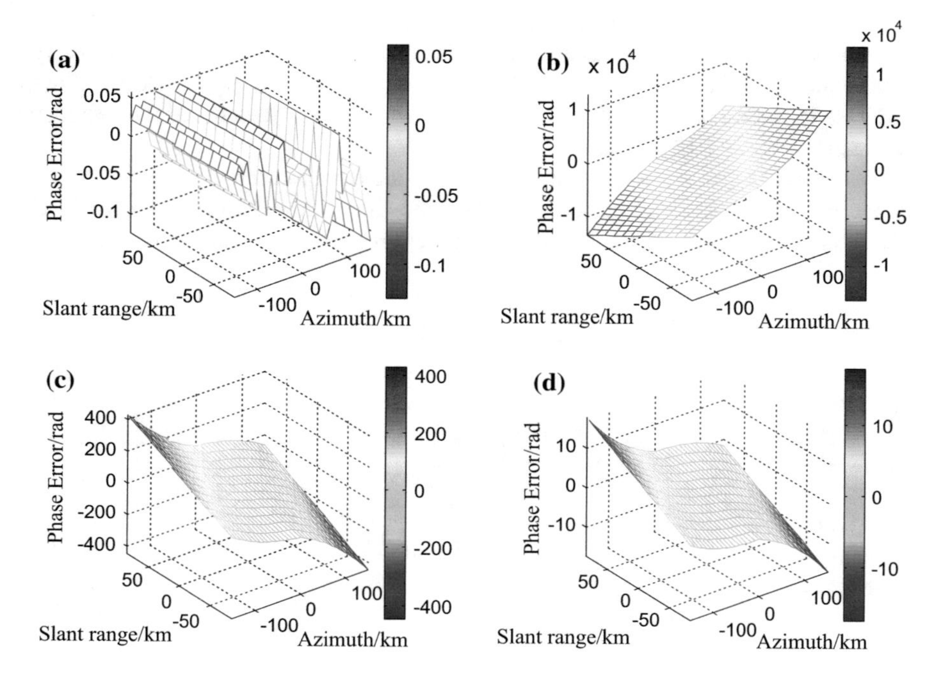

Fig. 3.11 Residual phase caused by 2D spatially variant: **a** k_1, **b** k_2, **c** k_3, **d** k_4

Note that k_1 is considered to be azimuth-invariant because the azimuth residual phase it causes will not affect the image quality.

(3) Small azimuth time-bandwidth product

As we all know, the expressions of the Doppler rate can be shown as:

$$f_{dr} = -\frac{2}{\lambda}k_2 = f_{dra} + f_{drv} \tag{3.71}$$

where f_{dra} is the acceleration-related component of the Doppler rate and f_{drv} is the velocity component of the Doppler rate. Their expressions are shown as

$$\begin{aligned} f_{dra} &= -\frac{\left(\vec{\mathbf{a}}_{s0} - \vec{\mathbf{a}}_{g0}\right)\left(\vec{\mathbf{r}}_{s0} - \vec{\mathbf{r}}_{g0}\right)^T + \left\|\vec{\mathbf{v}}_{s0} - \vec{\mathbf{v}}_{g0}\right\|^2}{\lambda\left\|\vec{\mathbf{r}}_{s0} - \vec{\mathbf{r}}_{g0}\right\|} \\ f_{drv} &= \frac{\left[\left(\vec{\mathbf{v}}_{s0} - \vec{\mathbf{v}}_{g0}\right)\left(\vec{\mathbf{r}}_{s0} - \vec{\mathbf{r}}_{g0}\right)^T\right]^2}{\lambda\left\|\vec{\mathbf{r}}_{s0} - \vec{\mathbf{r}}_{g0}\right\|^3} \end{aligned} \tag{3.72}$$

As shown in (3.71) and (3.72), Doppler rate of GEO SAR can be divided into two parts. At some special orbital positions, the sum of the two parts is zero or nearly zero, and the Doppler bandwidth is zero or nearly zero, too. In order to illustrate this phenomenon directly, a computer simulation was performed. The orbital and simulation parameters are shown in Fig. 3.1 and Table 3.5, respectively, and the simulation results are shown in Fig. 3.12.

Figure 3.12a shows the acceleration-related component f_{dra} and the velocity component f_{drv} of the Doppler rate. Figure 3.12b shows the Doppler rate f_{dr} of the GEO SAR, which is zero at the orbit positions where the arguments of the latitudes are approximately 51° and 129°.

When the Doppler rate is zero, the azimuth observation angle θ_ϕ between the target and the radar is still changing. Thus, the azimuth resolution can be obtained via the accumulation of the azimuth observation angle θ_ϕ. The work model is

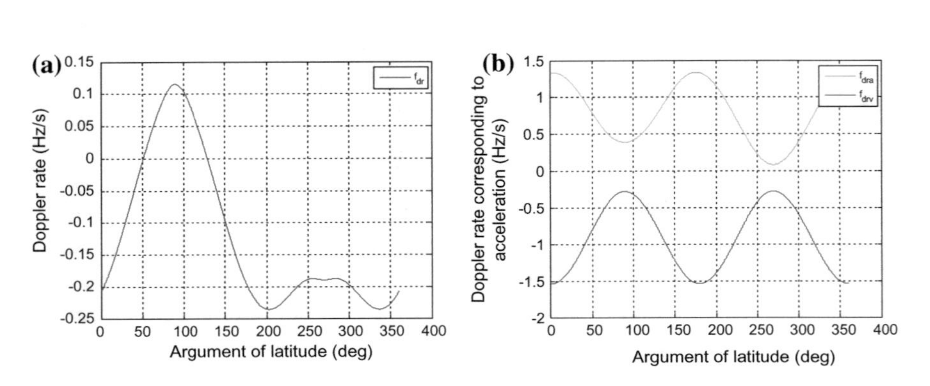

Fig. 3.12 Simulation results of: **a** Doppler rate, **b** f_{dra} and f_{drv}

similar to the spotlight SAR, and the scene is illuminated by the radar beam during the synthetic aperture time. The SAR imaging geometry is shown in Fig. 3.13.

The azimuth signal property at the zero Doppler rate position is shown in Fig. 3.14. The target is at the scene center. The original RCM of the target is significantly different from the traditional "far-near-far" RCM. The azimuth spectrum of the original signal is compressed significantly, as shown in Fig. 3.14b. The nonlinear Doppler frequency versus time relationship is shown in Fig. 3.14c. The Doppler bandwidth is only 0.04 Hz because of the zero Doppler rate, which is due to the acceleration component of the Doppler rate. However, the azimuth observation angle between the target and the radar, which determines the azimuth resolution, is changed about 0.38° for a 20-mazimuth resolution. Thus, GEO SAR is capable of imaging with a low bandwidth. The Doppler frequency is folded in the original signal, and one Doppler frequency corresponds to two different RCMs, as shown in Fig. 3.14d.

(4) Azimuth spatial variance of RCM

The same as the azimuth-variant slant range model coefficients, azimuth-variant RCM which affects RCMC also stems from the curved trajectory. Correspondingly, the azimuth-variant RCM of high squint mode is more severe than that of broadside mode because of increased trajectory curvature. However, the correction of azimuth-variant RCM, which should be performed before azimuth processing, and the compensation of the azimuth residual phase, which is usually performed during azimuth processing, cannot both be performed at the same time. Moreover, their effects on image quality are not exactly the same. Furthermore, the threshold values

Fig. 3.13 Geometry of the radar beam and the scene

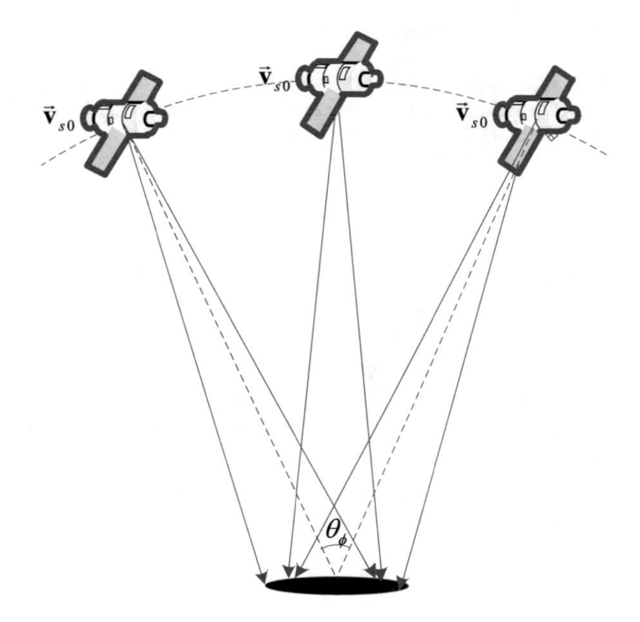

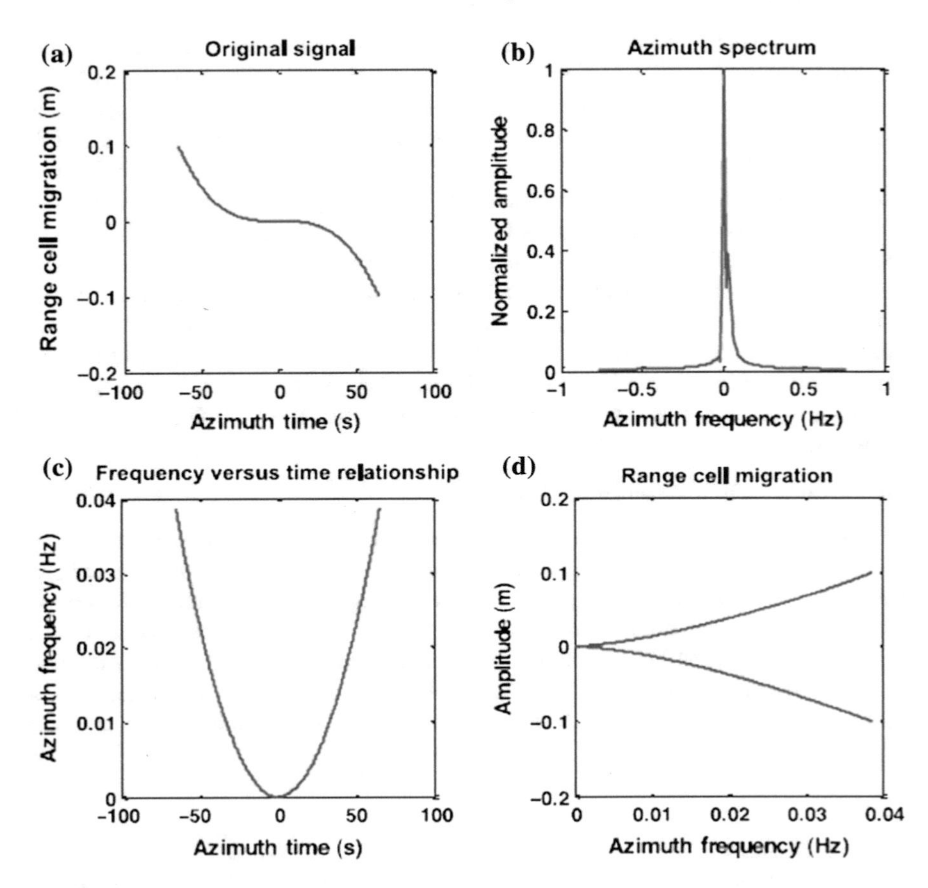

Fig. 3.14 Azimuth signal at the orbit position with a zero Doppler rate. **a** RCM of the original signal. **b** Azimuth spectrum. **c** Doppler frequency versus time relationship. **d** Relationship between the RCM and the azimuth frequency

of the errors they cause are also different. Therefore, the analysis and modeling of azimuth-variant RCM are presented separately in this section. It should be noted that the analysis in this section is also based on high squint mode, and the analysis results and corresponding model are both suitable for broadside mode.

Assuming that the range walk has been removed, a computer simulation was performed to illustrate the azimuth-variant RCM. The orbital and simulation parameters are shown in Fig. 3.1 and Table 3.5, respectively. The satellite is at the equator, the required SAT is 150 s and the simulation results are shown in Fig. 3.15.

According to Fig. 3.15, azimuth-variant RCM caused by k_{a21}, k_{a22} and k_{a31} are larger than half of the range cell. Therefore, the expression of azimuth-variant RCM can be written as

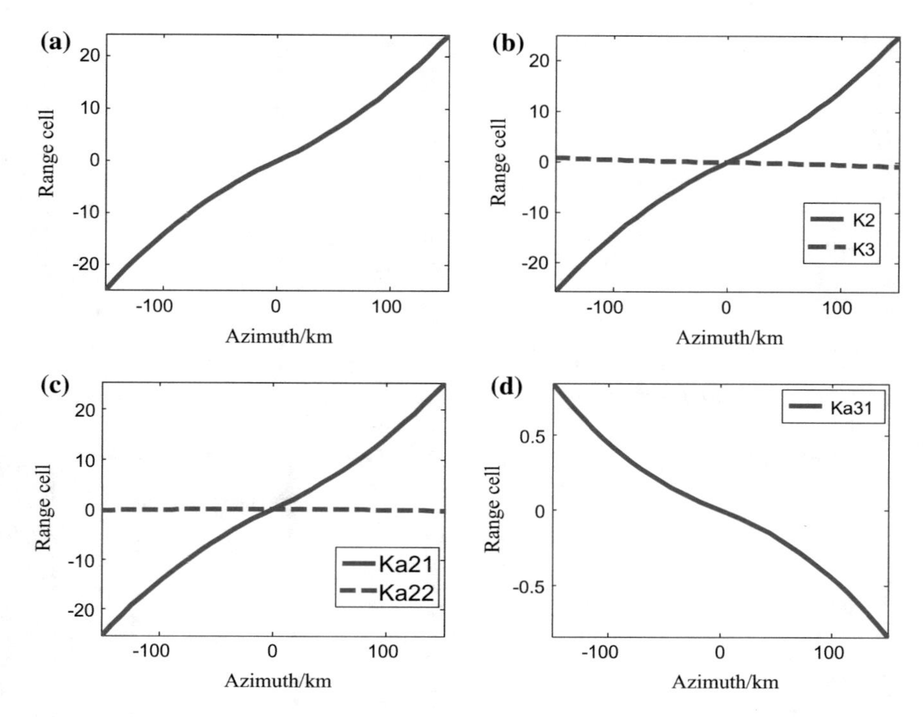

Fig. 3.15 **a** Azimuth-variant RCM, **b** Components of azimuth-variant RCM, **c** Azimuth-variant RCM caused by k_2, **d** Azimuth-variant RCM caused by k_3

$$RCM_{azv}\left(t_a, R, t_p\right) = \left(k_{a21}(R) \cdot t_p + k_{a22}(R) \cdot t_p^2\right) \cdot \left(t_a - t_p\right)^2 + \left(k_{a31}(R) \cdot t_p\right)\left(t_a - t_p\right)^3 \tag{3.73}$$

3.4.2 Derivation of Azimuth Compensation

According to the analysis in Sect. 3.4.1, the Doppler rate of echo signal is zero at some special orbital positions, the Doppler time-frequency relationship is nonlinear, and the azimuth spectrum of echo signal is folded correspondingly. Under this circumstance, if the acceleration-related component of the Doppler rate is compensated by a quadratic factor, the nonlinear Doppler time-frequency relationship can become linear, as shown in Fig. 3.16a. Thus, the frequency domain algorithms can then be applied to these positions. Therefore, a quadratic factor compensation should be performed for the original echo to introduce the frequency domain algorithms.

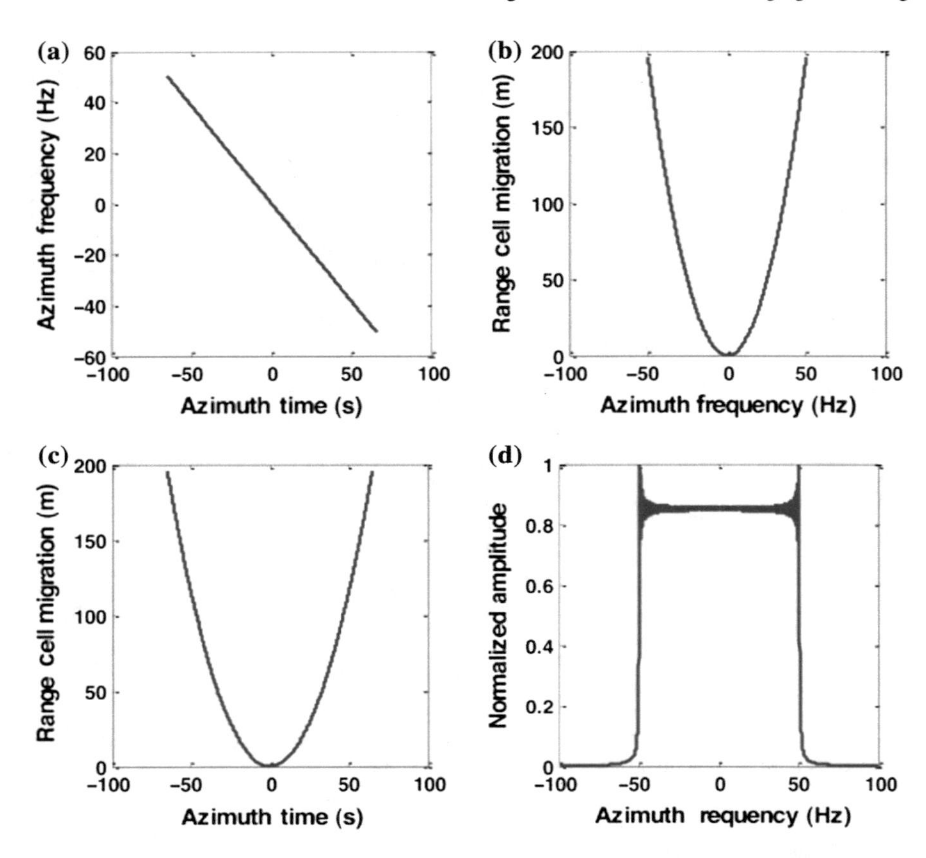

Fig. 3.16 Azimuth signal property after the quadratic factor compensation. **a** Doppler time-frequency relationship. **b** Relationship between the RCM and the azimuth frequency. **c** Relationship between the RCM and the azimuth time. **d** Azimuth spectrum

The Doppler rate can be changed by one compensation in 2D time domain and one compensation in range-frequency-azimuth-time domain which are called azimuth compensation. The principle of the azimuth compensation is shown in Fig. 3.17.

As shown in Fig. 3.17, azimuth compensation which consists of a time domain compensation and a frequency domain compensation equivalently changes the slant range histories of targets and further change the echo signal characteristics. Furthermore, the distributions of targets after azimuth compensation are also changed, which means the azimuth variance of slant range coefficients will also change. In addition, the azimuth compensation is performed before the azimuth processing; thus, the azimuth variant RCM can be weakened via this method.

Based on this conclusion, an azimuth polynomial compensation that can remove the range walk, unfold the folded azimuth spectrum and weaken the azimuth variance of slant range coefficients is proposed and derived in this section.

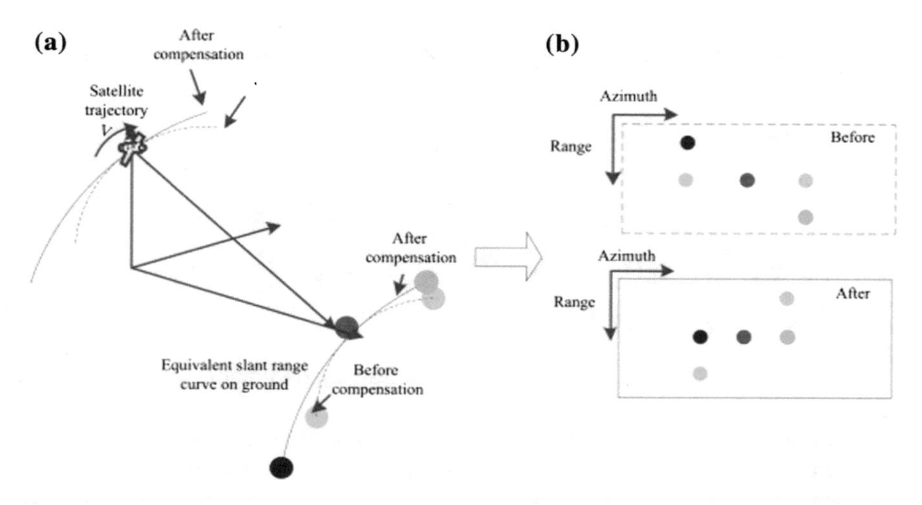

Fig. 3.17 Target distribution before and after azimuth compensation. **a** Equivalent slant range curve on the ground. **b** Result of the target distribution of the focus

To simplify the derivation, the linear compensation used to remove range walk is not performed here. For an arbitrary point target, its slant range history after azimuth polynomial compensation can be written as follows:

$$\begin{aligned} R_c\left(R_{pc}, t_{pc}\right) &= R_p + k_1\left(R_p\right)\left(t_a - t_p\right) + k_2\left(R_p, t_p\right)\left(t_a - t_p\right)^2 \\ &\quad + k_3\left(R_p, t_p\right)\left(t_a - t_p\right)^3 + k_4\left(R_p, t_p\right)\left(t_a - t_p\right)^4 - at_a^2 - bt_a^2 - dt_a^2 \\ &= R_{pc} + k_{1c}\left(R_{pc}\right)\left(t_a - t_{pc}\right) + k_{2c}\left(R_{pc}, t_{pc}\right)\left(t_a - t_{pc}\right)^2 \\ &\quad + k_{3c}\left(R_{pc}, t_{pc}\right)\left(t_a - t_{pc}\right)^3 + k_{4c}\left(R_{pc}, t_{pc}\right)\left(t_a - t_{pc}\right)^4 \end{aligned} \tag{3.74}$$

where R_c is the compensated slant range history, t_{pc} and R_{pc} are the compensated beam crossing time and corresponding instantaneous slant range, respectively, and $k_{1c}\left(R_{pc}\right)$ and $k_{2c}\left(R_{pc}, t_{pc}\right)-k_{4c}\left(R_{pc}, t_{pc}\right)$ are the compensated coefficients.

The expression for t_{pc} can be written as shown in (3.75). It should be noted that the influence of third and higher order azimuth compensation is ignored to obtain the analytic solution. In addition, for simplicity, the spatial variance of $k_2\left(R_p, t_p\right)$ is also ignored.

$$t_{pc} \approx \frac{k_2\left(R_p, t_p\right)}{k_2\left(R_p, t_p\right) - a} t_p \approx \frac{k_{20} - a}{k_{20}} t_p \tag{3.75}$$

Thus, the expressions of $k_2^{cp}\left(R_p^{cp}, t_p^{cp}\right) \sim k_4^{cp}\left(R_p^{cp}, t_p^{cp}\right)$ can be solved as follows:

$$\begin{aligned} k_{2c}\left(R_{pc}, t_{pc}\right) &= k_2\left(R_p, t_p\right) - 6k_{4c}\left(R_{pc}, t_{pc}\right)\left(\frac{a}{k_{20}-a}t_p\right)^2 \\ &\quad + 3k_{3c}\left(R_{pc}, t_{pc}\right)\left(\frac{a}{k_{20}-a}t_p\right) - a - 3bt_p - 6dt_p^2 \\ k_{3c}\left(R_{pc}, t_{pc}\right) &= k_3\left(R_p, t_p\right) + 4k_{4c}\left(R_{pc}, t_{pc}\right)\left(\frac{a}{k_{20}-a}t_p\right) - b - 4dt_p \\ k_{4c}\left(R_{pc}, t_{pc}\right) &= k_4\left(R_p, t_p\right) - d \end{aligned} \tag{3.76}$$

Because R_{pc} and $k_{1c}\left(R_{pc}\right)$ do not affect the following derivation, they are omitted here.

The analysis in Sect. 3.3.2 shows that most of the RCM come from k_{a21}, k_{a22} and k_{a31}; thus, the polynomial compensation should set the compensated azimuth variance coefficients k_{a21c}, k_{a22c} and k_{a31c} to zero to minimize the azimuth-variant RCM. Combining $t_p = (k_{20} - a)t_{pc}/k_{20}$ with (3.76), the derivatives of $k_{2c}\left(R_{pc}, t_{pc}\right)$ and $k_3^{cp}\left(R_p^{cp}, t_p^{cp}\right)$ with respect to t_{pc} can be obtained as follows:

$$\begin{aligned} k_{a21c} &= k_{a21} \cdot \left(\frac{k_{20}-a}{k_{20}}\right) + \frac{3k_{a30} \cdot a}{k_{20}} - 3b \\ k_{a22c} &= \frac{\left[\begin{array}{l} k_{a22}(k_{20}-a)^2 - 6d \cdot (k_{20}-a) \cdot (k_{20}+a) \\ \quad + 3k_{a31} \cdot (k_{20}-a) \cdot a + 6(k_{40}-d) \cdot a^2 \end{array}\right]}{k_{20}^2} \\ k_{a31c} &= k_{a31} \cdot \left(\frac{k_{20}-a}{k_{20}}\right) + 4(k_{40}-d) \cdot \frac{a}{k_{20}} - 4d \cdot \left(\frac{k_{20}-a}{k_{20}}\right) \end{aligned} \tag{3.77}$$

Then, the polynomial compensation factors can be obtained by setting (3.77) to 0:

$$\begin{aligned} a &= \frac{k_{20} \cdot (2k_{a22} - 3k_{a31})}{2 \cdot (k_{a22} + 6k_{40} - 3k_{a31})} \\ b &= \frac{a \cdot k_{30}}{k_{20}} + \frac{k_{a21}(k_{20}-a)}{3k_{20}} \\ d &= \frac{k_{a31}(k_{20}-a)}{4k_{20}} + \frac{ak_{40}}{k_{20}} \end{aligned} \tag{3.78}$$

Furthermore, there are several other kinds of compensation factors, and (3.78), which sets k_{a21c}, k_{a22c} as well as k_{a31c} to zero, is just the most commonly used one. In addition, there are also some other different kinds of compensation factors which have different expressions and effects compared with (3.78). Some kinds of them which are commonly used are shown as follows:

- Higher order compensation factors b and d equal 0, and derivation k_{a21c} equals 0:

$$a = \frac{k_{20}k_{a21}}{k_{a21} - 3k_{30}} \tag{3.79}$$

- Higher order compensation factor d equal 0, and derivations k_{a21c} and k_{a31c} equal 0:

$$\begin{aligned} a &= \frac{k_{a31} \cdot k_{a20}}{k_{a31} - 4k_{a40}} \\ b &= \frac{a \cdot k_{a30}}{k_{a20}} + \frac{k_{a21}(k_{a20} - a)}{3k_{a20}} \end{aligned} \tag{3.80}$$

Moreover, other kinds of compensation factors can be obtained in a similar way.

Furthermore, it should be noted that the quadratic compensation factor a cannot be equal to 0, for it unfolds the folded azimuth frequency spectrum at the orbital positions whose Doppler rates equal 0. Besides, for the orbital positions with zero Doppler rates, the analytical compensation factor a is zero, and it is incorrect obviously. Thus, a random quadratic factor is compensated for the expansion of the azimuth spectrum, and the optimal compensation factor can be obtained after calculating the new azimuth variance of the compensation echo.

3.4.3 Details of 2D NCSA Based on the Azimuth Compensation

Based on aforementioned analysis, in GEO SAR imaging processing, the range model coefficients have severe nonlinear 2D spatial variance, which means that the RDA, CSA, ωKA and SPECAN which don't consider the spatial variance or only consider the linear spatial variance of range model coefficients are not accurate enough. Thus, the NCSA which can deal with the nonlinear spatial variant range model coefficients should be adopted in GEO SAR imaging. Furthermore, as shown in 3.4.1 and 3.4.2, the azimuth polynomial compensation should also be utilized to remove the range walk and reduce the azimuth spatial variant RCM. The flow chart of the 2D NCSA based on azimuth compensation is shown as Fig. 3.18.

The 2D NCSA consists of two parts. One part is the azimuth compensation which includes time domain compensation and frequency domain compensation. The other part is 2D NCSA which consists of range NCSA and azimuth NCSA. In the following, these two parts will be introduced in detail.

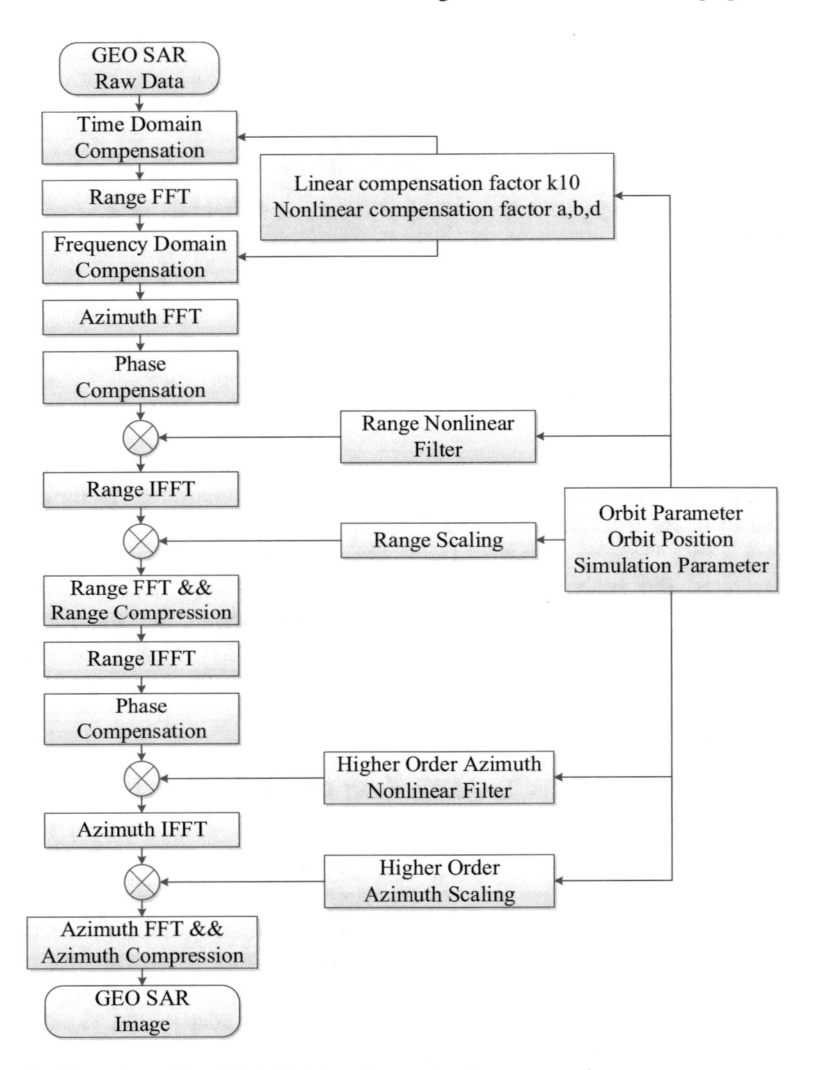

Fig. 3.18 Flowchart of the 2D NCSA based on azimuth compensation

A. Azimuth compensation

The compensation functions can be written as follows:

$$
\begin{aligned}
H_{cpt} &= \exp\left[j\tfrac{4\pi}{\lambda}\left(k_{10}t_a + at_a^2 + bt_a^3 + dt_a^4\right)\right] \\
H_{cpf} &= \exp\left[j\tfrac{4\pi}{c}\left(k_{10}t_a + at_a^2 + bt_a^3 + dt_a^4\right)f_r\right]
\end{aligned}
\tag{3.81}
$$

where H_{cpt} and H_{cpf} are the compensation functions in the time and frequency domains, respectively, k_{10} is the linear compensation factor, and its related compensation can remove the range walk. Here, a, b and d are the compensation factors, and their related compensation can weaken the azimuth variance of RCM. The expressions of the compensation factors are as (3.78), (3.80) or (3.81).

B. Range NCSA

Through range FFT and azimuth FFT, the 2D spectrum is obtained. In 2D frequency domain, the first step is to remove the migration phase $b(f_a)$ at reference point through compensation function shown in (31). Then, multiply a nonlinear frequency-modulated function shown in (32), this function can not only remove the effect of cubic phase, but also remove the residual cubic phase error caused by adjusting the spatial variance of FM rate.

$$H_1 = \exp(-j2\pi b(f_a) \cdot f_r), \tag{3.82}$$

$$H_2 = \exp\left(j\frac{2\pi}{3}Y(f_a)f_r^3\right), \tag{3.83}$$

where

$$Y(f_a) = \frac{\Delta k_s(f_a)(\alpha - 0.5)}{K_s^3(f_a, R_0)(\alpha - 1)} - \frac{3}{2\pi} \cdot \phi_3(f_a, R_0) \tag{3.84}$$

$$\alpha = \frac{M(f_{ref})}{M(f_a)} \tag{3.85}$$

f_{ref} is azimuth reference frequency and depends on reference target position.
After compensation with (3.82) and (3.83), 2D spectrum can be rewritten as:

$$\begin{aligned} S_2(f_r, f_a) = {} & A_r\left(\frac{f_r}{K_r}\right) \cdot A_a\left[f_a + \frac{2k_1}{c} \cdot (f_r + f_c)\right] \exp[j2\pi\phi_0(f_a, R)] \\ & \exp\left(-j\frac{4\pi R}{cM(f_a)}f_r\right) \exp\left(-j\pi\frac{f_r^2}{K_s(f_a, R)}\right) \exp\left[j\frac{2\pi}{3}Y_m(f_a)f_r^3\right] \end{aligned} \tag{3.86}$$

where

$$Y_m(f_a) = \frac{\Delta k_s(f_a)(\alpha - 0.5)}{K_s^3(f_a, R_0)(\alpha - 1)}. \tag{3.87}$$

Via range IFFT to (3.86), echo signal in RD domain can be written as

$$\begin{aligned} S_3(t_r,f_a) = a_r\left\{\frac{K_s(f_a,R)}{K_r}\left[t_r - \frac{2R}{CM(f_a)}\right]\right\}A_a(f_a)\exp[j2\pi\phi_0(f_a,R)] \\ \exp\left(j2\pi K_s(f_a,R)\left[t_r - \frac{2R}{cM(f_a)}\right]^2\right) \\ \exp\left(j\frac{2\pi}{3}Y_m(f_a)K_s^3(f_a,R)\left[t_r - \frac{2R}{cM(f_a)}\right]^3\right) \end{aligned} \tag{3.88}$$

In order to remove spatial variance of RCM and range FM rate, nonlinear chirp scaling operation is implemented in RD domain. After multiplying the compensation function (3.89) in RD domain, the spatial variance of target RCM in whole scene is removed. Thus, the residual RCM can be bulk processed in 2D frequency domain.

$$H_3 = \exp\left\{j\pi q_2[t_r - \tau(f_a,R_0)]^2\right\}\cdot\exp\left\{j\frac{2\pi}{3}q_3[t_r - \tau(f_a,R_0)]^3\right\} \tag{3.89}$$

where

$$\begin{aligned} q_2 &= K_s(f_a,R_0)(\alpha-1) \\ q_3 &= \frac{\Delta k_s(f_a)(\alpha-1)}{2} \end{aligned} \tag{3.90}$$

The detailed derivation of q_2 and q_3 are shown in Appendix A.

Multiplying (3.88) with (3.89), and implementing range FFT, the echo signal in 2D frequency domain can be further rewritten as:

$$\begin{aligned} S_4(f_r,f_a) = \; & A_r\left(\frac{f_r}{K_s(f_a,R_0)\alpha}\right)A_a(f_a)\exp[j2\pi\phi_0(f_a,R)]\exp(j\pi C_0) \\ & \exp\left(-j\pi\frac{f_r^2}{\alpha K_s(f_a,R_0)}\right)\exp\left\{j\frac{2\pi}{3}\frac{\left[Y_m(f_a)K_s^3(f_a,R_0)+q_3\right]}{\left[\alpha K_s(f_a,R_0)\right]^3}\cdot f_r^3\right\} \\ & \exp\left(-j\frac{4\pi R_0}{c}\cdot\left[\frac{1}{M(f_a)} - \frac{1}{M(f_{ref})}\right]\cdot f_r\right)\exp\left(-j\frac{4\pi R}{cM(f_{ref})}\cdot f_r\right) \end{aligned} \tag{3.91}$$

In (3.91),C_0 is the residual term after solving $Y_m(f_a)$, q_2 and q_3, and it can be expressed as:

$$\begin{aligned} C_0 = & K_s(f_a,R)\Delta\tau^2\left(\frac{1}{\alpha}-1\right)^2 + \frac{2}{3}Y_m(f_a)K_s^3(f_a,R)\Delta\tau^3\left(\frac{1}{\alpha}-1\right)^3 \\ & + q_2\left(\frac{\Delta\tau}{\alpha}\right) + \frac{2}{3}q_3\left(\frac{\Delta\tau}{\alpha}\right)^3 \end{aligned} \tag{3.92}$$

where

$$\Delta\tau = \tau(f_a, R) - \tau(f_a, R_0) \tag{3.93}$$

C_0 not only relates to azimuth Doppler frequency, but also changes along range direction, thus the compensation of C_0 must be implemented in RD domain.

After the nonlinear chirp scaling operation with (3.89), RCMC can be bulk processed in azimuth frequency domain. Therefore, the corresponding range compression and secondary range compression will be carried out in 2D frequency domain at the same time. The RCMC function can be written as

$$H_4 = \exp\left\{j\frac{4\pi R_0}{c}\left[\frac{1}{M(f_a)} - \frac{1}{M\left(f_{ref}\right)}\right]\cdot f_r\right\} \tag{3.94}$$

Furthermore, the range compression function and secondary range compression function can be expressed respectively as

$$H_5 = \exp\left\{j\pi\frac{f_r^2}{\alpha K_s(f_a, R_0)}\right\} \tag{3.95}$$

$$H_6 = \exp\left\{-j\frac{2\pi}{3}\frac{\left[Y_m(f_a)K_s^3(f_a, R_0) + q_3\right]}{\left[\alpha K_s(f_a, R_0)\right]^3}\cdot f_r^3\right\} \tag{3.96}$$

After processing with (3.94), (3.95) and (3.96), the 2D frequency domain expression can be further written as

$$\begin{aligned} S_5(f_r, f_a) =& A_r\left(\frac{f_r}{K_s(f_a, R_0)\alpha}\right)A_a(f_a)\exp[j2\pi\phi_0(f_a, R)] \\ & \exp\left(-j\frac{4\pi R}{cM\left(f_{ref}\right)}\cdot f_r\right)\exp(j\pi C_0) \end{aligned} \tag{3.97}$$

C. Azimuth NCSA

Before the azimuth NCSA processing, the residual phase should be compensated first, and the compensation function can be expressed as

$$H_7 = \exp[-j2\pi\phi_{RP}(R) - j\pi C_0] \tag{3.98}$$

Then, the signal can be written as:

$$S_2(t_r, f_a) = \sin c\left(t_r - \frac{2R}{c}\right)\exp\{j\phi_{az}(f_a, R)\} \tag{3.99}$$

where

$$\phi_{az}(f_a,R)=\pi\cdot\left\{\begin{array}{l}\left(\frac{k_1}{k_2}+\frac{3k_3k_1^2}{4k_2^3}+\frac{k_1^3\left(9k_3^2-4k_2k_4\right)}{8k_2^5}\right)f_a\\+\left(\frac{\lambda}{4k_2}+\frac{3k_3k_1\lambda}{8k_2^3}+\frac{3\lambda k_1^2\left(9k_3^2-4k_2k_4\right)}{32k_2^5}\right)f_a^2\\+\left(\frac{k_3\lambda^2}{16k_2^3}+\frac{k_1\lambda^2\left(9k_3^2-4k_2k_4\right)}{32k_2^5}\right)f_a^3\\+\left(\frac{\lambda^3\left(9k_3^2\lambda^3-4k_4k_2\right)}{256k_2^5}\right)f_a^4\end{array}\right\} \tag{3.100}$$

Because k_1 is very small and the azimuth variance of k_1 can be neglected, the phase related to k_1 is first compensated in the range-Doppler domain. The phase of the compensation function can be written as

$$\begin{aligned}H_9(f_a,R)=&\left(\frac{k_1}{k_2}+\frac{3k_3k_1^2}{4k_2^3}+\frac{k_1^3\left(9k_3^2-4k_2k_4\right)}{8k_2^5}\right)f_a\\&+\left(\frac{3k_3k_1\lambda}{8k_2^3}+\frac{3\lambda k_1^2\left(9k_3^2-4k_2k_4\right)}{32k_2^5}\right)f_a^2\\&+\left(\frac{k_1\lambda^2\left(9k_3^2-4k_2k_4\right)}{32k_2^5}\right)f_a^3\end{aligned} \tag{3.101}$$

After compensation, the azimuth phase can be written as

$$\begin{aligned}H_9(f_a,R)=&\left(\frac{k_1}{k_2}+\frac{3k_3k_1^2}{4k_2^3}+\frac{k_1^3\left(9k_3^2-4k_2k_4\right)}{8k_2^5}\right)f_a\\&+\left(\frac{3k_3k_1\lambda}{8k_2^3}+\frac{3\lambda k_1^2\left(9k_3^2-4k_2k_4\right)}{32k_2^5}\right)f_a^2\\&+\left(\frac{k_1\lambda^2\left(9k_3^2-4k_2k_4\right)}{32k_2^5}\right)f_a^3\end{aligned} \tag{3.102}$$

Next, expand the cubic and quartic terms in (3.102) with the Taylor series and compensate for their azimuth-invariant components. The azimuth phase after compensation can be written as follows:

$$\phi_{az3}(f_a,R)=\pi\left[\frac{\lambda}{4k_2}f_a^2+\left(a_{rt1}\cdot t_p+a_{rt2}\cdot t_p^2\right)\cdot f_a^3+a_{rt3}\cdot t_p\cdot f_a^4\right] \tag{3.103}$$

where

$$\begin{aligned}
a_{rt1} &= \frac{1}{16}\cdot\frac{\lambda^2(-3k_{30}k_{a21}+k_{a31}k_{20})}{k_{20}^5}\\
a_{rt2} &= \frac{1}{16}\cdot\frac{\lambda^2\left(-3k_{30}k_{a22}k_{20}+6k_{30}k_{a21}^2+k_{a32}k_{20}^2-3k_{a31}k_{a21}k_{20}\right)}{k_{20}^5}\\
a_{rt3} &= -\frac{1}{256}\frac{\lambda^3\left(45k_{a21}k_{30}-16k_{a21}k_{20}k_{40}-18k_{20}k_{30}k_{a31}+4k_{20c}^2k_{41}\right)}{k_{20}^5}
\end{aligned} \tag{3.104}$$

Then, multiply the signal using the nonlinear filter:

$$H_{ANF} = \exp\left[j\pi\left(p_3f_a^3 + p_4f_a^4 + p_5f_a^5\right)\right] \tag{3.105}$$

where p_3, p_4 and p_5 are the nonlinear phase coefficients. After nonlinear filtering, azimuth IFFT is performed to transform the signal to the 2D time domain, and azimuth nonlinear chirp scaling, whose function is shown in (3.106), is performed:

$$H_{ANCS} = \exp\left[j\pi\left(q_2t_a^2 + q_3t_a^3 + q_4t_a^4 + q_5t_a^5\right)\right] \tag{3.106}$$

where q_2, q_3, q_4 and q_5 are the nonlinear scaling factors.

For the sake of brevity, the derivations of the phase coefficients and scaling factors are shown in Appendix B, and their expressions are

$$\begin{cases}
q_2 = (2\beta-1)f_{dr0}\\
q_3 = \frac{(2\beta-1)f_{dr1}}{3}\\
q_4 = \frac{14f_{dr1}^2\beta-5f_{dr1}^2-4f_{dr0}f_{dr2}\beta+6f_{dr0}^4a_{rt1}-3f_{dr0}^4a_{rt1}}{12\cdot f_{dr0}}\\
q_5 = \frac{\left(\begin{array}{l}7f_{dr1}^3\beta+2f_{dr1}f_{dr0}^4a_{rt1}\beta+2f_{dr0}^6a_{rt3}\beta-f_{dr0}^5a_{rt2}\beta\\ -3f_{dr1}f_{dr0}f_{dr2}\beta-f_{dr0}^4f_{dr1}a_{rt1}-f_{dr0}^6a_{rt3}-2f_{dr1}^3\end{array}\right)}{5\cdot f_{dr0}^2}\\
p_3 = \frac{f_{dr1}(4\beta-1)}{3\cdot f_{dr0}^3(2\beta-1)}\\
p_4 = \frac{(4\beta-1)\cdot\left(3f_{dr0}^4a_{rt1}+5f_{dr1}^2\right)-4f_{dr2}f_{dr0}\beta}{12\cdot f_{dr0}^5(2\beta-1)}\\
p_5 = \frac{\left(\begin{array}{l}4f_{dr1}f_{dr0}^4a_{rt1}\beta+4f_{dr0}^6a_{rt3}\beta+9f_{dr1}^3\beta-f_{dr0}^5a_{rt2}\beta\\ -3f_{dr1}f_{dr2}f_{dr0}\beta-f_{dr0}^4f_{dr1}a_{rt1}-2f_{dr1}^3-f_{dr0}^6a_{rt3}\end{array}\right)}{5\cdot f_{dr0}^7(2\beta-1)}
\end{cases} \tag{3.107}$$

where $f_{drn} = -\frac{4k_{a2nc}}{\lambda}$(n = 0, 1, 2), and $\beta(\neq \frac{1}{2})$ is a constant that represents the linear scaling of the image introduced by azimuth NCS processing.

After azimuth nonlinear filtering and nonlinear chirp scaling, the azimuth variance of the focusing parameters has been removed. Thus, unified azimuth compression can be performed by

$$H_{MF} = \exp\left[j\pi\left(\begin{array}{c}\frac{1}{q_2+f_{dr0}}f_a^2 - \frac{p_3 f_{dr0}^3+q_3}{(q_2+f_{dr0})^3}f_a^3 \\ -\frac{p_4 f_{dr0}^4+q_4}{(q_2+f_{dr0})^4}f_a^4 - \frac{p_5 f_{dr0}^5+q_5}{(q_5+f_{dr0})^5}f_a^5\end{array}\right)\right] \tag{3.108}$$

Finally, azimuth IFFT is performed, and the well-focused image is

$$s(t_r, t_a) = \sin c\left(t_r - \frac{2R_{pc}}{c}\right) \cdot \sin c\left(t_a - \frac{t_{pc}}{2\beta}\right) \tag{3.109}$$

where t_{pc} and R_{pc} are the beam crossing time and the corresponding instantaneous slant range after optimal azimuth polynomial compensation.

3.5 Discussion

The proposed algorithm can be used for the entire orbit of GEO SAR, regardless of whether the ground squint angle and Doppler rate are zeroes. For work mode with non-zero ground squint angle, the linear azimuth compensation can remove the range walk and decouple the azimuth direction and range direction. For orbit positions with near-zero Doppler rates, the quadratic azimuth compensation can transform the nonlinear frequency versus time relationship into a linear relationship. Furthermore, the azimuth spatial variance of the focus parameters could be weakened by optimizing the coefficients of the compensation functions. Finally, the residual 2D spatial variance of the focus parameters can be eliminated by the 2D NCSA based on the accurate slant range model.

Although the proposed algorithm has many advantages and can overcome the difficulties existing in GEO SAR imaging, it also has two main disadvantages. The first is that the proposed algorithm will result in additional geometric distortion in the final GEO SAR images, which must be corrected, and the second is that the azimuth swath width that the proposed algorithm can handle will change with the argument of latitude. Thus, these two disadvantages will be discussed in this section to make the readers have a complete understanding of the proposed algorithm.

A. Geometric distortion

Because of the azimuth polynomial compensation and the azimuth NCSA, there is a tremendous amount of geometric distortion in the GEO SAR image obtained by the proposed algorithm. This geometric distortion will have a negative impact on the interpretation of SAR image; therefore, the analysis of geometric distortion is significant. The distortion of azimuth NCSA involves simple linear scaling; consequently, a discussion is omitted in this paper. However, the distortion caused by polynomial compensation is complex; thus, it will be introduced in detail.

After polynomial compensation, the beam center crossing time and corresponding instantaneous slant range both change in complex ways. Considering the analysis in [11] and the influence of range walk correction, their changes can be expressed as

$$\begin{cases} R_{pc} = R_p - k_{10} \cdot t_p - a \cdot t_p^2 - b \cdot t_p^3 - d \cdot t_p^4 \\ + k_{1c}\left(R_{pc}\right) \cdot \left(\frac{a}{k_{20}-a} t_p\right) - k_{2c}\left(R_{pc}, t_{pc}\right) \cdot \left(\frac{a}{k_{20}-a} t_p\right)^2 \\ + k_{3c}\left(R_{pc}, t_{pc}\right) \cdot \left(\frac{a}{k_{20}-a} t_p\right)^3 - k_{4c}\left(R_{pc}, t_{pc}\right) \cdot \left(\frac{a}{k_{20}-a} t_p\right)^4 \\ t_{pc} = \frac{k_{20}}{k_{20}-a} t_p + \frac{k_{10}}{2k_{20}} \end{cases} \tag{3.110}$$

Then, geometric correction can be realized based on the function shown in (3.110), and well-focused GEO SAR images without geometric distortion can be obtained.

B. Azimuth swath width

The algorithm proposed in this chapter can solve the azimuth space variance effectively and enlarge the azimuth swath width of high squint GEO SAR. However, the azimuth swath width that the proposed algorithm can handle is not constant: it depends on the wavelength, orbital parameters and orbital positions. Commonly, when the wavelength and orbital parameters are decided, a greater argument of latitude (AOG) will result in a longer SAT and increase the trajectory curvature. Consequently, the imaging processing becomes more difficult and the azimuth swath width decreases.

Because the effect of the azimuth residual phase can be mitigated by the azimuth NCSA used in the algorithm, the key factor that restricts the azimuth swath width is azimuth-variant RCM. Therefore, the theoretical maximum azimuth swath width can be obtained by

$$\max\left[\frac{2f_s}{c} \cdot \left| \begin{array}{r} \left(k_{a21c} t_{pc} + k_{a22c}\left(t_{pc}\right)^2\right) \cdot \left(t_a - t_{pc}\right)^2 \\ + \left(k_{a31c} t_{pc}\right) \cdot \left(t_a - t_{pc}\right)^3 \end{array} \right| \right] \leq \frac{1}{2} \tag{3.111}$$

Theoretically, k_{a21c}, k_{a22c}, k_{a31c} and the corresponding azimuth-variant RCM should equal zero after optimal azimuth polynomial compensation. However, because of the approximation used in (3.75), k_{a21c}, k_{a22c}, k_{a31c} and the corresponding azimuth-variant RCM cannot strictly equal zero. Moreover, other factors such as the 2D spatially variant range-azimuth coupling will also restrict the available azimuth swath width. Thus, the practical available azimuth swath width will be less than the result obtained from (3.111).

3.5.1 Computer Simulation

A. Simulation for orbital position with 0 Doppler rate

To verify the effectiveness of the 2D NCSA based on azimuth compensation at the orbital positions with Doppler rate of 0, computer simulations based on a target array which consists of 11 × 11(ground range × azimuth) points and SAR raw data are performed. The synthetic aperture time is 130 s, the azimuth resolution is 20 m, and the orbital parameters as well as simulation parameters are shown in Tables 3.1 and 3.4, respectively. Both the target array and the SAR raw data have a scene size of 400 km × 400 km. The imaging result of the target array is shown in Fig. 3.19. The 2D contour maps of six targets are shown in Figs 3.20 and 3.21. Four targets are on the edge, one target is at the center, and the last target is chosen randomly. The evaluation results for the six target points are shown in Table 3.6. The focus results of the SAR raw data are shown in Fig. 3.21. As shown in the imaging results and

Table 3.1 GEO SAR Orbital Parameters

Parameter (unit)	Value
Semi major axis (km)	42,164
Inclination (°)	50
Longitude of ascending node (°)	108E
Eccentricity	0
Argument of perigee (°)	90

Table 3.2 Traditional BP computational budget

Operation	FLOP
Pulse compression	$5N_aN_r\log_2 N_r + 6N_aN_r$ $+ 5N_aN_{up}N_r\log_2(N_{up}N_r)$
Target slant range calculation	$11N_aN_xN_y$
Reverse point selection	$3N_aN_xN_y$
Phase compensation	$6N_aN_xN_y$
Accumulation	$2N_aN_xN_y$

Table 3.3 Fast BP computational budget of pulse compression

Operation	FLOP
Pulse compression	$5N_aN_r\log_2 N_r + 6N_aN_r$ $+ 5N_aN_{up}N_r\log_2(N_{up}N_r)$
Imaging processing based on sub-apertures	$\frac{N_a}{M}\left[\begin{array}{l}22NM^2 + 5NM\log_2(NM)\\ + 5M_{up}NM\log_2(M_{up}NM) + 11N^2\end{array}\right]$
Sub-apertures based SAR images accumulation	$8\frac{N_a}{M}N^2$

Table 3.4 Simulation parameters for BP algorithms

Parameter (unit)	Value	Parameter (unit)	Value
Wavelength (m)	0.24	PRF (Hz)	200
SAT(s)	130	Bandwidth (MHz)	20
Sampling rate (MHz)	24	Pulse width (μs)	500
Squint angle (°)	0	Ground squint angle (°)	0

Table 3.5 Simulation parameters

Parameters (unit)	Value	Parameters (unit)	Value
Wavelength (m)	0.24	PRF (Hz)	200
Elevation angle (°)	4.5	Incident angle (°)	30
Bandwidth (MHz)	22	Sample frequency (MHz)	26.4
Squint angle (°)	3.8961	Ground squint Angle (°)	60
Ground range resolution (m)	20	Azimuth resolution (m)	20

Table 3.6 Evaluation results for a zero Doppler rate

Target	A	B	C	D	E	F
Range PSLR (dB)	−13.36	−13.27	−13.29	−13.26	−13.26	−13.28
Range ISLR (dB)	−10.05	−10.04	−10.07	−10.01	−10.03	−10.06
Azimuth PSLR (dB)	−13.0	−13.09	−13.25	−13.1	−13.26	−13.1
Azimuth ISLR (dB)	−12.16	−11.39	−11.99	−11.93	−10.37	−11.98

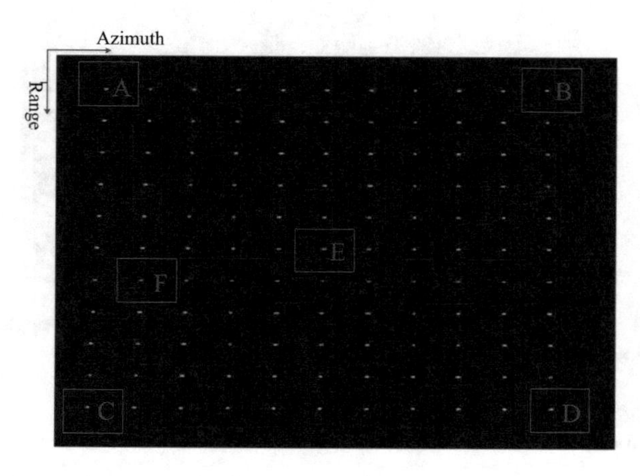

Fig. 3.19 Imaging results at an orbital position with a zero Doppler rate using the 2D NCS algorithm (400 km ×400 km)

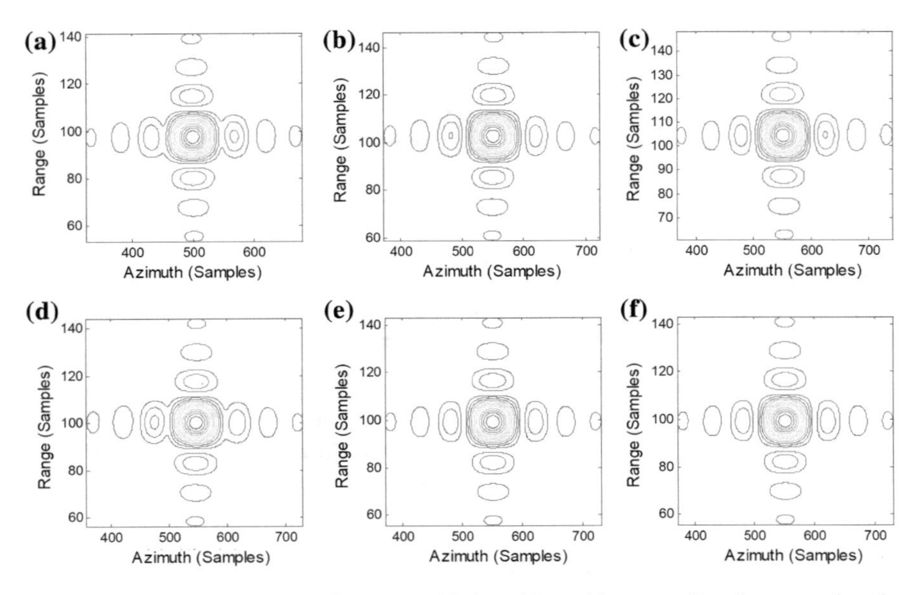

Fig. 3.20 GEO SAR image results at an orbital position with a zero Doppler rate using the proposed algorithm: **a** Target A; **b** Target B; **c** Target C; **d** Target D; **e** Target E; and **f** Target F

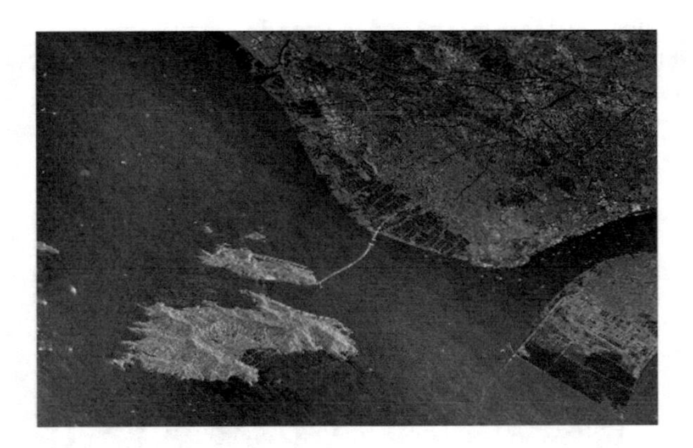

Fig. 3.21 SAR raw data simulation result of a target area at a zero Doppler rate position (400 km ×400 km)

Table 3.7 Evaluation results for equator with ground squint angle of 60°

Target	A	B	C	D	E
Range PSLR (dB)	−13.15	−13.36	−13.26	−13.05	−13.16
Range ISLR (dB)	−10.44	−10.61	−10.42	−10.74	−10.68
Azimuth PSLR (dB)	−13.31	−13.61	−13.31	−13.54	−13.21
Azimuth ISLR (dB)	−10.06	−10.39	−10.09	−10.40	−10.06

evaluation results, well focused GEO SAR images at an orbital position with a zero Doppler rate can be achieved by using the proposed GEO SAR imaging algorithm.

B. Simulation for equator with ground squint angle of 60°

To validate the effectiveness of the 2D NCSA based on azimuth compensation in high squint conditions, computer simulations based on a target array which consists of 11 × 11(ground range × azimuth) points and SAR raw data are performed. The synthetic aperture time is 150 s, the azimuth resolution is 20 m, and the orbital parameters as well as simulation parameters are shown in Fig. 3.1 and Table 3.5, respectively. Both the target array and the SAR raw data have a scene size of 400 km × 400 km. The imaging result of the target array is shown in Fig. 3.22. The 2D contour maps of five targets are shown in Fig. 3.24. Four targets are on the edge, and one target is at the center. The evaluation results for the five target points

Fig. 3.22 Imaging results at the equator with ground squint angle of 60° (300 km ×300 km)

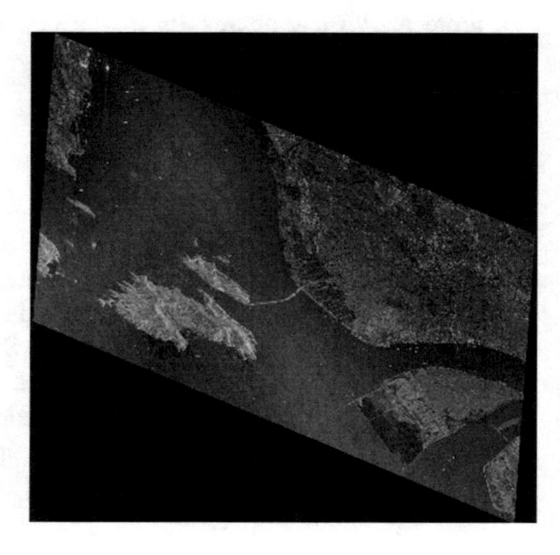

Fig. 3.23 SAR raw data simulation result at the equator with ground squint angle of 60° (300 km ×300 km)

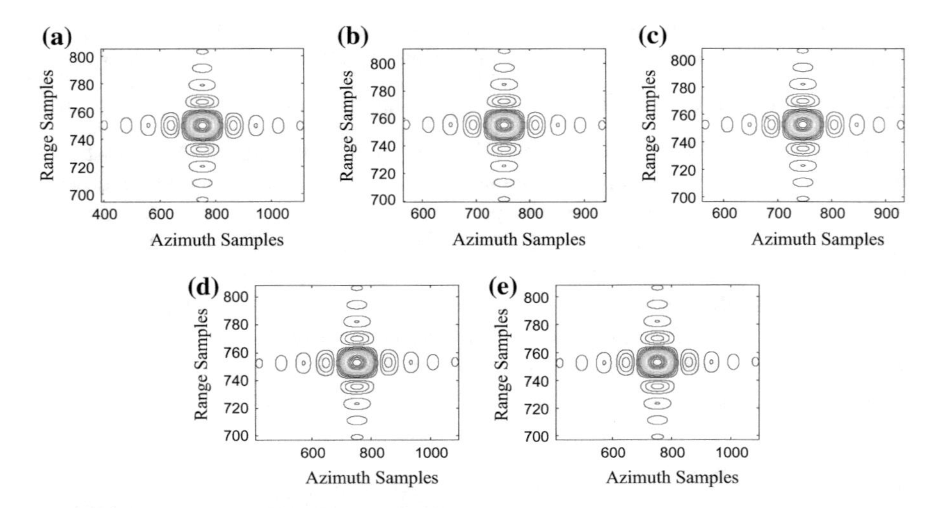

Fig. 3.24 Contour maps of selected targets obtained by proposed algorithm: **a** A, **b** B, **c** C, **d** D, **e** E

Table 3.8 Evaluation results for equator with ground squint angle of 60°

Target	A	B	C	D	E
Range PSLR (dB)	−13.15	−13.36	−13.26	−13.05	−13.16
Range ISLR (dB)	−10.44	−10.61	−10.42	−10.74	−10.68
Azimuth PSLR (dB)	−13.31	−13.61	−13.31	−13.54	−13.21
Azimuth ISLR (dB)	−10.06	−10.39	−10.09	−10.40	−10.06

are listed in Table 3.7. The focus results of the SAR scene target is shown in Fig. 3.23. As shown in the imaging results and evaluation results, well focused GEO SAR images with ground angle of 60° can be achieved by using the proposed GEO SAR imaging algorithm.

Figure 3.22 Imaging results at the equator with ground squint angle of 60° (300 km × 300 km).

3.6 Summary

This chapter introduces the key issues in GEO SAR imaging, including signal model, time domain algorithm and frequency domain algorithm.

Based on the characteristics of "non-stop-and-go" assumption, long SAT, curved trajectory and large imaging scene in GEO SAR system, this chapter elaborates the GEO SAR signal model and obtains the accurate GEO SAR slant range model, 2D spectrum as well as 2D spatially variant model of the slant range model coefficients.

Moreover, this chapter introduces the application of time domain algorithm, which can be applied to any condition, in GEO SAR imaging processing. The principle and processing flow of the traditional and fast BP imaging algorithms are introduced, respectively in this section, and the imaging results of traditional and fast BP imaging algorithms are given and compared in Sect. 3.3.

Although the time domain algorithm is adequately precise and suitable for SAR imaging with any kind of condition, its tremendous computational budget is an inevitably issue in the engineering of GEO SAR because of its long synthetic aperture time and large scene size. Therefore, this chapter introduces the 2D NCS algorithm in details which is more suitable for GEO SAR imaging whose focus parameters are 2D spatially variant. In this chapter, the difficulties in GEO SAR frequency domain imaging are analyzed. Moreover, the improved 2D NCS algorithm based on azimuth compensation is proposed and verified, and well-focused GEO SAR images are obtained.

Appendix A: Derivation of the Range NCS Factor

After introducing a nonlinear FM function in 2D frequency domain, via the range IFFT operation, the range-Doppler expression can be written as

$$\begin{aligned} S_1(t_r,f_a) = & \exp\left[j2\pi\phi_{az}(f_a,R)\right]\exp[j2\pi\phi_{RP}(R)] \\ & \exp\left\{j\pi K_s(f_a,R)\left[t_r-\frac{2R}{cM(f_a)}\right]^2\right\} \\ & \exp\left\{j\frac{2\pi}{3}Y_m(f_a)K_s(f_a,R)\left[t_r-\frac{2R}{cM(f_a)}\right]^3\right\} \end{aligned} \tag{3.112}$$

Here, we ignore the effect of amplitude, and H_1 has been compensated in 2D frequency domain in the derivation of (3.112). The nonlinear chirp scaling operation is implemented in range-Doppler domain and the chirp scaling function can be expressed as

$$H_3 = \exp\left\{j\pi q_2[t_r-\tau(f_a,R_0)]^2\right\}\exp\left\{j\frac{2\pi}{3}q_3[t_r-\tau(f_a,R_0)]^3\right\}. \tag{3.113}$$

After multiplying (3.112) by (3.113), we have

$$\begin{aligned} S_1(t_r,f_a) = & \exp\left[j2\pi\phi_{az}(f_a,R)\right]\exp[j2\pi\phi_{RP}(R)] \\ & \exp\left\{j\pi\left[\begin{array}{l} C_0+2C_1[t_r-\tau_s(f_a,R)]+C_2[t_r-\tau_s(f_a,R)]^2 \\ +\frac{2}{3}C_3[t_r-\tau_s(f_a,R)]^3 \end{array}\right]\right\} \end{aligned} \tag{3.114}$$

where

$$\begin{aligned}
C_1 &= \left[K_s(f_a, R_0)\left(\frac{1}{\alpha} - 1\right) + \frac{q_2}{\alpha}\right]\Delta\tau \\
&+ \left[\Delta k_s(f_a)\left(\frac{1}{\alpha} - 1\right) + Y_m(f_a)K_s^3(f_a, R_0)\left(\frac{1}{\alpha} - 1\right)^2 + \frac{q_3}{\alpha^2}\right]\Delta\tau^2 + \cdots \\
C_2 &= [K_s(f_a, R_0) + q_2] \\
&+ \left[\Delta k_s(f_a) + 2Y_m(f_a)K_s^3(f_a, R_0) + \frac{2q_3}{\alpha^2}\right]\Delta\tau + \cdots \\
C_3 &= Y_m(f_a)K_s^3(f_a, R_0) + q_3 + \cdots
\end{aligned} \tag{3.115}$$

In order to adjust the spatial variance, we should define the following equation to exist.

$$\begin{aligned}
&K_s(f_a, R_0)\left(\frac{1}{\alpha} - 1\right) + \frac{q_2}{\alpha} = 0 \\
&\Delta k_s(f_a)\left(\frac{1}{\alpha} - 1\right) + Y_m(f_a)K_s^3(f_a, R_0)\left(\frac{1}{\alpha} - 1\right)^2 + \frac{q_3}{\alpha^2} = 0 \\
&\Delta k_s(f_a) + 2Y_m(f_a)K_s^3(f_a, R_0) + \frac{2q_3}{\alpha^2} = 0
\end{aligned} \tag{3.116}$$

Solving (3.116), we can obtain

$$\begin{aligned}
q_2 &= K_s(f_a, R_0)(\alpha - 1) \\
q_3 &= \frac{\Delta k_s(f_a)(\alpha - 1)}{2} \\
Y_m(f_a) &= \frac{\Delta k_s(f_a)(\alpha - 0.5)}{K_s^3(f_a, R_0)(\alpha - 1)}
\end{aligned} \tag{3.117}$$

Appendix B: Derivation of the Azimuth NCS Factor

After azimuth nonlinear filtering, the signal phase can be written as

$$\phi_{az4}(f_a, R) = \frac{\lambda}{4k_{2c}}f_a^2 + \left(p_3 + a_{rt1}t_p + a_{rt2}t_p^2\right)f_a^3 + (p_4 + a_{rt3}t_p)f_a^4 + p_5f_a^5 \tag{3.118}$$

Then, azimuth IFFT is performed to transform the signal into the 2D time domain. For an arbitrary point target, its signal phase in the time domain can be written as follows:

$$\phi_{az5}(t_a,R)=\pi\left[\begin{array}{l}-\dfrac{4k_{2c}\left(t_a-t_p\right)^2}{\lambda}-\dfrac{64\left(p_3+a_{rt1}t_p+a_{rt2}t_p^2\right)k_2^3\left(t_a-t_p\right)^3}{\lambda^3}\\+\dfrac{256\left(p_4+a_{rt3}t_p\right)k_2^4\left(t_a-t_p\right)^4}{\lambda^4}-\dfrac{1024p_5k_2^5\left(t_a-t_p\right)^5}{\lambda^5}\end{array}\right] \tag{3.119}$$

Next, azimuth nonlinear chirp scaling and azimuth FFT are conducted, and the azimuth phase in the range-Doppler domain can be obtained by

$$\phi_{az5}(f_a,R)=\pi\left[P_0+P_1f_a+P_2f_a^2+P_3f_a^3+P_4f_a^4+P_5f_a^5\right], \tag{3.120}$$

where P_m(m = 0, 1, 2, 3, 4, 5) are the mth-order coefficients. In fact, the P_n(n = 1, 2, 3, 4) are dependent on t_{pc}, which is the beam crossing time after azimuth compensation, and the expressions of P_1-P_5 can be written as shown below:

$$\begin{aligned}P_1&=P_{a11}\cdot t_p+P_{a12}\cdot t_p^2\\&=-\frac{2\pi f_{dr0}}{q_2+f_{dr0}}t_p+\frac{\pi\left(-2q_2^2f_{dr1}-2q_2f_{dr1}f_{dr0}+3p_3q_2^2f_{dr0}^3+3q_3f_{dr0}^2\right)}{\left(q_2+f_{dr0}\right)^3}t_p^2\end{aligned} \tag{3.121}$$

$$\begin{aligned}P_2&=P_{20}+P_{a21}\cdot t_p+P_{a22}\cdot t_p^2\\&=-\frac{\pi}{q_2+f_{dr0}}-\frac{\pi\left(-q_2f_{dr1}+3p_3q_2f_{dr0}^3-f_{dr0}f_{dr1}+3q_3f_{dr0}\right)}{\left(q_2+f_{dr0}\right)^3}t_p\\&\quad\pi\left(\begin{array}{l}9p_3q_2^2f_{dr0}^2f_{dr1}+3q_2^2a_{rt1}f_{dr0}^3+3q_2a_{rt1}f_{dr0}^4-3q_2q_3f_{dr1}\\-2q_2f_{dr0}f_{dr2}-q_2^2f_{dr2}-f_{dr0}^2f_{dr2}+q_2f_{dr1}^2+f_{dr0}f_{dr1}^2\\-\dfrac{+q_3f_{dr0}f_{dr1}-6p_4q_2^2f_{dr0}^4-6q_4f_{dr0}^2}{6(q_2+f_{dr0})^4t_p^2}\end{array}\right)\end{aligned} \tag{3.122}$$

$$
\begin{aligned}
P_3 &= P_{30} + P_{a31} \cdot t_p + P_{a32} \cdot t_p^2 \\
&= \frac{\pi\left(p_4 f_{dr0}^4 + q_4\right)}{\left(q_2 + f_{dr0}\right)^3} + \pi\left(\frac{-4p_4 q_2 f_{dr0}^4 + 4q_4 f_{dr0} + q_2 a_{rt1} f_{dr0}^3 + a_{rt1} f_{dr0}^4 + 3p_3 q_2 f_{dr0}^2 f_{dr1} - 3q_3 f_{dr1}\left(q_2 + f_{dr0}\right)^4}{}\right) t_p \\
&+ \frac{\pi\left(\begin{array}{l} -4q_2^2 a_{rt3} f_{dr0}^4 + 3p_3 q_2 f_{dr0}^3 f_{dr2} - 3q_2 q_3 f_{dr2} + q_2^2 a_{rt2} f_{dr0}^3 \\ -3q_3 f_{dr0} f_{dr2} + a_{rt2} f_{dr0}^5 + 6q_3 f_{dr1}^2 - 4q_2 a_{rt3} f_{dr0}^5 + 3p_3 q_2^2 f_{dr0}^2 f_{dr1} \\ -12q_4 f_{dr0} f_{dr1} + 2q_2 a_{rt2} f_{dr0}^4 + 10p_5 q_2^2 f_{dr5}^5 + 3q_2^2 a_{rt1} f_{dr0}^2 f_{dr1} \\ -3p_3 q_2 f_{dr0}^2 f_{dr1}^2 + 4q_2 q_4 f_{dr1} + 3q_2 a_{rt1} f_{dr0}^3 f_{dr1} - 16p_4 q_2^2 f_{dr0}^3 f_{dr1} \\ +10q_5 f_{dr0}^2 + 3p_3 q_2^2 f_{dr0} f_{dr1}^2 \end{array}\right)}{\left(q_2 + f_{dr0}\right)^5} t_p^2
\end{aligned}
\tag{3.123}
$$

$$
\begin{aligned}
P_4 &= P_{40} + P_{a41} \cdot t_p \\
&= \frac{\pi\left(p_4 f_{dr0}^4 + q_4\right)}{\left(q_2 + f_{dr0}\right)^4} + \pi \frac{\left(\begin{array}{l} -4q_4 f_{dr1} - 5p_5 q_2 f_{dr0}^5 + 5q_5 f_{dr0} \\ + q_2 a_{rt3} f_{dr0}^4 + a_{rt3} f_{dr0}^5 + 4p_4 q_2 f_{dr0}^3 f_{dr1} \end{array}\right)}{\left(q_2 + f_{dr0}\right)^5} t_p
\end{aligned}
\tag{3.124}
$$

$$
P_5 = \frac{\pi\left(p_5 f_{dr0}^4 + q_5\right)}{\left(q_2 + f_{dr0}\right)^5} \tag{3.125}
$$

where P_{i0}(i = 1, 2, 3, 4) are the scene center coefficients, and P_{aij}(i = 1, 2, 3, 4) are the jth-order derivatives of P_i in the azimuth direction. It should be noted that P_{a11} and P_{a12} will result in azimuth distortion and that other azimuth derivatives will result in residual phase, which will degrade the image quality.

Thus, to remove the azimuth distortion and the residual phase, the azimuth derivatives should satisfy

$$
P_{aij} = \begin{cases} -\frac{\pi}{\beta} & i = j = 1 \\ 0 & other \end{cases} \tag{3.126}
$$

Finally, the solutions of phase coefficients and scaling factors can be obtained as follows:

$$\begin{cases} q_2 = (2\beta - 1)f_{dr0} \\ q_3 = \frac{(2\beta-1)f_{dr1}}{3} \\ q_4 = \frac{14f_{dr1}^2\beta - 5f_{dr1}^2 - 4f_{dr0}f_{dr2}\beta + 6f_{dr0}^4 a_{rt1} - 3f_{dr0}^4 a_{rt1}}{12 \cdot f_{dr0}} \\ q_5 = \frac{\begin{pmatrix} 7f_{dr1}^3\beta + 2f_{dr1}f_{dr0}^4 a_{rt1}\beta + 2f_{dr0}^6 a_{rt3}\beta - f_{dr0}^5 a_{rt2}\beta \\ -3f_{dr1}f_{dr0}f_{dr2}\beta - f_{dr0}^4 f_{dr1}a_{rt1} - f_{dr0}^6 a_{rt3} - 2f_{dr1}^3 \end{pmatrix}}{5 \cdot f_{dr0}^2} \\ p_3 = \frac{f_{dr1}(4\beta-1)}{3 \cdot f_{dr0}^3(2\beta-1)} \\ p_4 = \frac{(4\beta-1)\cdot\left(3f_{dr0}^4 a_{rt1} + 5f_{dr1}^2\right) - 4f_{dr2}f_{dr0}\beta}{12 \cdot f_{dr0}^5(2\beta-1)} \\ p_5 = \frac{\begin{pmatrix} 4f_{dr1}f_{dr0}^4 a_{rt1}\beta + 4f_{dr0}^6 a_{rt3}\beta + 9f_{dr1}^3\beta - f_{dr0}^5 a_{rt2}\beta \\ -3f_{dr1}f_{dr2}f_{dr0}\beta - f_{dr0}^4 f_{dr1}a_{rt1} - 2f_{dr1}^3 - f_{dr0}^6 a_{rt3} \end{pmatrix}}{5 \cdot f_{dr0}^7(2\beta-1)} \end{cases} \tag{3.127}$$

After nonlinear filtering and nonlinear chirp scaling, azimuth distortion and residual phase have been removed, and the azimuth phase is unified as

$$\phi_{az6} = \pi \begin{pmatrix} -\frac{1}{q_2 + f_{dr0}} f_a^2 + \frac{p_3 f_{dr0}^3 + q_3}{(q_2 + f_{dr0})^3} f_a^3 \\ + \frac{p_4 f_{dr0}^4 + q_4}{(q_2 + f_{dr0})^4} f_a^4 + \frac{p_5 f_{dr0}^5 + q_5}{(q_5 + f_{dr0})^5} f_a^5 \end{pmatrix} \tag{3.128}$$

References

1. AIAA (1976) A digital system to produce imagery from SAR data. Syst Design Driven Sens (AIAA). AIAA J
2. Wu C, Liu KY, Jin M (1982) Modeling and a correlation algorithm for spaceborne SAR aignals. IEEE Trans Aerospace Electron Syst AES 18(5):563–575
3. Cumming I, Bennett JR (1979) Digital processing of Seasat SAR data. In: Acoustics, speech, and signal processing, IEEE international conference on ICASSP. pp 710–718
4. Runge H, Bamler RA (1992) Novel high precision SAR focussing algorithm based on chirp scaling. In: Geoscience and remote sensing symposium, 1992. IGARSS '92, international. pp 372–375
5. Cumming I, Wong F, Raney KA (1992) SAR processing algorithm with no interpolation. In: Geoscience and remote sensing symposium, 1992, IGARSS '92. International. pp 376–379
6. Raney RK, Runge H, Bamler R, Cumming IG, Wong FH (1994) Precision SAR processing using chirp scaling. IEEE Trans Geosci Remote Sens 32(4):786–799
7. Hellsten H, Andersson LE (1987) An inverse method for the processing of synthetic aperture radar data. Inverse Prob 3(1):111–124
8. Sack M, Ito MR, Cumming IG (1985) Application of efficient linear FM matched filtering algorithms to synthetic aperture radar processing. Commun Radar Signal Processing IEEE Proc F 132(1):45–57
9. Soumekh M (1999) Synthetic aperture radar signal processing with MATLAB algorithms

10. Yegulalp AF (1999) Fast backprojection algorithm for synthetic aperture radar. In: radar conference, the record of the, 1999. pp 60–65
11. Li Z, Li C, Yu Z, Zhou J, Chen J (2011) Back projection algorithm for high resolution GEO-SAR image formation. In: Geoscience and remote sensing symposium. pp 336–339
12. Zhao B, Qi X, Song H, Wang R, Mo Y, Zheng S (2013) An accurate range model based on the fourth-order doppler parameters for geosynchronous SAR. IEEE Geosci Remote Sens Lett 11(1):205–209
13. Liu Z, Hu C, Zeng T (2011) Improved secondary range compression focusing method in GEO SAR. In: IEEE international conference on acoustics, speech and signal processing. pp 1373–1376
14. Liu Z, Yao D, Long T (2011) An accurate focusing method in GEO SAR. In: Radar conference. pp 237–241
15. Liu Z, Long T, Hu C, Zeng T (2012) An improved wide swath imaging algorithm based on series reversion in GEO SAR. In: Geoscience and remote sensing symposium. pp 2078–2081
16. Hu C, Long T, Liu Z, Zeng T, Tian Y (2014) An improved frequency domain focusing method in geosynchronous SAR. IEEE Trans Geosci Remote Sens 52(9):5514–5528
17. Yang W, Zhu Y, Liu F, Hu C (2010) Modified range migration algorithm in GEO SAR system. In: European conference on synthetic aperture radar. pp 1–4
18. Liu F, Hu C, Zeng T, Long T, Jin L (2010) A novel range migration algorithm of GEO SAR echo data. In: Geoscience and remote sensing symposium. pp 4656–4659
19. Hu C, Long T, Zeng T, Liu F, Liu Z (2011) The accurate focusing and resolution analysis method in geosynchronous SAR. IEEE Trans Geosci Remote Sens 49(10):3548–3563
20. Hu C, Liu F, Yang W, Zeng T, Long T (2010) Modification of slant range model and imaging processing in GEO SAR. In: Geoscience and remote sensing symposium. pp 4679–4682
21. Hu C, Liu Z, Long T (2012) An improved CS algorithm based on the curved trajectory in geosynchronous SAR. IEEE J Sel Topics Appl Earth Observations Remote Sens 5(3): 795–808
22. Li D, Wu M, Sun Z, He F, Dong Z (2015) Modeling and processing of two-dimensional spatial-variant geosynchronous SAR data. IEEE J Sel Topics Appl Earth Observations Remote Sens 8(8):3999–4009
23. Sun GC, Xing M, Wang Y, Yang J, Bao Z (2014) A 2D space-variant chirp scaling algorithm based on the rcm equalization and subband synthesis to process geosynchronous SAR Data. IEEE Trans Geosci Remote Sens 52(8):4868–4880
24. Bao M, Xing MD, Li YC (2012) Chirp scaling algorithm for GEO SAR based on fourth-order range equation. Electron Lett 48(1):41–42
25. Zeng T, Yang W, Ding Z, Liu D, Long T (2013) A refined two-dimensional nonlinear chirp scaling algorithm for geosynchronous earth orbit SAR. Prog Electromagnet Res 143(143): 19–46
26. Hu C, Long T, Tian Y (2013) An improved nonlinear chirp scaling algorithm based on curved trajectory in geosynchronous SAR. Prog Electromagnet Res 135(1):481–513
27. Ding Z, Shu B, Yin W, Zeng T, Long T (2016) A modified frequency domain algorithm based on optimal azimuth quadratic factor compensation for geosynchronous SAR Imaging. IEEE J Sel Topic Appl Earth Observations Remote Sens 9(3):1119–1131
28. Chen Q, Li D, He F, Liang D (2013) Exact spectrum of non-linear chirp scaling and its application in geosynchronous synthetic aperture radar imaging. J Eng 1
29. Hu B, Jiang Y, Zhang S, Yeo TS (2015) Focusing of geosynchronous SAR with nonlinear chirp scaling algorithm. Electron Lett 51(15):1195–1197
30. Chen J, Sun GC, Wang Y, Xing M, Li Z, Zhang Q, Liu L, Dai C (2016) A TSVD-NCS algorithm in range-doppler domain for geosynchronous synthetic aperture radar. IEEE Geosci Remote Sens Lett 13(11):1631–1635
31. Zhang T, Ding Z, Tian W, Zeng T, Yin W (2017) A 2-D Nonlinear Chirp Scaling Algorithm for High Squint GEO SAR Imaging Based on Optimal Azimuth Polynomial Compensation. IEEE J Sel Topics Appl Earth Observations Remote Sens 10(12):5724–5735

Chapter 4
Analysis of Temporal-Spatial Variant Atmospheric Effects on GEO SAR

Abstract Due to the ultra-long integration time and large coverage characteristics of geosynchronous synthetic aperture radar (GEO SAR), the atmospheric frozen model in the traditional low Earth orbit SAR (LEO SAR) imaging fails in GEO SAR. The temporal-spatial variation of troposphere and ionosphere should be taken into account for the GEO SAR imaging. Based on the accurate GEO SAR signal model, the two-dimensional spectrum of GEO SAR signal in the context of temporal-spatial variant troposphere and background ionosphere are derived, and then the two-dimensional image shift and defocusing are investigated. The boundary conditions of relevant effects are analyzed and summarized which are related to the status of troposphere and background ionosphere, the GEO SAR imaging geometry and the integration time dependent on the resolution requirement. GEO SAR is also sensitive to ionospheric scintillation which causes the amplitude and phase fluctuations of signals. The corresponding degradation will have a different pattern from LEO SAR. The azimuth point spread function considering the scintillation sampling model is constructed. Then, based on the measurable statistical parameters of ionospheric scintillation, performance is quantitatively analyzed. The analysis suggests that in GEO SAR imaging the azimuth integrated side lobe ratio deteriorates severely, while the degradations of the azimuth resolution and azimuth peak-to-sidelobe ratio are negligible when scintillation occurs.

4.1 Introduction

Atmosphere is wrapped outside the Earth which can be divided into the troposphere, stratosphere, ionosphere and magnetosphere from bottom to top based on the height. For analyzing the radio wave propagation in SAR, only the troposphere

---

T. Long et al., *Geosynchronous SAR: System and Signal Processing*,
https://doi.org/10.1007/978-981-10-7254-3_4

and ionosphere should be considered. The troposphere is distributed within the height scope of 0–12 km; and the ionosphere is distributed from 60 to 1000 km.

Troposphere is non-dispersive and thus the focusing in range will not be influenced. But the variations of the troposphere status within the integration time of SAR will induce phase errors and degrade the focusing [1]. Furthermore, the turbulence in troposphere will cause the amplitude and phase fluctuations of signals [2], which can induce the image shift and defocusing.

In comparison, ionosphere has an important impact on the spaceborne SAR imaging, and would bring about phase errors to the SAR signal traversing therein [3], and further lead to image shift, resolution degradation and defocusing [4–10]. The Faraday rotation of the ionosphere might cause attenuations [11], and the ionospheric scintillation might give rise to random fluctuations of the SAR signal amplitude and phase, and thus affect the focusing effects [12].

Currently, the atmospheric analysis is allusion to the LEO SAR satellites. As the synthetic aperture time of LEO SAR is short (approximately 1–2 s), the above analyses were carried out under the assumption of the ionosphere frozen model [5]. However, GEO SAR has an integration time up to the order of a few hundred seconds, thus troposphere and ionosphere within the aperture time can no longer satisfy the frozen model assumption, and it should also take the temporal-variation background ionosphere into consideration. Additionally, owing to the large coverage of GEO SAR, the status of troposphere and ionosphere in different locations within the imaging scene are varying, i.e. the spatial-variation characteristics. Therefore, in GEO SAR, the temporal-spatial variations of troposphere and ionosphere turn out to be serious, which has become the critical issue constraining the implementation of GEO SAR.

4.1.1 Troposphere

Troposphere is located in the bottom layer of atmosphere and distributed from the ground to the upper air which is 12 km far. The tropopause height is varying for different latitudes: it is about 18 km near the equator, 12 km in mid-latitude region, and 7 km in polar region. The tropopause height is also related to the season. In summer it is higher than that in winter.

For troposphere, three of the most important factors are temperature (T), press (P) and humidity (e_w), which decrease as the height rises. The influence of troposphere on radio wave propagation is depicted by the reflective index n generally, whose value is among 1.00026–1.00046. For convenience, it usually uses reflectivity N to replace the reflective index n. Their relationship is

$$N = (n - 1) \times 10^6 \tag{4.1}$$

The reflectivity of troposphere can be represented by the three meteorological parameters: temperature, press and humidity. Its definition is

$$N = N_d + N_w = \frac{77.6}{T}\left(P + 4810\frac{e_w}{T}\right) \tag{4.2}$$

where $N_d = 77.6 \cdot P/T$ is the dry item which has no relationship with humidity and $N_w = 3.73 \times 10^5 \cdot e_w/T^2$ is humid item.

When radio wave propagates in the troposphere, the propagation speed is different from that in vacuum. Due to the heterogeneous reflectivity, its propagation path will bend. Therefore, when spaceborne SAR signal passes through the troposphere, it will generate time-delay errors and curvature errors, and resultantly introduce phase errors.

We assume that the relative geometry between the spaceborne SAR and the troposphere is shown as Fig. 4.1, O is the Earth center; R is the Earth radius; the target P is located in the height of h_p from the ground; the curve, from spaceborne SAR, through the puncture points P'' and P' to the target P, shows the actual propagation path in troposphere; the straight line from spaceborne SAR to the target P shows the ideal path of the signal along the straight line; P'' is the puncture point between the propagation path of spaceborne SAR and the top of troposphere; P' shows a certain point along the actual signal path; θ_1 is the elevation angle at P; θ_2 is the elevation angle at P'; $h_{p'}$ is the height of P'; h_{up} is the tropopause height.

The radio wave propagation error in troposphere can be calculated by the ray tracing method. Based on the Snell's law, under the condition of hierarchical

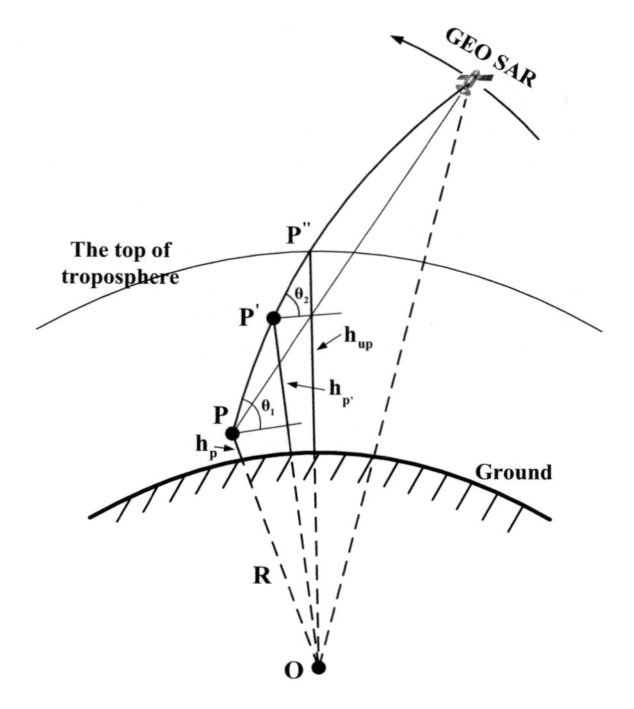

Fig. 4.1 Sketch map of relative geometry of spaceborne SAR and the troposphere

hypothesis, the product of reflectivity, geocentric distance and cosine value of elevation angle is identical in different heights as

$$n_p\left(h_p + R\right)\cos\theta_1 = n_{p''}\left(h_{p'} + R\right)\cos\theta_2 = A \tag{4.3}$$

The real distance is used to substitute the actual propagation distance of satellite signal in troposphere. It is equal to the sum of three parts: the propagation distance of satellite signal along the straight line, the time-delay errors and the curvature errors. Based on the geometry in Fig. 4.1, the real distance of the propagation path in troposphere is

$$R_{real} = \int_{r_p}^{r_{p''}} n \cdot \csc\theta \cdot dr \tag{4.4}$$

where, $r_p = h_p + R$ and $r_{p''} = h_{up} + R$ are the distances from P and P'' to the Earth center.

For SAR signals passing through troposphere, the total error is

$$\begin{aligned}\Delta r &= R_{real} - R_{str} \\ &= \underbrace{\int_{r_p}^{r_{p''}} (n-1)\csc\theta dr}_{\text{time delay errors}} + \underbrace{\left(\int_{r_p}^{r_{p''}} \csc\theta dr - R_{str}\right)}_{\text{curvature errors}}\end{aligned} \tag{4.5}$$

In (4.5), R_{str} represents the length of the SAR signal propagation path along the assumed straight line in troposphere. The first item represents the time-delay error caused by the slow-down of signal propagation velocity; the second item represents the curvature error caused by the bending of the signal propagation path.

Based on analysis of the real data, the total errors of troposphere can reach the meter level, wherein the curvature error is generally below 0.1 m and occupies only a small proportion. Therefore, in subsequent analysis, we neglect the influence of the curvature error and only consider time delay errors. Thus the propagation error of the troposphere can be simplified as

$$\Delta r = R_{real} - R_{str} = \int_{r_p}^{r_{p''}} (n-1)\csc\theta dr \tag{4.6}$$

Based on Snell's law, (4.6) can be simplified as

$$\Delta r = \int_{r_p}^{r_{p''}} \frac{nr(n-1)}{\sqrt{n^2r^2 - A^2}} \cdot dr \tag{4.7}$$

For the further simplification, (4.7) can be re-expressed as

$$\begin{aligned}\Delta r &= \int_{h_p}^{h_{up}} \frac{n(n-1)\left(R+h_{p'}\right)}{\sqrt{n^2\left(R+h_{p'}\right)^2 - n^2\left(R+h_{p'}\right)^2\cos^2\theta_1}} \cdot dh \\ &= \int_{h_p}^{h_{up}} \frac{n(h)-1}{\sin\theta_1} \cdot dh \\ &= \csc\theta_1 \cdot \int_{h_p}^{h_{up}} (n(h)-1) \cdot dh\end{aligned} \tag{4.8}$$

Based on the relationship in (4.1), (4.8) can be re-written as

$$\Delta r = \csc\theta_1 \cdot \frac{\int_{h_p}^{h_{up}} N(h) \cdot dh}{10^6} \tag{4.9}$$

It can be seen from (4.9), the tropospheric time-delay errors in GEO SAR is dependent on the integral of reflectivity and the elevation angle of GEO SAR propagation path. Considering the two-way propagation, the phase errors introduced by tropospheric effects can be expressed as $\Delta\phi_{trop}$ which will influence the imaging

$$\Delta\phi_{trop} = \frac{4\pi \cdot \Delta r}{\lambda} = \frac{4\pi \cdot \csc\theta_1 \cdot \int_{h_p}^{h_{up}} N(h) \cdot dh}{10^6 \cdot \lambda} \tag{4.10}$$

4.1.2 Ionosphere

The ionosphere is constituted of the ionized layers (D, E and F layers) with lots of electrons and electrically charged molecules and atoms, which distributes at the upper of the atmosphere (from about 60 km to more than 1000 km) [9]. D and E layers are the related lower layers, distributing from 60 to 150 km. The F layer lies at an altitude range of 150 km to more than 400 km. The free electrons in the ionosphere will cause variations of ionospheric reflectivity and further induce the change of propagation velocity of radio wave passing through the ionosphere. The ionospheric reflectivity n_{iono} can be calculated as

$$\mathrm{n_{iono}} \approx \left(1 - \frac{1}{2}\frac{e_l^2}{4\pi^2 m\varepsilon_0}\frac{n_e}{f^2}\right) = 1 - K\frac{n_e}{f^2} \tag{4.11}$$

where $K = 40.28\,\mathrm{m}^3/\mathrm{s}^2$, n_e is electron density of the ionosphere; e_l is elementary charge; m is electronic mass; ε_0 is permittivity of vacuum; f is carrier frequency.

Due to the different reflectivity between the ionosphere and the vacuum, the propagation velocity of radio wave in the ionosphere is relatively slower than that in vacuum. Therefore, time delay will be induced when the signal passes through the ionosphere. The one-way signal propagation time-delay caused by the ionosphere can be obtained through the path integral of reflectivity.

$$\Delta t = \frac{\int (n_{iono} - 1) \cdot ds}{c} \tag{4.12}$$

where c is light speed in vacuum; s shows the signal propagation path in ionosphere.

Substitute (4.11) into (4.12), and simplify to obtain

$$\Delta t = -\frac{40.28 \cdot TEC}{c \cdot f^2} \tag{4.13}$$

wherein, define $TEC = \int n_e \cdot ds$, which is the total electron content along the signal propagation path.

Considering the two-way propagation of the SAR signal, the phase errors can be expressed as $\Delta\phi_{iono}$ which will influence focusing

$$\Delta\phi_{iono} = -\frac{4\pi(\Delta t \cdot c)}{\lambda} \approx -\frac{2\pi \cdot 80.6 \cdot TEC}{c \cdot f}. \tag{4.14}$$

Including the background ionosphere, radio wave is also affected by the ionospheric scintillation. The ionospheric scintillation forms in the F layer which is the densest ionization layer during the day. It is strongly influenced by variations of the Sun, Earth's magnetic field and the unstable electron density, which will make the ionosphere have lots of irregularities. Generally, ionospheric scintillation often happens at the equatorial zone from the early evening after sunset to the midnight (local time 18:00 to 24:00) and can occur at the polar zones at any time in a day [13]. Ionospheric scintillation activity also relates to season, longitude and solar activities [9, 13, 14].

According to the scales of the ionospheric irregularities, they are divided into the large-scale irregularities (larger than 10 km) and the small-scale irregularities (smaller than 10 km) [15]. The large-scale irregularities correspond to the low frequency variation of background ionosphere, which give rise to the effects of the dispersion and the time delay in traversing signals [6]. Small-scale irregularities are mainly brought by the ionospheric plasma bubbles inside the ionosphere and will cause the random fluctuations of the phase and amplitude of traversed signals. These random errors will degrade focusing quality, or even lead to defocusing.

4.1.3 Summary

GEO SAR has the characteristics of long integration time and large wide swath. Resultantly, the temporal-spatial variations of atmosphere should be considered. For GEO SAR, the induced phase errors will be accumulated within the integration time, degrading the focusing. Meantime, the spatial variations will cause image distortions. In this chapter, the influences of troposphere and ionosphere are considered. For the troposphere, it is non-dispersive and thus the relevant conclusion can be applied to other wavebands SAR system under long-time accumulation. In comparison, ionosphere is dispersive. So the influence is related to the carrier frequency. Therefore, the influences of different wavebands need to be analyzed respectively. But the analytical method in this chapter can be fully applied to GEO SAR in other wavebands or SAR systems under the long-time accumulation.

4.2 Tropospheric Influences

4.2.1 Signal Model Considering Time-Varying Troposphere

The troposphere will introduce phase errors to the SAR signals and influence the imaging. Here taking the point target as an example, GEO SAR signal model is constructed under the influence of time-varying troposphere and its influence on imaging is analyzed. The time-varying meteorological parameters should be considered within the long integration time in GEO SAR. The time-delay error caused by time-varying troposphere can be expressed by the high-order polynomial approximation

$$\Delta r(t_a) = \Delta r_0 + q_1 \cdot t_a + q_2 \cdot t_a^2 + q_3 \cdot t_a^3 + q_4 \cdot t_a^4 + \cdots \tag{4.15}$$

In (4.15), Δr_0 is the constant part of $\Delta r(t_a)$, $q_i (i = 1 \cdots n)$ is the all-order polynomial coefficients of $\Delta r(t_a)$ on time. The phase errors can be re-written as

$$\Delta\phi_{trop}(t_a) = \frac{4\pi}{\lambda} \cdot \Delta r(t_a) = \frac{4\pi}{\lambda} \cdot \left[\Delta r_0 + q_1 \cdot t_a + q_2 \cdot t_a^2 + q_3 \cdot t_a^3 + q_4 \cdot t_a^4 + \cdots\right] \tag{4.16}$$

The GEO SAR geometry is shown in Fig. 4.2. The coordinate system O-XYZ is the scene coordinate system, and $\vec{P}(x, y, 0)$ is the position of Target P in the scene coordinate system, $r(t_a)$ is the instantaneous slant range from the GEO SAR to Target P at the time of t_a, while $\vec{V}_s$ is the satellite velocity.

The echo signal of GEO SAR is expressed as

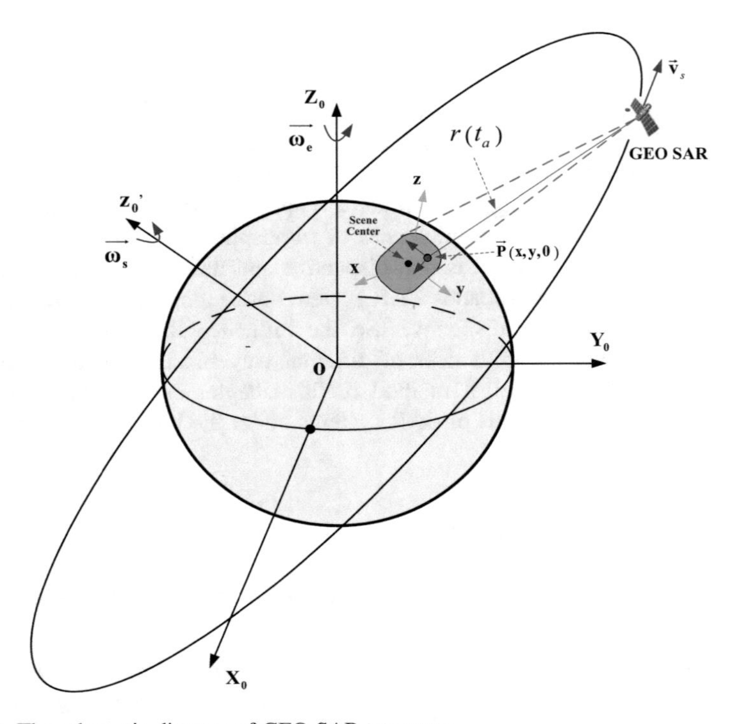

Fig. 4.2 The schematic diagram of GEO SAR geometry

$$s(t_a, t) = A_r(t) \cdot A_a(t_a) \cdot \exp\left[j\pi k_r\left(t - \frac{2 \cdot r(t_a)}{c}\right)^2\right] \cdot \exp\left[-j4\pi\frac{r(t_a)}{\lambda}\right] \tag{4.17}$$

where, $A_r(t)$ and $A_a(t_a)$ are the range and azimuth envelope function respectively, k_r is the chirp rate, and λ is the signal wavelength. Through the Fourier Transform (FT), the echo signal can be transformed into the range frequency domain:

$$S(f_r, t_a) = A_r(f_r) \cdot A_a(t_a) \cdot \exp\left[-j\frac{4\pi(f_r + f_0) \cdot r(t_a)}{c}\right] \cdot \exp\left(-j\frac{\pi f_r^2}{k_r}\right) \tag{4.18}$$

Considering the accurate signal model of GEO SAR, the signal model under the influences of troposphere is

$$\begin{aligned} S(f_r, t_a) = {} & A_r(f_r) \cdot A_a(t_a) \cdot \exp\left[-j\frac{4\pi(f_r + f_0) \cdot r(t_a)}{c}\right] \cdot \exp\left(-j\frac{\pi f_r^2}{k_r}\right) \\ & \cdot \exp\left[j \cdot \frac{4\pi}{\lambda} \cdot \left[\Delta r_0 + q_1 \cdot t_a + q_2 \cdot t_a^2 + q_3 \cdot t_a^3 + q_4 \cdot t_a^4 + \cdots\right]\right] \end{aligned} \tag{4.19}$$

Then (4.19) can be modified as

$$S(f_r, t_a) = A_r(f_r) \cdot A_a(t_a) \cdot \exp\left[-j\frac{4\pi(f_r + f_0) \cdot r(t_a)}{c}\right] \cdot \exp\left(-j\frac{\pi f_r^2}{k_r}\right)$$
$$\cdot \underbrace{\exp\left[j \cdot \frac{4\pi}{\lambda} \cdot \Delta r_0\right]}_{\Delta\phi_{tropo}} \cdot \underbrace{\exp\left[j \cdot \frac{4\pi}{\lambda} \cdot \left[q_1 \cdot t_a + q_2 \cdot t_a^2 + q_3 \cdot t_a^3 + q_4 \cdot t_a^4 + \cdots\right]\right]}_{\Delta\phi_{\Delta tropo}} \quad (4.20)$$

where $\Delta\phi_{tropo}$ shows the constant part and $\Delta\phi_{\Delta tropo}$ is the time-varying part of the phase errors.

4.2.2 Theoretical Analysis of Influences on Focusing

Troposphere is non-dispersive medium and thus its influences will be coincidence for different frequency signals. The imaging in the range direction will not be affected. However, the time-varying troposphere will cause different time delays among different PRTs and influence imaging in the azimuth direction. Therefore, this section only analyzes the azimuth focusing of GEO SAR. Through (4.20), the azimuth signal under the influence of $\Delta\phi_{\Delta tropo}$ in (4.20) is

$$\begin{aligned} s(t_a) = rect\left(\frac{t_a}{T_a}\right) \quad & \exp\left(j\pi f_{dr} \cdot t_a^2\right) \\ & \exp\left[-j \cdot \frac{4\pi}{\lambda} \cdot \left(q_1 \cdot t_a + q_2 \cdot t_a^2 + q_3 \cdot t_a^3 + q_4 \cdot t_a^4 + \cdots\right)\right] \end{aligned} \quad (4.21)$$

The azimuth signal spectrum can be obtained as

$$\begin{aligned} \phi_a(f_a) = A \cdot & \underbrace{\exp\left(-j \cdot \frac{2\pi \cdot 2 \cdot q_1}{\lambda \cdot f_{dr}} \cdot f_a\right)}_{\phi_{a1}} \\ & \cdot \underbrace{\exp\left[-j \cdot \pi \Big/ \left(f_{dr} - \frac{4 \cdot q_2}{\lambda}\right) \cdot f_a^2\right]}_{\phi_{a2}} \cdot \underbrace{\exp\left(-j \cdot \frac{2\pi \cdot 2 \cdot q_3}{\lambda \cdot f_{dr}^3} \cdot f_a^3\right)}_{\phi_{a3}} \end{aligned} \quad (4.22)$$

In (4.22), the phase error caused by the linear part q_1 is denoted as ϕ_{a1}; the phase errors caused by the non-linear parts q_2 and q_3 are denoted as ϕ_{a2} and ϕ_{a3}, respectively. The phase ϕ_{a1} will induce azimuth image offset which is

$$\Delta L_a = v_{nadir} \cdot \frac{2 \cdot q_1}{\lambda f_{dr}} \quad (4.23)$$

The azimuth image offset is related to the signal wavelength and the linear change rate of the troposphere delay with respect to the time when GEO SAR satellite orbit configuration is fixed (namely f_{dr} remains unchanged). The smaller the signal wavelength or the larger the linear change rate q_1 is, the larger the azimuth image offset is.

The quadratic azimuth phase error ϕ_{a2} will cause the main lobe widening and the sidelobe rising, leading to defocusing in azimuth. Considering (4.22), ϕ_{a2} can be expressed as

$$\phi_{a2} = \exp\left[-j \cdot \pi f_{dr}^2 \Big/ \left(4\left(f_{dr} - \frac{4q_2}{\lambda}\right)\right) \cdot T_a^2\right] \tag{4.24}$$

which is related to the signal wavelength, the second-order change rate of the troposphere delay along the time, the integration time.

In comparison, asymmetric sidelobes will arise by the influences of the cubic azimuth phase error ϕ_{a3}

$$\phi_{a3} = \exp\left(-j \cdot \frac{\pi q_3}{2\lambda} \cdot T_a^3\right) \tag{4.25}$$

It can be seen that the smaller the signal wavelength, the larger the third-order change rate of the troposphere delay along the time, or the longer the integration time is, the larger the cubic phase error ϕ_{a3} is.

4.2.3 Simulation

4.2.3.1 Influences on Echoes

For troposphere, the two-way signal delay caused by the troposphere can be expressed as

$$\Delta r = 2\csc\theta_1 \cdot \frac{\int_{h_p}^{h_{up}} N(h) \cdot dh}{10^6} \tag{4.26}$$

Therefore, the troposphere delay is closely related to the reflectivity of troposphere and the elevation angle of the signal propagation path. In order to simulate the signal delay caused by troposphere, it needs to obtain the tropospheric reflectivity.

There exist several tropospheric reflectivity models including the linear model, the exponential model, the subsection model, the Saastamoinen model, the Hopfield model, etc. In this section, the Hopfield model is adopted as its high accuracy and generality. It is expressed as

$$\begin{cases} & N = N_d + N_w \\ \text{if } h \leq H_i, & N_i(h) = \frac{N_{i0}}{(H_i - h_0)^4}(H_i - h)^4, \quad i = d, w \\ \text{if } h > H_i, & N_i = 0, \quad i = d, w \end{cases} \tag{4.27}$$

where h_0 is the height from the ground, N_d and N_w are the dry and the humid items of the reflectivity respectively, N_{d0} and N_{w0} are the dry item and the humid item of the reflectivity at the ground level respectively whose specific expressions are

$$\begin{aligned} N_{d0} &= 77.6 \cdot \frac{\mathrm{P}_0}{(\mathbf{T_0} + 273.15)} \\ N_{w0} &= 3.37 \times 10^3 \cdot \frac{RH \times a}{(\mathbf{T_0} + 273.15)^2} \cdot \exp\left(\frac{b \cdot \mathbf{T_0}}{c + \mathbf{T_0}}\right) \end{aligned} \tag{4.28}$$

where $\mathbf{T_0}$ is the temperature at the ground in degree, P_0 is the pressure at the ground whose mean value is 1013.25 hPa, RH is the humidity at the ground, and a, b, c are the conversion constants whose specific values are shown in Table 4.1.

In (4.27), H_d and H_w are the equivalent heights of the dry and humid items of reflectivity respectively. H_w is varying along with time, location and season. Its value range is from 10 to 14 km and the typical value is 12 km. The equivalent height of the dry item H_d is related to the temperature at the ground. Its specific value is

$$H_d = 40.082 + 0.148 \cdot \mathrm{T}_0 \tag{4.29}$$

When the pressure takes the averaged value, the temperature is 20° and the relative humidity is 50%, the reflectivity based on the Hopfield model is shown in Fig. 4.3.

The two-way time delay and the phase errors caused by the troposphere can be calculated by using the ray tracing method. The results are as shown in Fig. 4.4.

From Fig. 4.4, the two-way time delay is below 3.62 m and the corresponding phase errors are below 94.73 rads within the integration time of 1000 s. The errors are minimal at the center of the aperture and maximal at the ends which are the results of the varying signal propagation path within the aperture.

Table 4.1 Humidity conversion constants

Constant coefficient	$\mathbf{T_0} \geq 20\,°\mathrm{C}$	$\mathbf{T_0} < 20\,°\mathrm{C}$
a	6.1121	6.1115
b	17.502	22.452
c	240.97	272.55

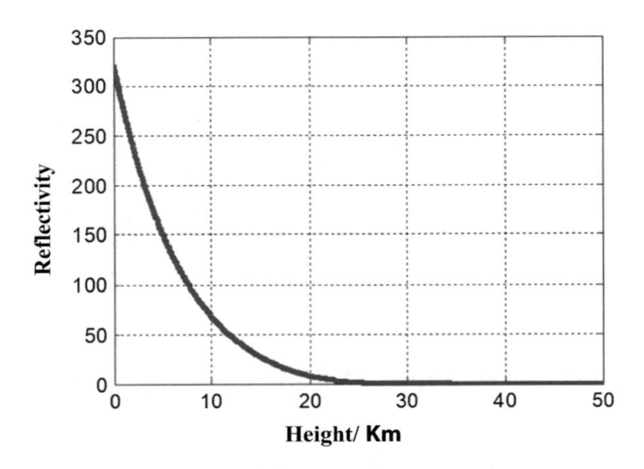

Fig. 4.3 The varying troposphere reflectivity corresponding to the height

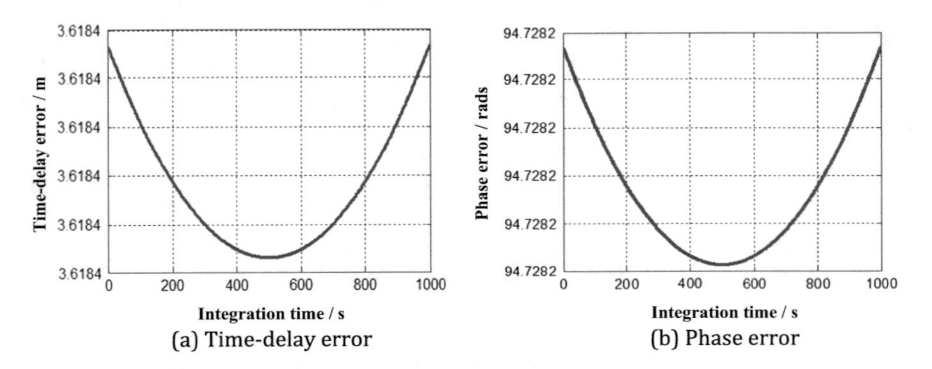

Fig. 4.4 GEO SAR signal delay and phase error caused by troposphere

4.2.4 Influences on Focusing

In this section, we will analyze influences of troposphere on GEO SAR imaging by the real data. China National Satellite Meteorological Center releases historic meteorological data and a part of real-time meteorological data, including historical data of atmospheric reflectivity profile which is released by Fengyun-3C satellite (FY-3C). The time interval is usually 2–5 min. The atmospheric reflectivity profile data consist of atmospheric reflectivity, exact time and the satellite position.

The selected data were acquired from 18:28 to 18:40 on May 27, 2015. The interval is 2 min. The time delay corresponding to the real reflectivity data is calculated by the ray tracing method, which is shown as red "+" in Fig. 4.5. Then the Lagrange's interpolation is used to obtain the time delay at each PRT.

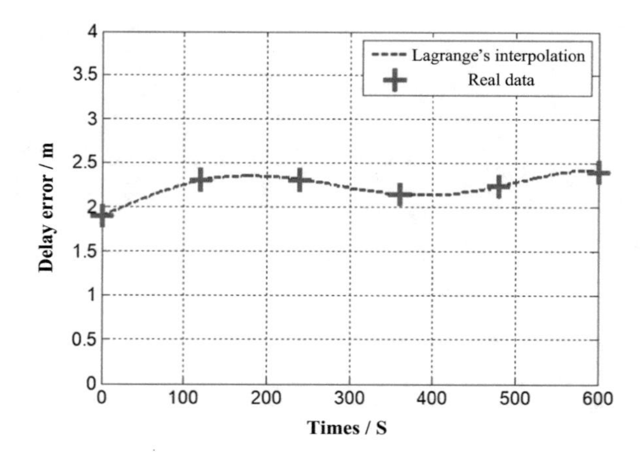

Fig. 4.5 Troposphere signal delay based on the FY-3C reflectivity profile data

As FY-3C runs in the low Earth orbit, the signal time delay obtained only represents the time delay in the FY-3C signal propagation path. It cannot be directly used in GEO SAR as the different propagation paths. So the equivalent processing is as follows. Firstly, the proper orbit trajectory is selected for GEO SAR and FY-3C satellite, obeying that their trajectories are approximately parallel. Then the troposphere puncture points (joint between signal ray and tropopause) along signal propagation paths of GEO SAR satellite and FY-3C satellite during the same period are calculated according to orbit parameters of GEO SAR and FY-3C satellite. Finally, the changing rates of the reflectivity are calculated based on the FY-3C data and then employed in GEO SAR. More details can be found in the equivalent processing for the background ionosphere monitoring experiment in Chap. 5.

Here we use the 100s' data for the equivalent processing to obtain the equivalent troposphere time delay error in GEO SAR. The all-order change rates of troposphere are obtained through (4.15), which is shown as Table 4.2.

The equivalent troposphere time delays in GEO SAR as shown in Table 4.2 are used for imaging simulation. Based on the aforementioned theoretical analysis, it can be known that the tropospheric influences are closely related to the integration time. So in the simulation, the integration times are set as 300, 500 and 900 s respectively. The signal frequency is 1.25 GHz and the bandwidth is 20 MHz. Based on (4.23), the azimuth offsets caused by troposphere are only 0.86 m. The GEO SAR focusing results are shown as Fig. 4.6. From these results, the troposphere has little influence on GEO SAR imaging and will not cause defocusing. The detailed assessments are listed in Table 4.3.

Table 4.2 Changing rates of GEO SAR troposphere time delay along with the time

Item	Δr_0/m	q_1/m/S	q_2/m/S^2	q_3/m/S^3
Value	2.21	2.52×10^{-4}	2.71×10^{-7}	1.64×10^{-13}

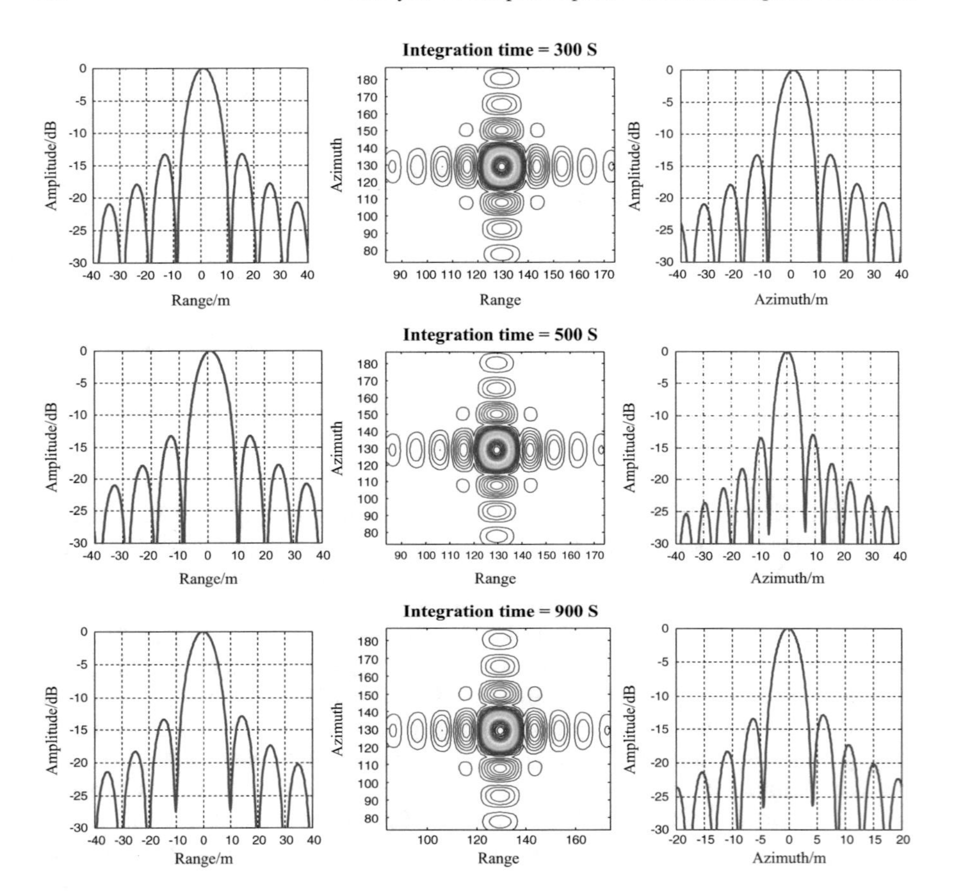

Fig. 4.6 GEO SAR imaging results influenced by troposphere under different integration times

Table 4.3 Assessment of the imaging results influenced by troposphere

Integration time/s	Range PSLR/dB	Azimuth PSLR/dB	Azimuth offset/m
300	−13.26	−13.21	0.86
500	−13.25	−13.22	0.86
900	−13.27	−13.18	0.86

4.3 Background Ionospheric Influences

4.3.1 Background Ionosphere Models

The ionospheric analysis in the traditional LEO SAR was carried out in the context of ionosphere frozen assumptions, namely, the temporal-spatial variation (TSV) of the ionosphere was very small and can be regarded as constant. However, in

GEO SAR, the ionosphere exhibits the characteristics of temporal and spatial variations. First of all, due to the ultra-long aperture time (up to hundreds to thousands of seconds) in GEO SAR, the ionospheric TEC will no longer be constant, but will tend to change within the aperture time. In contrast, the aperture time of LEO SAR is usually 1–2 s, thus the ionospheric variation of GEO SAR within the aperture time is much more severe than that of LEO SAR. Figure 4.7 gives the variations in TEC within 1 and 1000 s calculated by using the Klobuchar model [16], which is less than 0.01 TECU within 1 s and exceeds 2 TECU within 1000 s.

Secondly, because of the large wide swath in GEO SAR, the ionosphere at different positions within the imaging area will also be subject to distinction. It is related to the length and the total electron contents (TEC) along the path that the GEO SAR signal traverses the ionosphere. Thus, it should take into consideration the effects of the TSV characteristics of background ionosphere on imaging.

Considering the temporal-variant $TEC(t_a)$ with respect to the slow time t_a within the long integration time in GEO SAR, the phase errors in (4.14) can be modified as

$$\varphi_{iono}(t_a) = -2\pi \cdot 80.6 \cdot TEC(t_a)/cf \tag{4.30}$$

The temporal variant ionosphere will induce the linear, quadratic or higher orders of phase errors, and be sure to result in the image drifts and defocusing. Based on the Taylor expansion, $TEC(t_a)$ can be modelled as the all-order derivatives of the slow time t_a. The expanded $TEC(t_a)$ is

$$TEC(t_a) = TEC_0 + k_1 \cdot t_a + k_2 \cdot t_a^2 + k_3 \cdot t_a^3 + .. \tag{4.31}$$

where TEC_0 is the constant part which is not variant with the time, and k_i is the ith temporal derivatives.

In nature, the ionosphere distributes along the altitudes between 50 and 1000 km and it can be represented by a layer at the height of 350 km. When the satellite transmits and receives signals, the line-of-sight will intercept this ionospheric layer and the intercept point is named as the puncture point. When the satellite runs along with its own trajectory, the puncture point also forms a trajectory called the puncture path. This gives

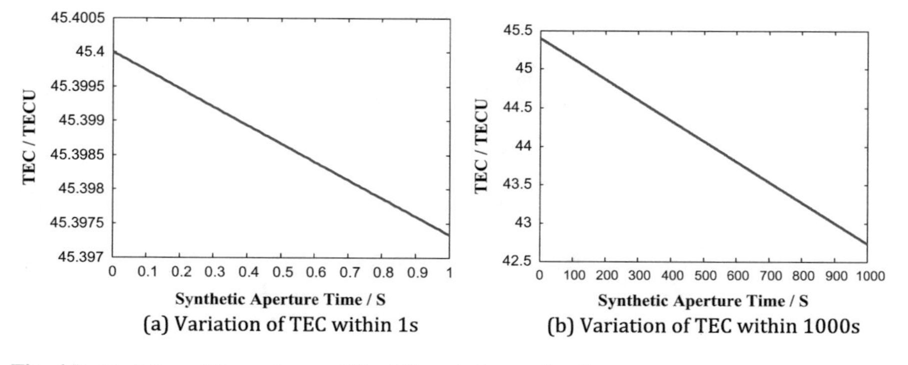

Fig. 4.7 Variation of ionosphere within different observation time spans

an approximation about the description of the trans-ionosphere propagation and the relative movement between GEO SAR and the ionosphere. The *TEC* values are also various at different puncture points where the propagation signals traverse the ionosphere. So considering the spatial variance, the phase errors can be modified as

$$\begin{aligned}\varphi_{iono}(t_a, P) = & -\frac{2\pi \cdot 80.6}{cf} \\ & \cdot \left(TEC_0(P) + k_1(P)t_a + k_2(P)t_a^2 + k_3(P)t_a^3 + \ldots\right)\end{aligned} \tag{4.32}$$

where P indicates the position of the puncture point. Therefore the phase errors induced by background ionosphere in GEO SAR are temporal and spatial variant.

4.3.2 Time-Frequency Signal Model

Considering the phase errors in (4.32), the GEO SAR signal model can be modified as [35]

$$\begin{aligned}S(f_r, t_a) = & A_r(f_r) \cdot A_a(t_a) \cdot \exp\left(-j\frac{\pi f_r^2}{k_r}\right) \\ & \cdot \exp\left[-j\frac{4\pi(f_r + f_0)r(t_a)}{c}\right] \cdot \exp\left[-j\frac{2\pi \cdot 80.6 \cdot TEC(t_a, P)}{c(f_r + f_0)}\right].\end{aligned} \tag{4.33}$$

Then the accurate two-dimensional spectrum of (4.33) can be expressed as follows through the specific derivation process in Appendix 1.

$$\begin{aligned}S(f_a, f_r) = & \sigma A_a(f_a)A_r(f_r) \cdot \exp\left(-j \cdot \pi\frac{f_r^2}{\beta}\right) \\ & \times \exp\left(-j \cdot 2\pi \cdot \frac{2(f_r + f_0)}{c} \cdot \left(r_0 + \frac{40.3 \cdot TEC_0}{(f_0 + f_r)^2}\right)\right) \\ & \times \exp\left(j \cdot 2\pi \cdot \frac{1}{4 \cdot \left(q_2 + \frac{40.3 \cdot k_2}{f_0^2}\right)} \cdot \frac{c}{2(f_r + f_0)}\right. \\ & \left. \cdot \left(f_a + \frac{2 \cdot (f_r + f_c)}{c} \cdot \left(q_1 + \frac{40.3 \cdot k_1}{f_0^2}\right)\right)^2\right) \\ & \times \exp\left(j \cdot 2\pi \cdot \frac{q_3 + \frac{40.3 \cdot k_3}{f_0^2}}{8 \cdot \left(q_2 + \frac{40.3 \cdot k_2}{f_0^2}\right)^3} \cdot \left(\frac{c}{2(f_r + f_0)}\right)^2\right. \\ & \left. \cdot \left(f_a + \frac{2 \cdot (f_r + f_c)}{c} \cdot \left(q_1 + \frac{40.3 \cdot k_1}{f_0^2}\right)\right)^3\right).\end{aligned} \tag{4.34}$$

4.3.3 Influences on Focusing

The background ionosphere can influence the range and azimuth focusing in GEO SAR. Specifically, the effects on range focusing include image shift due to the group delay, and defocusing caused by dispersion. In comparison, the effects on azimuth focusing also include image shift and defocusing but are caused by the changes of the Doppler history due to the varying ionospheric time delays. The detailed influences can be derived from the two-dimensional spectrum in (4.34) which can be rewritten as

$$\begin{aligned} S(f_a,f_r) = \sigma A_a(f_a)A_r(f_r) \cdot \exp(j\cdot\phi_r(f_r)) \cdot \exp(j\cdot\phi_a(f_a)) \\ \cdot \exp(j\cdot\phi_{RCM}(f_r,f_a)) \cdot \exp(j\cdot\phi_{SRC}(f_r,f_a)) \cdot \exp(j\cdot\phi_{residual}) \end{aligned} \quad (4.35)$$

where, $\exp(j\cdot\phi_r(f_r))$ is the range compression term, $\exp(j\cdot\phi_a(f_a))$ is the azimuth compression term, $\exp(j\cdot\phi_{RCM}(f_r,f_a))$ is the range cell migration term, $\exp(j\cdot\phi_{SRC}(f_r,f_a))$ is the secondary range compression term, and $\exp(j\cdot\phi_{residual})$ is the residual phase term. Their specific expressions are listed as (4.36)–(4.40) where $M_i(k_i) = q_i + \frac{40.3\cdot k_i}{f_0^2}$, $i = 1,2,3$.

The range compression term $\exp(j\cdot\phi_r(f_r))$ is expressed as

$$\begin{aligned} \phi_r(f_r) = & -\frac{4\pi r_0}{c}\cdot f_r + \frac{2\pi}{c}\cdot\left(\frac{80.6\cdot TEC_0}{f_0^2} + \frac{M_1(k_1)^3\cdot M_3(k_3)}{4\cdot M_2(k_2)^3} + \frac{M_1(k_1)^2}{2\cdot M_2(k_2)}\right)\cdot f_r \\ & -\pi\frac{f_r^2}{\beta} + \frac{2\pi}{cf_0}\cdot\left(-\frac{80.6\cdot TEC_0}{f_0^2} + \frac{M_1(k_1)^3\cdot M_3(k_3)}{16\cdot M_2(k_2)^3}\right)\cdot f_r^2 \\ & + \frac{2\pi}{cf_0^2}\cdot\left(\frac{80.6\cdot TEC_0}{f_0^2} + \frac{M_1(k_1)^3\cdot M_3(k_3)}{16\cdot M_2(k_2)^3}\right)\cdot f_r^3 \end{aligned} \quad (4.36)$$

The azimuth compression term $\exp(j\cdot\phi_a(f_a))$ is specifically expressed as

$$\begin{aligned} \phi_a(f_a) = & \pi\cdot\frac{M_1(k_1)}{M_2(k_2)}\cdot\left(1 + \frac{3M_1(k_1)\cdot M_3(k_3)}{4M_2(k_2)^2}\right)\cdot f_a \\ & + \frac{\pi c}{f_0}\cdot\frac{1}{M_2(k_2)}\cdot\left(\frac{1}{4} + \frac{3M_1(k_1)\cdot M_3(k_3)}{8M_2(k_2)^2}\right)\cdot f_a^2 + \frac{\pi c^2}{16f_0^2}\cdot\frac{M_3(k_3)}{M_2(k_2)^3}\cdot f_a^3 \end{aligned} \quad (4.37)$$

The range cell migration term $\exp(j\cdot\phi_{RCM}(f_r,f_a))$ is specifically expressed as

$$\phi_{RCM}(f_r,f_a) = -2\pi f_r \cdot \left(\left(\frac{c}{8f_0^2 M_2(k_2)} + \frac{3c \cdot M_1(k_1) \cdot M_3(k_3)}{16 f_0^2 M_2(k_2)^3} \right) \cdot f_a^2 + \frac{c^2 \cdot M_3(k_3)}{16 f_0^3 M_2(k_2)^3} \cdot f_a^3 \right) \tag{4.38}$$

The secondary range compression term $\exp(j \cdot \phi_{SRC}(f_r,f_a))$ is specifically expressed as

$$\phi_{SRC}(f_r,f_a) = 2\pi \begin{pmatrix} \left(\frac{M_1(k_1)^2 \cdot M_3(k_3)}{16 f_0^2 \cdot M_2(k_2)^3} \cdot f_a + \left(\frac{c}{8f_0^3 \cdot M_2(k_2)} + \frac{3c \cdot M_1(k_1) \cdot M_3(k_3)}{16 f_0^3 \cdot M_2(k_2)^3} \right) \right. \\ \left. \cdot f_a^2 + \frac{3c^2 \cdot M_3(k_3)}{32 f_0^4 \cdot M_2(k_2)^3} \cdot f_a^3 \right) \cdot f_r^2 \\ - \left(\left(\frac{c}{8f_0^4 \cdot M_2(k_2)} + \frac{3c \cdot M_1(k_1) \cdot M_3(k_3)}{16 f_0^4 \cdot M_2(k_2)^3} \right) \right. \\ \left. \cdot f_a^2 + \frac{c^2 \cdot M_3(k_3)}{8 f_0^5 \cdot M_2(k_2)^3} \cdot f_a^3 \right) \cdot f_r^3 \end{pmatrix} \tag{4.39}$$

The residual phase term $\exp(j \cdot \phi_{residual})$ is specifically expressed as

$$\begin{aligned} \phi_{residual} = & -\frac{2\pi}{c} \cdot \left(2R_0 \cdot f_0 + \frac{80.6 \cdot TEC_0}{f_0} \right) \\ & + \frac{\pi f_0 \cdot M_1(k_1)^2}{c \cdot M_2(k_2)} \cdot \left(1 + \frac{\pi f_0 \cdot M_1(k_1) \cdot M_3(k_3)}{2 \cdot M_2(k_2)^2} \right) \end{aligned} \tag{4.40}$$

It can be seen that the aforementioned phase terms are all related to the status of TEC, i.e. the values of TEC_0, k_1, k_2 and k_3. In order to compare the influences of various terms, the real TEC data was used for analysis. The U.S. Total Electron Content (US-TEC) data at 140° W and 35° N on June 29, 2014 at UTC 18: 00 was applied. The preprocessed data is listed in Table 4.4.

It can also be indicated that, $\phi_r(f_r)$ is only in correlation with f_r, which will affect the range imaging; $\phi_a(f_a)$ is only in correlation with f_a, $\phi_{RCM}(f_r,f_a)$ and $\phi_{SRC}(f_r,f_a)$ are related with f_r and f_a simultaneously, which will affect the azimuth imaging; but $\phi_{residual}$ has nothing to do with the both, and it does not affect the imaging results.

The terms $\phi_a(f_a)$, $\phi_{RCM}(f_r,f_a)$ and $\phi_{SRC}(f_r,f_a)$ have various influences on the azimuth imaging. Without loss of generality, considering the off-nadir angle of

Table 4.4 US-TEC data at UTC 18: 00 on June 29, 2014 at 1400 W and 350 N

Item	TEC_0 (TECU)	k_1 (TECU/S)	k_2 (TECU/S^2)	k_3 (TECU/S^3)
Value	35.4	0.0072	6.89×10^{-6}	8.35×10^{-12}

4.65°, the resolution of range-5 m corresponds to 67.1 MHz of signal bandwidth. The phase errors corresponding to the three terms under different integration time are shown in Fig. 4.8.

As can be seen from Fig. 4.8, the azimuth compression term plays an important role in the azimuth focusing. In contrast, the effects of the range cell migration term and the secondary range compression term are very limited, thus it just takes the effects of the azimuth compression term $\exp(j \cdot \phi_a(f_a))$ into account in terms of azimuth aspects and just takes the effects of the range compression term $\exp(j \cdot \phi_r(f_r))$ into account in terms of range aspects, subsequently.

4.3.4 *Effects on Range Focusing*

The effects of background ionosphere on range focusing are mainly correlated with the term $\exp(j \cdot \phi_r(f_r))$, and its specific analytical expression is

$$\begin{aligned}\phi_r(f_r) = {} & \frac{2\pi}{c} \cdot \left(\frac{80.6 \cdot TEC_0}{f_0^2} + \frac{\left(q_1 + \frac{40.3k_1}{f_0^2}\right)^3 \left(q_3 + \frac{40.3k_3}{f_0^2}\right)}{4 \cdot \left(q_2 + \frac{40.3k_2}{f_0^2}\right)^3} + \frac{\left(q_1 + \frac{40.3k_1}{f_0^2}\right)^2}{2 \cdot \left(q_2 + \frac{40.3k_2}{f_0^2}\right)} \right) \\ & \cdot f_r - \frac{4\pi r_0}{c} \cdot f_r + \frac{2\pi}{cf_0} \cdot \left(-\frac{80.6 \cdot TEC_0}{f_0^2} + \frac{\left(q_1 + \frac{40.3k_1}{f_0^2}\right)^3 \left(q_3 + \frac{40.3k_3}{f_0^2}\right)}{16 \cdot \left(q_2 + \frac{40.3k_2}{f_0^2}\right)^3} \right) \\ & \cdot f_r^2 - \pi \frac{f_r^2}{\beta} + \frac{2\pi}{cf_0^2} \cdot \left(\frac{80.6 \cdot TEC_0}{f_0^2} + \frac{\left(q_1 + \frac{40.3k_1}{f_0^2}\right)^3 \left(q_3 + \frac{40.3k_3}{f_0^2}\right)}{16 \cdot \left(q_2 + \frac{40.3k_2}{f_0^2}\right)^3} \right) \cdot f_r^3\end{aligned} \tag{4.41}$$

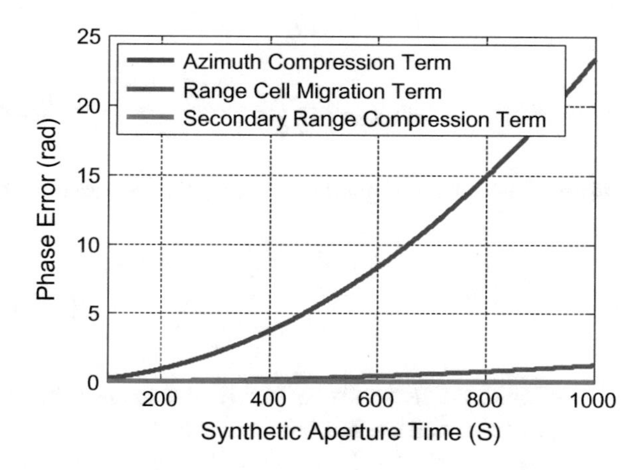

Fig. 4.8 Comparison of phase errors caused by the azimuth compression term, range cell migration term and secondary range compression term

The following part will probe into the effects of background ionosphere on range focusing in combination with (4.41).

(1) *Image Shift*

The background ionosphere can induce the group delay, resulting in the image shift along the range direction. According to the linear term of (4.41), it is able to calculate the image shift caused by the background ionosphere as follows

$$\Delta L_r = c \cdot \tau_{gd} = \frac{80.6 \cdot TEC_0}{f_0^2} + \left(\frac{\left(q_1 + \frac{40.3k_1}{f_0^2}\right)^3 \left(q_3 + \frac{40.3k_3}{f_0^2}\right)}{4 \cdot \left(q_2 + \frac{40.3k_2}{f_0^2}\right)^3} + \frac{\left(q_1 + \frac{40.3k_1}{f_0^2}\right)^2}{2 \cdot \left(q_2 + \frac{40.3k_2}{f_0^2}\right)} - \frac{q_1^3 \cdot q_3}{4 \cdot q_2^3} - \frac{q_1^2}{2 \cdot q_2} \right) \tag{4.42}$$

Therefore, the range image shift is dependent on both the constant component TEC_0 and the changing TEC at various orders. In addition, the carrier frequency will also affect the quantity of range image shift.

(2) *Image Defocusing*

GEO SAR signal will be stretched or compressed when traversing the ionosphere as the dispersion phenomenon. Thus it will induce range quadratic phase errors (QPE) and range cubical phase errors (CPE). The QPE will give rise to the broadening of the main lobe, the arising of the sidelobe and the degradation of the range image quality. The CPE would give rise to the asymmetrical sidelobes. The corresponding range QPE and CPE are written as

$$\phi_{range2} = \frac{2\pi}{cf_0} \cdot \left(-\frac{80.6 \cdot TEC_0}{f_0^2} + P(k_i, q_i) \right) \cdot B^2 \tag{4.43}$$

$$\phi_{range3} = \frac{2\pi}{cf_0^2} \cdot \left(\frac{80.6 \cdot TEC_0}{f_0^2} + P(k_i, q_i) \right) \cdot B^3 \tag{4.44}$$

where B is the signal bandwidth, and the expression of $P(k_i, q_i)$ is written as

$$P(k_i, q_i) = \frac{\left(q_1 + \frac{40.3k_1}{f_0^2}\right)^3 \left(q_3 + \frac{40.3k_3}{f_0^2}\right)}{16 \cdot \left(q_2 + \frac{40.3k_2}{f_0^2}\right)^3} - \frac{q_1^3 \cdot q_3}{16 \cdot q_2^3}, \quad i = 1, 2, 3 \tag{4.45}$$

As indicated in (4.43) to (4.45), the range QPE and CPE introduced by the ionospheric dispersion are both correlated with TEC_0 and the changing TEC over time at various orders. In addition, the signal frequency and signal bandwidth will also have a significant impact to the range QPE and CPE.

4.3.5 Effects on Azimuth Focusing

The integration time of GEO SAR can be up to hundreds or even thousands of seconds, and changes in *TEC* within the integration time will cause changes of the Doppler rates, and thus will result in azimuth shift and defocusing in GEO SAR.

In view of $\exp(j \cdot \phi_a(f_a))$ that may affect the azimuth imaging, the specific expression is

$$\begin{aligned}\phi_a(f_a) = {} & \pi \cdot \frac{q_1 + \frac{40.3k_1}{f_0^2}}{q_2 + \frac{40.3k_2}{f_0^2}} \cdot \left(1 + \frac{3\left(q_1 + \frac{40.3k_1}{f_0^2}\right)\left(q_3 + \frac{40.3k_3}{f_0^2}\right)}{4\left(q_2 + \frac{40.3k_2}{f_0^2}\right)^2}\right) \cdot f_a \\ & + \frac{\pi c}{f_0} \cdot \frac{1}{\left(q_2 + \frac{40.3k_2}{f_0^2}\right)} \cdot \left(\frac{1}{4} + \frac{3\left(q_1 + \frac{40.3k_1}{f_0^2}\right)\left(q_3 + \frac{40.3k_3}{f_0^2}\right)}{8\left(q_2 + \frac{40.3k_2}{f_0^2}\right)^2}\right) \cdot f_a^2 \\ & + \frac{\pi c^2}{16 f_0^2} \cdot \frac{\left(q_3 + \frac{40.3k_3}{f_0^2}\right)}{\left(q_2 + \frac{40.3k_2}{f_0^2}\right)^3} \cdot f_a^3\end{aligned} \tag{4.46}$$

The linear phase error will cause the azimuth shift, while the QPE and CPE will cause the azimuth defocusing.

(1) *Image Shift*

The linear changing components of TEC over time will lead to linear changes in the slant range, resulting in the changes of the Doppler centroid frequency, thereby causing the shift of image along the azimuth which is calculated as

$$\Delta L_a = v_{nadir} \cdot \frac{\pi}{2} \cdot \left(\begin{array}{c} \frac{q_1 + \frac{40.3k_1}{f_0^2}}{q_2 + \frac{40.3k_2}{f_0^2}} \cdot \left(1 + \frac{3\left(q_1 + \frac{40.3k_1}{f_0^2}\right)\left(q_3 + \frac{40.3k_3}{f_0^2}\right)}{4\left(q_2 + \frac{40.3k_2}{f_0^2}\right)^2}\right) \\ - \frac{q_1}{q_2} \cdot \left(1 + \frac{3 \cdot q_1 q_3}{4 \cdot q_2^2}\right) \end{array}\right) \tag{4.47}$$

where v_{nadir} is the Nadir-point velocity. Thus, when the GEO SAR satellite orbit is fixed, the azimuth shift is in correlation with the signal frequency and change rate of *TEC* over time at various orders.

(2) *Image Defocusing*

The azimuth QPE and CPE will cause azimuth defocusing. The azimuth signal bandwidth can be expressed as $B_a = f_{dr} \cdot T_a$, wherein, f_{dr} is the Doppler chirp rate, and T_a is the integration time. The azimuth QPE is written as

$$\phi_{azimuth2} = \frac{\pi c}{f_0} \cdot \left(\begin{array}{l} \frac{1}{\left(q_2 + \frac{40.3k_2}{f_0^2}\right)} \cdot \left(\frac{1}{4} + \frac{3\left(q_1 + \frac{40.3k_1}{f_0^2}\right)\left(q_3 + \frac{40.3k_3}{f_0^2}\right)}{8\left(q_2 + \frac{40.3k_2}{f_0^2}\right)^2} \right) \\ - \frac{1}{q_2} \cdot \left(\frac{1}{4} + \frac{3 \cdot q_1 q_3}{8 \cdot q_2^2} \right) \end{array} \right) \cdot (f_{dr} \cdot T_a)^2 \quad (4.48)$$

The phase error may cause the main lobe broadening, the sidelobe arising and the azimuth defocusing. Similarly, the azimuth CPE will bring about asymmetric sidelobes. It can be calculated as

$$\phi_{azimuth3} = \frac{\pi c^2}{16 f_0^2} \cdot \left(\frac{\left(q_3 + \frac{40.3k_3}{f_0^2}\right)}{\left(q_2 + \frac{40.3k_2}{f_0^2}\right)^3} - \frac{q_3}{q_2^3} \right) \cdot (f_{dr} \cdot T_a)^3 \quad (4.49)$$

As indicated in (4.48) and (4.49), the changing rates of TEC over time at various orders will cause the azimuth defocusing. In addition, the signal frequency and the synthetic aperture time will also affect the azimuth focusing.

4.3.6 Performance Analysis and the Changing TEC Boundaries

This part will give the boundary conditions of related various ionospheric parameters required for the GEO SAR imaging in the context of the background ionosphere, including the threshold values for the absolute TEC and the changing rates of TEC at various orders.

4.3.6.1 Requirements for Absolute TEC

Based on the above analysis, the absolute TEC and the changing rate of TEC will cause impacts on the range imaging. However, since the values of the change rates of TEC at various orders are much smaller than that of TEC_0, TEC_0 is decisive and critical for range focusing. Therefore, the range QPE and CPE in (4.43) and (4.44) can be simplified as

$$\phi_{range2} \approx -\frac{2\pi}{cf_0} \cdot \frac{80.6 \cdot TEC_0}{f_0^2} \cdot B^2 \quad (4.50)$$

$$\phi_{range3} \approx \frac{2\pi}{cf_0^2} \cdot \frac{80.6 \cdot TEC_0}{f_0^2} \cdot B^3 \tag{4.51}$$

Assuming the required resolution is 20 m, the signal bandwidth can be calculated based on different incident angles. The required signal bandwidths at different arguments of latitude are the same (Fig. 4.9).

Due to the classical SAR theory, the QPE of the signal should be above $\pi/4$ rads if it influences SAR focusing. Therefore, to make sure that the range can be focused, TEC must meet the following conditions

$$TEC \leq \frac{c \cdot f_0^3}{161.2 \cdot B^2} \tag{4.52}$$

Similarly, when the CPE of signal reaches $\pi/8$ rads, it will also cause the image defocusing. Hence, TEC should also satisfy the following conditions in order to avoid the defocusing in the range direction

$$TEC \leq \frac{c \cdot f_0^4}{161.2 \cdot B^3} \tag{4.53}$$

As shown in Fig. 4.10, in the case of the QPE and CPE up to the extent affecting the range focusing, TEC should be up to 301 TECU and 10,810 TECU respectively, which are obviously unrealistic. Therefore, in respect of the current GEO SAR parameters, the background ionosphere will not affect GEO SAR focusing in range.

4.3.6.2 Requirements for the Changing Rates of TEC

The changes in *TEC* over time will cause azimuth defocusing when the azimuth QPE and CPE exceeds $\pi/4$ and $\pi/8$ rads, respectively. Below, it will be given the thresholds of TEC change rates at various orders.

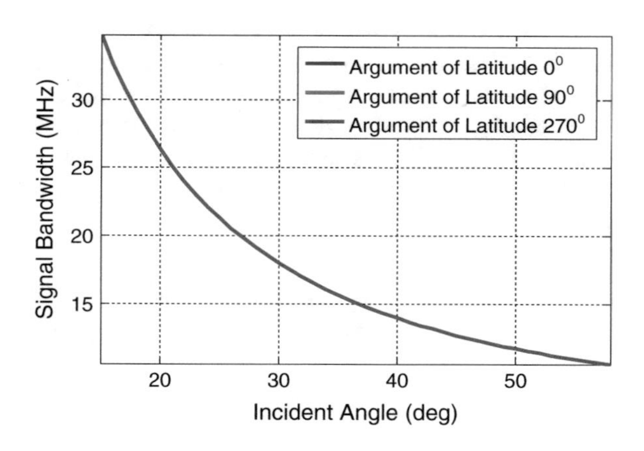

Fig. 4.9 Signal bandwidth required for the resolution of 20 m at different incident angles and arguments of latitude

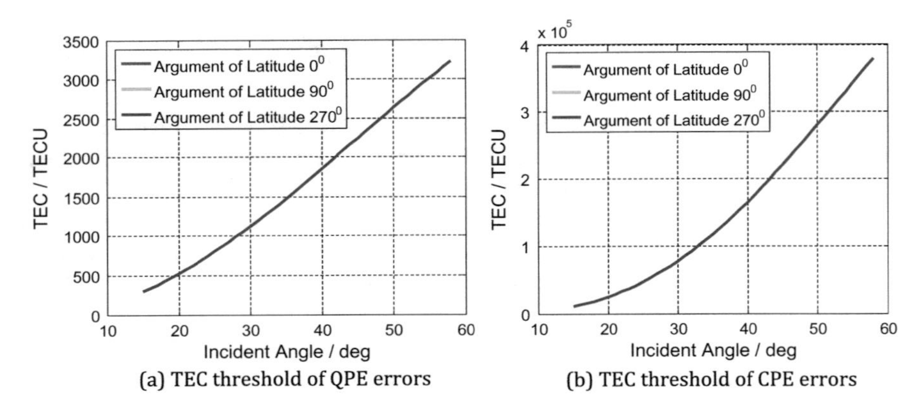

Fig. 4.10 TEC thresholds of the range QPE and CPE

In the case of well-focusing in azimuth, the azimuth QPE in (4.48) should be below $\pi/4$ rads. Moreover, when the second-order and third-order change rates of *TEC* are 0, k_1 will gain the maximum value. Therefore, the first-order change rate of *TEC* needs to satisfy the following conditions

$$k_1 \leq \frac{2 \cdot f_0^3 \cdot q_2^3}{3 \cdot 40.3 \cdot c \cdot f_{dr}^2 \cdot T_a^2 \cdot q_3} \tag{4.54}$$

Similarly, it is derived that the second-order and third-order change rates of *TEC* are required to meet the following conditions respectively

$$k_2 \leq \frac{f_0^3 \cdot q_2^2}{40.3 \cdot \left(f_0 \cdot q_2 + c \cdot f_{dr}^2 \cdot T_a^2\right)} \tag{4.55}$$

$$k_3 \leq \frac{2 \cdot f_0^3 \cdot q_2^3}{3 \cdot 40.3 \cdot c \cdot f_{dr}^2 \cdot T_a^2 \cdot q_1} \tag{4.56}$$

In the view of (4.54), (4.55) and (4.56), when the resolution is up to 20 m, the thresholds for the first-order, second-order and third-order change rate of *TEC* corresponding to the azimuth QPE are illustrated in Fig. 4.11.

Similarly, in the case of well-focusing in azimuth, the azimuth CPE in (4.49) cannot exceed $\pi/8$ rads. It can be elicited that thresholds for the second-order and third-order change rates of *TEC* corresponding to the azimuth CPE is written as follows

$$k_2 \leq \frac{f_0^2 \cdot q_2}{40.3} \cdot \left(f_{dr} \cdot T_a \cdot \sqrt[3]{\frac{c^2 q_3}{2 f_0^2 q_2^3 + c^2 q_3 f_{dr}^3 T_a^3}} - 1\right) \tag{4.57}$$

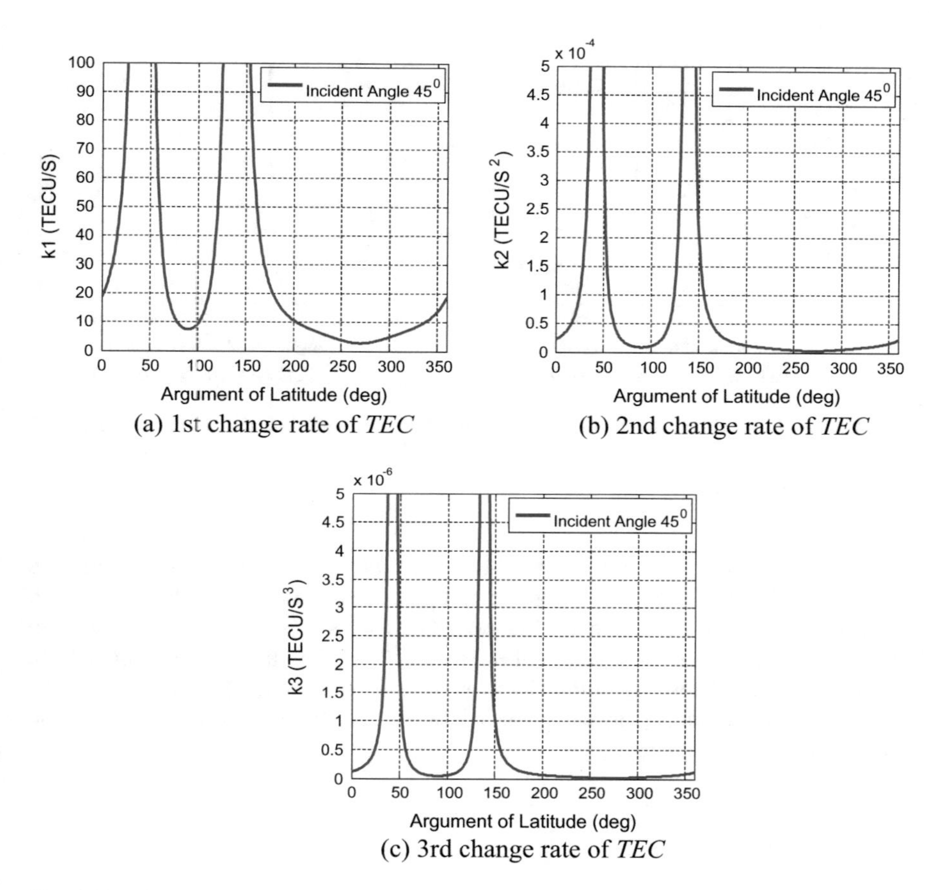

(a) 1st change rate of *TEC*

(b) 2nd change rate of *TEC*

(c) 3rd change rate of *TEC*

Fig. 4.11 Change rate threshold of TEC corresponding to the resolution of 20 m (azimuth QPE)

$$k_3 \leq \frac{2 \cdot f_0^4 \cdot q_2^3}{40.3 \cdot c^2 \cdot f_{dr}^3 \cdot T_a^3} \tag{4.58}$$

In the view of (4.57) and (4.58), when the resolution is up to 20 m, thresholds for the second-order and third-order change rate of *TEC* corresponding to the azimuth CPE are illustrated in Fig. 4.12.

It is revealed in accordance with the analysis results in Figs. 4.11 and 4.12, at the resolution of 20 m, the first-order change rate of *TEC* should be at least up to 2.8 *TECU/S* before influencing the GEO SAR azimuth focusing, however, which is out of reach in the nature. Therefore, the first-order change rate of *TEC* will not affect the GEO SAR azimuth focusing. In contrast, the second-order and third-order change rates of *TEC* should be at least up to 3.38×10^{-6} TECU/S^2 and 1.29×10^{-8} TECU/S^3 respectively before influencing the azimuth focusing. But the detailed thresholds will be also dependent on the GEO SAR geometry.

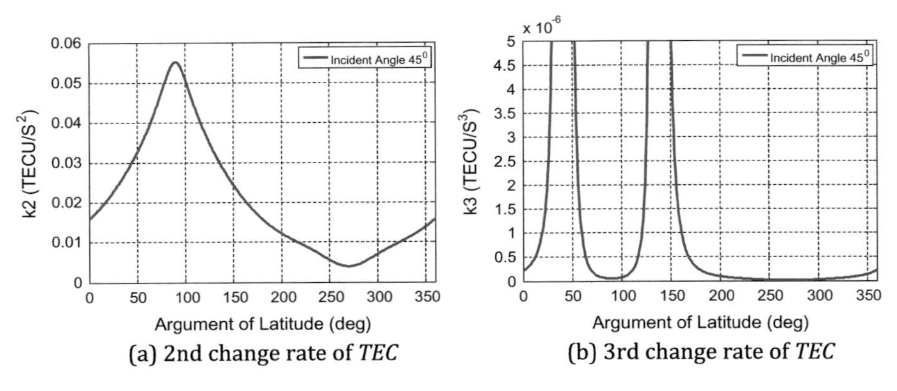

Fig. 4.12 Change rate thresholds of TEC corresponding to the resolution of 20 m (azimuth CPE)

4.3.7 Simulations

In order to analyze the ionospheric effects on GEO SAR, it needs to acquire the *TEC* status along the GEO SAR signal propagation path within the integration time. Nonetheless, due to the absence of in-orbit GEO SAR satellite, it is impossible to accurately measure the *TEC* values along the GEO SAR signal propagation path. In this section, we propose an approach to equivalently construct *TEC* values along the GEO SAR propagation path using the Klobuchar model. Based on this approach, the influences are analyzed.

4.3.7.1 TSV TEC Data

The Beidou (BD) Navigation Satellite System has an orbital altitude of 36,000 km. Its orbital parameters are similar to GEO SAR and listed in Table 4.5. Its nadir-point track has the shape of "figure-8" (as shown in Fig. 4.13). The information of the *TEC* values is embedded in the head message of BD signals. So the TEC values retrieved from BD signals can be employed in analyzing ionospheric influences on GEO SAR.

The ionosphere is regarded as a thin layer at the altitude of 350 km, and the intersection between the satellite signal and the ionosphere is called as the puncture point (IPP). As shown in Fig. 4.13, when the BD IGSO satellite and GEO SAR are simultaneously illuminating Beijing, both of their IPPs have almost the same latitude and longitude within the integration time of 200 s.

Moreover, as the BD IGSO has similar carrier frequency and signal bandwidth as GEO SAR, the approximately identical errors will be induced in case of traversing the ionosphere. Thus, the BD IGSO satellite can be served as the equivalently alternative satellite to validate the ionospheric effects on GEO SAR.

The BD IGSO signals broadcast the D1 navigation message, which contains the basic navigation information of the satellite, and it consists of eight ionospheric

Table 4.5 BD IGSO satellite orbital parameters

Items	Value	Items	Value
Eccentricity	0.0025	Inclination	56.4°
Argument of perigee	185°	Right ascension of ascending node	85°
Mean anomaly	279°	Semi-major axes	42,164 km

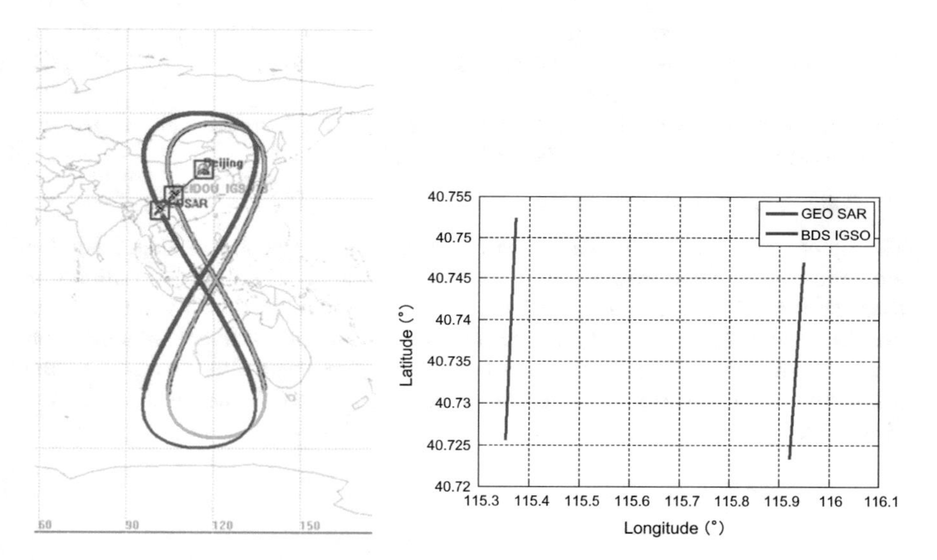

Fig. 4.13 Schematic diagrams of the IGSO and GEO SAR nadir-point track (left) and the IPPs' latitude and longitude (right)

delay model parameters that can represent the current ionospheric information ($\alpha_1 \sim \alpha_4$, $\beta_1 \sim \beta_4$,). Then in combination with the coordinates of the satellite and the receiver, the *TEC* values along the BD IGSO satellite signal propagation path can be calculated. The specific calculation method is detailed in Appendix 2.

4.3.7.2 Point Target

This part will analyze the effects of the background ionosphere on GEO SAR imaging based on the retrieved *TEC* data retrieved from the BD IGSO satellite. The simulations include single point target, point array targets and area target.

Southern China is located in the lower-latitude region where the influences of the ionosphere are significant. It can be characterized by relatively great ionospheric strength and diurnal variations, which can be used as an ideal test observation area. Therefore, the test site was selected in Zhuhai City (113.5°E, 22.3°N). The messages of BD IGSO-5 satellite were collected and then the *TEC* values were calculated based on the Klobuchar model. The duration of the collection was 900 s and the *TEC* values were shown in Fig. 4.14.

Then the changes of *TEC* data over time are fitted and the changing rates are summarized in Table 4.6.

(1) *Single-point target simulation results*

In the simulation, the signal frequency was L-band (1.25 GHz), the signal bandwidth was 20 MHz, and the argument of latitude in simulation was 90°. In addition, different integration times were selected for analysis, including 300, 500 and 700 s. The imaging results were shown in Fig. 4.15. When the integration time exceeded 500 s, defocusing in azimuth is obvious. There exists no degradation in range. The corresponding evaluations are listed in Table 4.7.

(2) *Point array targets simulation*

The point array targets simulation can simultaneously test the influences of the temporal-variation and spatial-variation of the ionosphere. The spatial-variation was embodied in the diversity of ionospheric *TEC* at different locations within the large imaging scene. The spatial-variant ionosphere will impose different influences on the different point targets within the large scene, and will also give rise to the distinction on the shifts of different point targets, and further lead to image distortion.

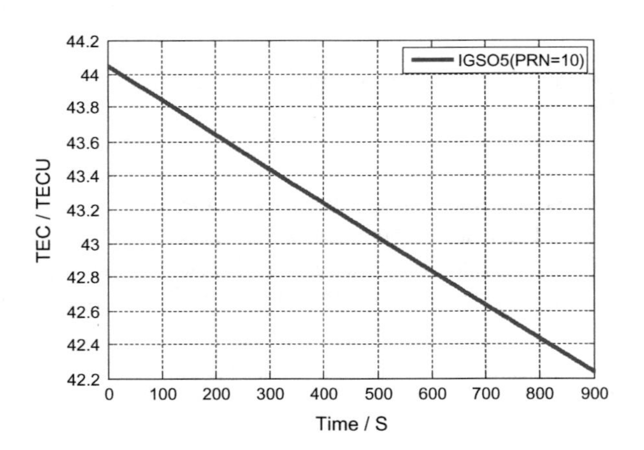

Fig. 4.14 *TEC* values along the IGSO-5 satellite propagation path

Table 4.6 Changing rates of *TEC* data of IGSO-5 satellite over time

Mean/ TECU	First-order change rate (TECU/S)	Second-order change rate ($TECU/S^2$)	Third-order change rate ($TECU/S^3$)
43.13	0.0021	9.24×10^{-7}	7.32×10^{-12}

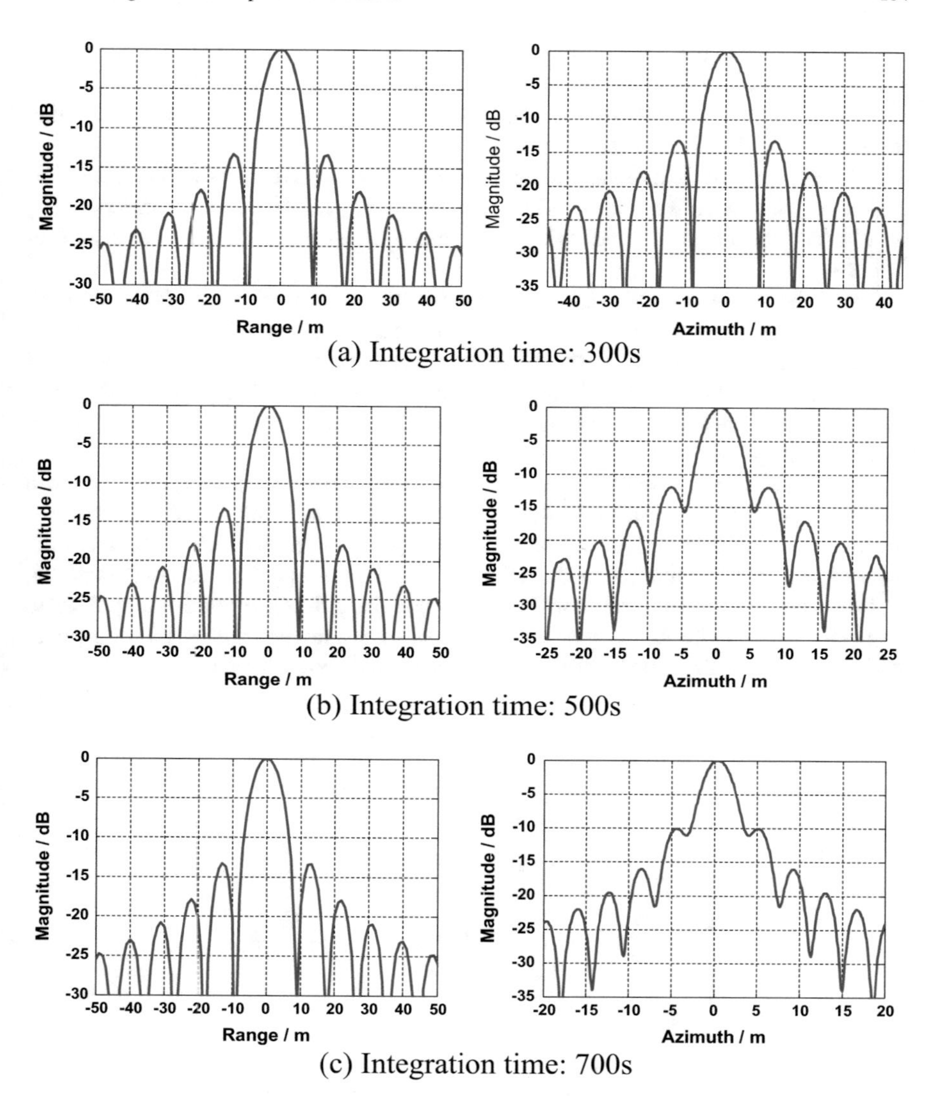

Fig. 4.15 Two-dimensional profiles of point target simulation under various integration times (Left: Range Profile; Right: Azimuth Profile)

Table 4.7 Evaluation of point target simulation under various integration times

Integration time/S	Range PSLR/dB	Azimuth PSLR/dB
300	−13.32	−13.27
500	−13.28	**−12.12**
700	−13.28	**−9.03**

With respect to the analysis in this section, a total of five point targets within the scene of 500 km × 500 km were taken into account. The point targets are set as shown in Fig. 4.16. The distance between the center target and the other edge targets is both 250 km in range and azimuth.

By adopting the Klobuchar model, the *TEC* data at five positions can be simulated as shown in Fig. 4.17. Then the mean of *TEC* and the corresponding first-order, second-order and third-order change rates are derived as listed in Table 4.8.

Then the retrieved *TEC* data at five positions were used to perform imaging analysis. Here the integration time was 300 s. Simulation results are illustrated in Fig. 4.18. In this simulation, the background ionosphere data listed in Table 4.8 would not affect the GEO SAR focusing, but would induce image shift. Since the *TEC* changing rates at those five positions are different, the two-dimensional image shift at each point varied.

As shown in Fig. 4.18, with regard to the image along the range, the minimum shift was point C (19.5 m), and the maximum shift was point D (25 m), thus the difference in the shifts of those two points along the range was 5.5 m. Meanwhile, with regard to the image along the azimuth, the minimum shift was point B (9 m), and the maximum shift was point C (16 m), thus the difference in the shifts of those two points along the azimuth was 7 m. Thus, the spatial-variation of ionosphere will cause image distortion. But it can be corrected through image co-registration techniques.

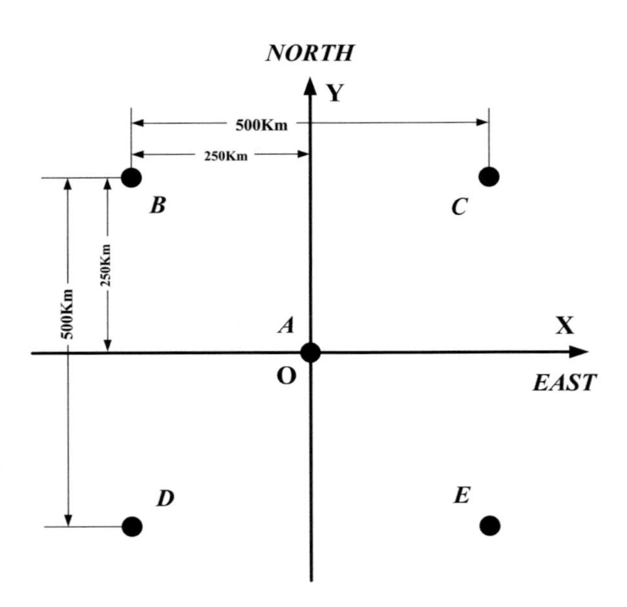

Fig. 4.16 Schematic diagram for the settings of five point targets

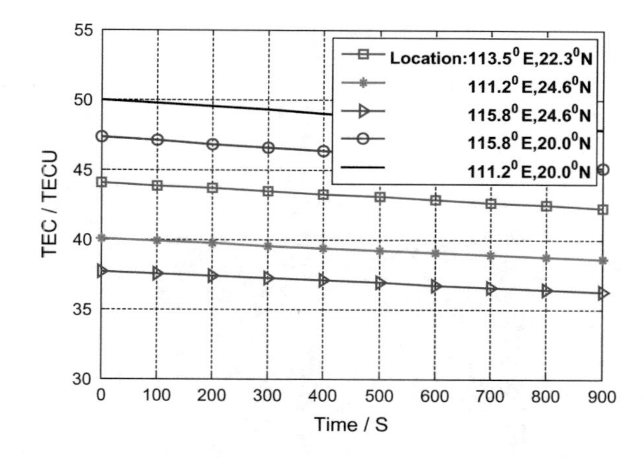

Fig. 4.17 *TEC* at five positions on the signal propagation path

Table 4.8 Changes in *TEC* data at five positions

Target position	Mean (*TEC*)	First-order change rate (*TECU/S*)	Second-order change rate (*TECU/S*^{2})	Third-order change rate (*TECU/S*^{3})
A (113.5°E, 22.3°N)	43.13	0.0021	9.24×10^{-7}	7.32×10^{-12}
B (111.2°E, 24.6°N)	39.06	0.0019	5.32×10^{-7}	4.54×10^{-12}
C (115.8°E, 24.6°N)	37.68	0.0033	9.89×10^{-7}	6.49×10^{-12}
D (111.2°E, 20.0°N)	48.85	0.0024	1.04×10^{-6}	8.48×10^{-12}
E (115.8°E, 20.0°N)	45.97	0.0030	8.36×10^{-7}	7.65×10^{-12}

4.3.7.3 Area Target

In order to observe the effects of the background ionosphere on the imaging of the real scene, the intensity of the real SAR images (China Yangtze River Three Gorges Dam and partial Three Gorges Reservoir, sourced from the TerraSAR-X) were regarded as the target Radar Cross Section (RCS) information, and performed echo simulation by using the GEO SAR orbit and system parameters [17]. The employed ionospheric TEC data were retrieved from Fig. 4.17 and Table 4.8. The integration time was 700 s, and the signal bandwidth was 20 MHz.

The imaging results are in Fig. 4.19 which show the comparison of the imaging results in the absence and the presence of ionospheric effects. Due to the influence of the ionosphere, the pixels in Fig. 4.19b appeared defocusing, and the image was blurred. For the sake of discerning the specific defocusing circumstances, the Three

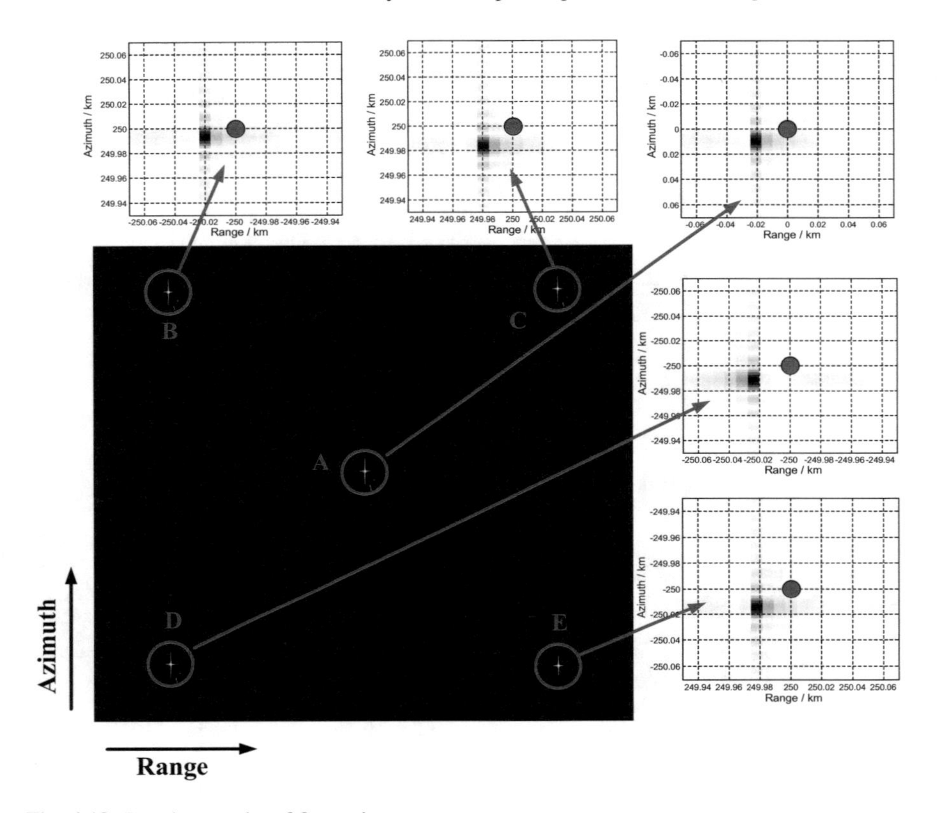

Fig. 4.18 Imaging results of five point targets

Gorges Dam and its nearby zones were enlarged to observe the imaging results. In Zone A, the Three Gorges Dam, ship locks and the surrounding mountains are clearly visible. In Zone B, due to the ionospheric effects, the Three Gorges Dam becomes obscure, and the double-line five-grade ship lock in the north side has been mixed into a pond, while the mountain chains on both sides of the dam also become no longer clear.

4.4 Ionospheric Scintillation Influences

4.4.1 Characteristics and Modelling Ionospheric Scintillation in GEO SAR

Ionospheric scintillation can cause the random fluctuations of signals which can be described by statistical parameters, probability distributions and power spectra [18]. The intensity of the ionospheric scintillation signal can be described by two

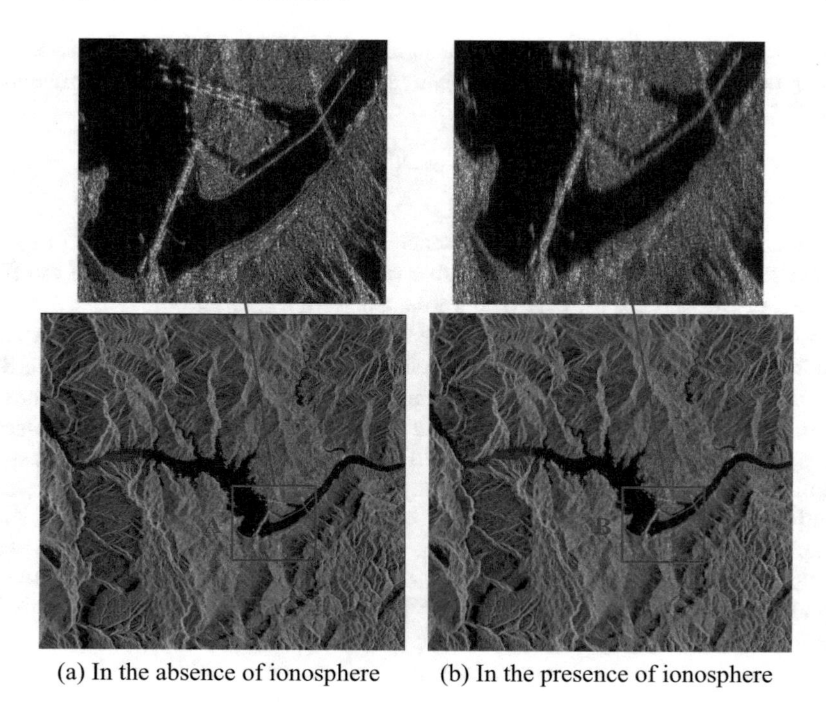

(a) In the absence of ionosphere (b) In the presence of ionosphere

Fig. 4.19 Area target imaging results and the zooms

statistical indexes: (1) S_4; (2) σ_φ. S_4 is defined by the normalized standard deviation of amplitude. σ_φ is defined by the standard deviation of the phase. Customarily, according to the value of S_4, ionospheric scintillation is cataloged as: weak scintillation ($0.1 < S_4 < 0.3$); moderate scintillation ($0.3 \leq S_4 \leq 0.5$); strong scintillation ($S_4 > 0.5$) [19]. In [20], aimed at phase scintillation, it also mentioned that $\sigma_\varphi < 0.3$ rad was considered as the weak phase scintillation, $\sigma_\varphi > 0.3$ rad for the strong phase scintillation. S_4 and σ_φ can be obtained by the commercial ionospheric scintillation monitors. Statistically, the amplitude scintillation and the phase scintillation fit Nakagami-m distribution and normal distribution, respectively [18].

Ionospheric scintillation power spectra are very important to describe the characteristics of ionospheric scintillation as well. The amplitude scintillation power spectrum has three parts: the low frequency, the high frequency roll-off and the noise floor [19]. The low frequency zone corresponds to the impacts of the large-scale ionospheric irregularities and some random noise [18, 21, 22]. The large-scale ionospheric structures do not contribute to diffraction effects. They mainly cause the effects of time delay, dispersion of the SAR signal, Doppler shift and azimuthal shifts in SAR images [8]. The high frequency roll-off part shows a power-law behavior, which corresponds to the impacts of the small-scale irregularities and contains the detailed information of the ionospheric scintillation [21]. The noise floor only contains some meaningless information brought by the high frequency noise. Fresnel frequency is a significant transition frequency between the low frequency zone and the high frequency roll-off zone, which can be expressed as [19, 23]

$$f_F = \frac{|\mathbf{V_f}|}{\sqrt{2\lambda z}} \tag{4.59}$$

where λ is the wavelength, z is the distance from the IPPs to the receiver. $|\cdot|$ stands for the norm of a vector. $\mathbf{V_f}$ is the relative velocity between the moving of the IPPs of the satellite and the moving of the ionospheric irregularity.

Similarly, the phase scintillation power spectrum has a power-law behavior zone which lies over the medium frequency range and relates the scintillation [18, 22]. The low frequency zone of the phase scintillation power spectrum corresponds to the impacts of the Doppler shift, the oscillator drift and the effects of the large-scale ionospheric irregularities [3, 18, 24, 25]. Unlike a pure diffraction process in amplitude scintillation, Fresnel frequency is not the cut-off frequency in phase scintillation power spectrum to separate the low frequency zone and the power-law zone. The cut-off frequency of the phase scintillation f_c is determined by the correlation time of the signal which is defined as the time lag at which a scintillation signal's autocorrelation function falls off by a factor e^{-1} [22, 26]. As the autocorrelation function highly depends on the power spectrum (inverse Fourier transform), the relationship between the correlation time and f_c under different scintillation intensities and spectral indexes is shown in Fig. 4.20 which is obtained by employing scintillation sampling model and 30 Monte Carlo experiments. In the simulation, scintillation indexes include S_4 and σ_φ (rad), p_A is the spectral index of the amplitude power spectrum, and p_Φ is the spectral index of the phase power spectrum. Fresnel frequency is assumed to be 0.1 Hz.

Based on the above description, we focus on the frequency zones with the power-law behavior in the ionospheric scintillation power spectra. Thus, we can describe the amplitude and phase scintillation power spectra by (Fig. 4.21).

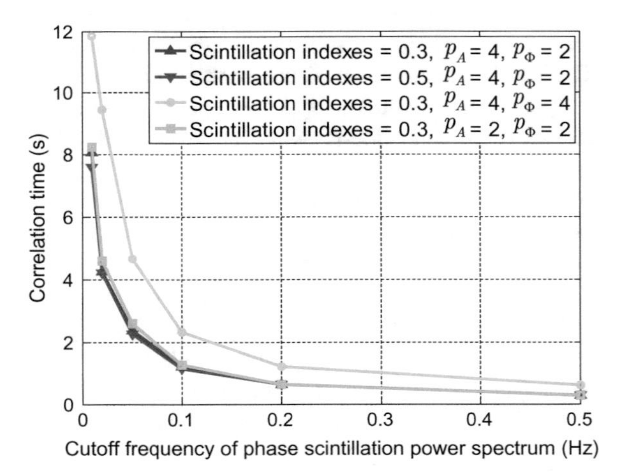

Fig. 4.20 Relationship between the correlation time and the cut-off frequency of the phase scintillation power spectrum under different scintillation intensities and spectral indexes

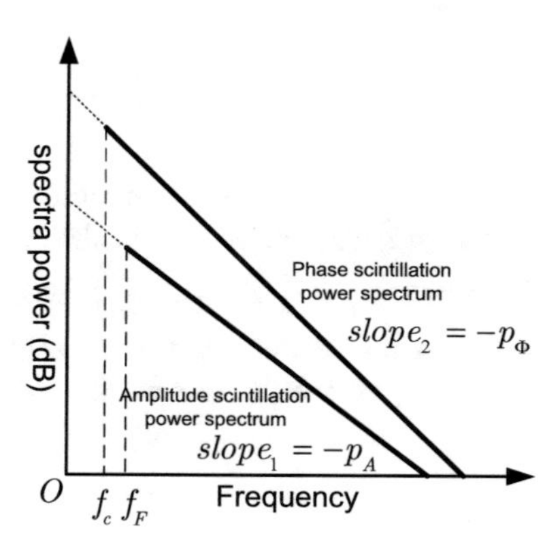

Fig. 4.21 Sketchmap of amplitude and phase scintillation power spectra

$$\begin{cases} S_\Phi(f) = T_\Phi f^{-p_\Phi}, f > f_c \\ S_A(f) = T_A f^{-p_A}, f > f_F \end{cases} \tag{4.60}$$

where $S_A(f)$ is the amplitude scintillation power spectra, $S_\Phi(f)$ is the phase scintillation power spectra, T_Φ and T_A are strength parameters corresponding to the power at $f = 1\,\text{Hz}$, p_Φ and p_A are the spectral indexes [18]. The spectral indexes depend on many factors, including the height and the size of ionospheric irregularities and the scintillation intensity, but they do not have a specified pattern [27]. Based on the previous studies by Li, Beniguel, et al. [18, 19, 22, 28], we select $p_A = 3.6, p_\Phi = 2.8$, which has a high probability to be detected in the real data set, in the following analysis and simulation.

In the GEO SAR case, assuming working in a 'figure 8' inclined GEO SAR orbit, the incidence angle is 45° and the height of the ionosphere is 350 km, the IPPs velocity of GEO SAR is about 14–41 m/s. Aimed at the ionospheric scintillation in the low latitude zone, the velocity of the moving of the ionospheric irregularities is about 50 m/s to more than 100 m/s with a mean value of approximately 75 m/s [19, 28]. Thus, we assume the IPPs velocity of the GEO SAR and the moving velocity of the ionospheric irregularities are 28 and 75 m/s, respectively. $|\mathbf{V_f}|$ has a scope of 47–103 m/s. The determined f_F is about 0.096–0.211 Hz. Only when the IPPs velocity of the GEO SAR and the moving velocity of the ionospheric irregularities are almost the same both in values and directions, f_F will be very small. Generally, as the moving velocity of the ionospheric irregularities is several times larger than the IPPs velocity of the GEO SAR, f_F is usually larger than 0.1 Hz. Meantime, ionospheric scintillation is a fast variation and the correlation time of the scintillation signal is limited less than 1–2 s [26, 29, 30]. With respect to weak and moderate scintillation, the spectral index of phase scintillation is near 2 [22]. Thus, according to Fig. 4.20, a cut-off frequency of 0.1 Hz is proper to obtain the scintillation signal which can satisfy the requirement of the correlation time.

Assuming $\varphi_{\Delta N}(\kappa)$ is the spatial spectrum in the direction of IPP's moving, where $\kappa = 2\pi f/|\mathbf{V_f}|$ is the spatial wavenumber, f is the azimuth frequency. The phase scintillation power spectrum $S_\Phi(f)$ is expressed as [22]

$$S_{\Phi}(f) = \frac{2\pi r_e^2 \lambda^2 L \sec\theta}{|\mathbf{V_f}|} \varphi_{\Delta N}\left(\frac{2\pi f}{|\mathbf{V_f}|}\right) \tag{4.61}$$

where r_e is classical electron radius, L is the layer thickness, θ is the incidence angle of the propagation vector on the ionospheric layer.

When $\frac{2\pi f}{|\mathbf{V_f}|} \geq \frac{2\pi}{L_o}$, $\varphi_{\Delta N}\left(\frac{2\pi f}{|\mathbf{V_f}|}\right)$ will show a power-law behavior. According to relationship between the outer scale size of the irregularities L_o and the cut-off frequency of the phase scintillation f_c [31], we have

$$L_o \leq \frac{|\mathbf{V_f}|}{f_c} \tag{4.62}$$

The boundary of the outer scale size of the irregularities to produce scintillation with different $|\mathbf{V_f}|$ is shown in Fig. 4.22. With the 0.1 Hz cut-off frequency of the phase scintillation, the ionospheric irregularities with the outer scale size smaller than about 1 km are the effective size scope of the irregularities to give rise to the typical phase scintillation. The larger irregularities will produce the relative slower phase variation to the signal.

Moreover, we have the relationship between the amplitude and phase scintillation power spectra and the statistical indexes by

$$\begin{cases} \int_{f_c}^{\infty} S_{\Phi}(f)df = \sigma_{\varphi}^2 \\ \int_{f_F}^{\infty} S_A(f)df = \sigma_A^2 \end{cases} \tag{4.63}$$

where σ_A^2 is the variance of amplitude scintillation and it can be expressed by S_4 as (see the Appendix 3 for details)

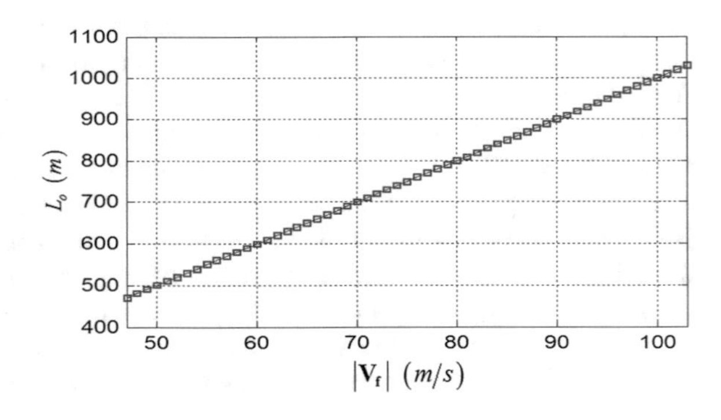

Fig. 4.22 Boundary of the outer scale size of the irregularities to produce scintillation with different $|\mathbf{V_f}|$ (the cut-off frequency of the phase scintillation is 0.1 Hz)

$$\sigma_A^2 = \Omega\left(1 - S_4^2\left(\frac{\Gamma\left(\frac{1}{S_4^2} + \frac{1}{2}\right)}{\Gamma\left(\frac{1}{S_4^2}\right)}\right)^2\right) \quad (4.64)$$

where $\Gamma(\cdot)$ is gamma function and Ω is the expectation of the ionospheric intensity scintillation.

4.4.2 GEO SAR Signal Model Considering Ionospheric Scintillation

Without the loss of generality, considering the compromise between the efficiency and accuracy of the algorithm, the SPECAN algorithm in is utilized to form an accurate GEO SAR signal model in the presence of ionospheric scintillation. GEO SAR acquisition geometry considering ionospheric scintillation is shown in Fig. 4.23. The black solid line is the orbit of the GEO SAR and the red solid line represents one synthetic aperture of the GEO SAR. The yellow line shows the nadir-point trajectory of the GEO SAR. The green region represents the ionosphere and the red region is the area where the ionospheric scintillation occurs. GEO SAR signal will traverse through the ionospheric scintillation region twice and both its amplitude and the phase will be impacted. $A_{\Delta down}(u)$ and $A_{\Delta up}(u)$ are the down-path and up-path amplitude fluctuations caused by ionospheric scintillation, respectively. $\varphi_{\Delta down}(u)$ and $\varphi_{\Delta up}(u)$ are the corresponding down-path and up-path phase fluctuations, respectively. u represents the azimuth slow time.

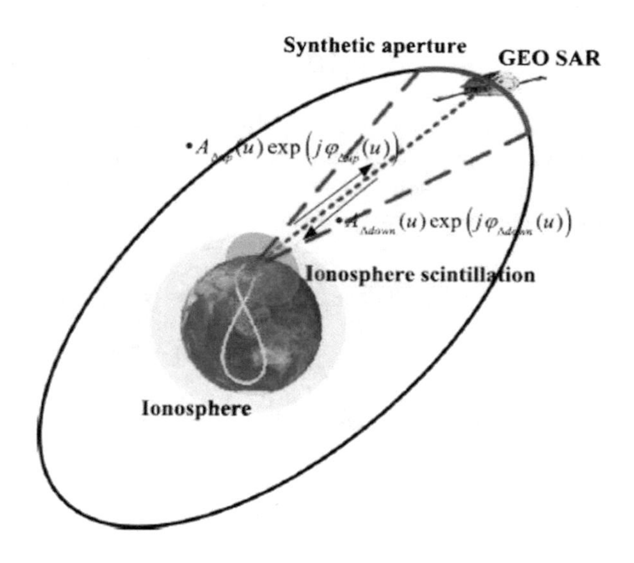

Fig. 4.23 GEO SAR acquisition geometry in the presence of ionospheric scintillation

Considering the two-way impacts of phase and amplitude fluctuations of the ionospheric scintillation on GEO SAR and the assumption of reciprocity, signal model $S_{oe}(t,u)$ in the presence of ionospheric scintillation is written as [34]

$$\begin{aligned} S_{oe}(t,u) = & A_{\Delta}^{2}(u) \cdot \exp(2j\varphi_{\Delta}(u)) \\ & \cdot \left[\rho_p W(t-\tau_A(u),u)\exp(j2\pi f_0(t-\tau_A(u)))\exp\left(j\pi k_r(t-\tau_A(u))^2\right)\right] \end{aligned} \tag{4.65}$$

where t represents the fast time in range direction, ρ_p is the normalized backscatter coefficient, $W(\cdot)$ is the envelop of the received GEO SAR signal, k_r is the range chirp rate, $\tau_A(u)$ represents the time delay in range direction. $A_{\Delta}(u)$ and $\varphi_{\Delta}(u)$ are the amplitude and phase fluctuations. In (4.65), the expression inside $[\cdot]$ represents the ideal GEO SAR signal without ionospheric scintillation.

After match-filtering in range and azimuth dechirp, two-dimensional imaged result is obtained after FFT. The imaging result in the azimuth frequency domain can be shown as

$$S_{oe}(\bar{\tau},f) = \int \left\{\bar{W}_{r,a}(\bar{\tau},u) \cdot \left[A_{\Delta}^{2}(u) \cdot \exp(2j\varphi_{\Delta}(u))\right]\right\}\exp(-j2\pi fu)du \tag{4.66}$$

where $S_{oe}(\bar{\tau},f)$ represents the target after imaging, $\bar{W}_{r,a}(\bar{\tau},u)$ is the envelop of the signal, $\bar{\tau}$ is the time delay related to the position of the target in range, f is the azimuth Doppler frequency.

When the ionospheric scintillation is weak ($0.1 < S_4 < 0.3$, $\sigma_{\varphi} < 0.3$ rad), $\varphi_{\Delta}(u) < 0.5$ rad is validated with a probability of more than 90%, by the approximation of the second-order Taylor expansion, we have

$$\exp(j2\varphi_{\Delta}(u)) \approx 1 + j2\varphi_{\Delta}(u) - 2\varphi_{\Delta}(u)^2 \tag{4.67}$$

Otherwise, if the phase scintillation is very severe, the σ_{φ} will be very large (with a low probability [20]), in which case the target will be entirely defocused and it will be no meaning to analysis the imaging quality of the target. Then the signal (4.66) can be re-written as

$$\begin{aligned} S_{oe}(\bar{\tau},f) = & \int \bar{W}_{r,a}(\bar{\tau},u)\left\{1 + \left[A_{\Delta}^{2}(u) - 1\right]\right\} \\ & \cdot \left[1 + j2\varphi_{\Delta}(u) - 2\varphi_{\Delta}(u)^2\right]\exp(-j2\pi fu)du \end{aligned} \tag{4.68}$$

Expanding (4.68), we obtain

$$S_{oe}(\bar{\tau},f) = \underbrace{S_o(\bar{\tau},f)}_{\text{Ideal output}} \underbrace{+S_{A^2d}(\bar{\tau},f)}_{\text{Distorted term of intensity scintillation}} \underbrace{+j2S_{\varphi d}(\bar{\tau},f)}_{\text{Distorted term of phase scintillation}}$$
$$\underbrace{+j2S_{A^2\varphi d}(\bar{\tau},f)}_{\text{Jointly distorted term of the intensity and phase scintillation}} \underbrace{-2S_{A^2\varphi^2 d}(\bar{\tau},f)}_{\text{Jointly distorted term of the intensity and square phase scintillation}}$$
$$\underbrace{-2S_{\varphi^2 d}(\bar{\tau},f)}_{\text{Distorted term of the square phase scintillation}} \tag{4.69}$$

where

$$\begin{cases} S_o(\bar{\tau},f) = \int \bar{W}_{r,a}(\bar{\tau},u)\exp(-j2\pi fu)du \\ S_{A^2d}(\bar{\tau},f) = \int \bar{W}_{r,a}(\bar{\tau},u)\left[A_\Delta^2(u)-1\right]\exp(-j2\pi fu)du \\ S_{\varphi d}(\bar{\tau},f) = \int \bar{W}_{r,a}(\bar{\tau},u)\varphi_\Delta(u)\exp(-j2\pi fu)du \\ S_{A^2\varphi d}(\bar{\tau},f) = \int \bar{W}_{r,a}(\bar{\tau},u)\left[A_\Delta^2(u)-1\right]\varphi_\Delta(u)\exp(-j2\pi fu)du \\ S_{A^2\varphi^2 d}(\bar{\tau},f) = \int \bar{W}_{r,a}(\bar{\tau},u)\left[A_\Delta^2(u)-1\right]\varphi_\Delta(u)^2\exp(-j2\pi fu)du \\ S_{\varphi^2 d}(\bar{\tau},f) = \int \bar{W}_{r,a}(\bar{\tau},u)\varphi_\Delta(u)^2\exp(-j2\pi fu)du \end{cases} \tag{4.70}$$

$S_o(\bar{\tau},f)$ is the ideal output term, $S_{A^2d}(\bar{\tau},f)$ is the distorted term of intensity scintillation, $S_{\varphi d}(\bar{\tau},f)$ is the distorted term of phase scintillation, $S_{A^2\varphi d}(\bar{\tau},f)$ is the jointly distorted term of the intensity (the square of the amplitude) and phase scintillation, $S_{A^2\varphi^2 d}(\bar{\tau},f)$ is the jointly distorted term of the intensity and squared phase scintillation and $S_{\varphi^2 d}(\bar{\tau},f)$ is the distorted term of the squared phase scintillation component.

Since the joint terms of amplitude scintillation and phase scintillation is small in the case of a weak ionospheric scintillation, (4.69) can be simplified as

$$S_{oe}(\bar{\tau},f) = S_o(\bar{\tau},f) + S_{A^2d}(\bar{\tau},f) + j2S_{\varphi d}(\bar{\tau},f) - 2S_{\varphi^2 d}(\bar{\tau},f) \tag{4.71}$$

Based on the convolution operation theory, the Fourier transform of the product of two signals in time domain is equal to the convolution of the spectra of the two signals in frequency domain. Therefore, (4.71) is re-expressed as

$$\begin{aligned} S_{oe}(\bar{\tau},f) = {} & S_o(\bar{\tau},f) + S_o(\bar{\tau},f)\otimes S_{A^2}(\bar{\tau},f) \\ & + j2\left[S_o(\bar{\tau},f)\otimes S_\varphi(\bar{\tau},f)\right] - 2\left[S_o(\bar{\tau},f)\otimes S_{\varphi^2}(\bar{\tau},f)\right] \end{aligned} \tag{4.72}$$

where $\otimes$ represents the convolution operation, $S_{A^2}(\bar{\tau},f)$ is the intensity scintillation spectrum, $S_{\varphi}(\bar{\tau},f)$ is the phase scintillation spectrum, $S_{\varphi^2}(\bar{\tau},f)$ is the spectrum of the square of the phase scintillation.

Compared with the integration time in LEO SAR, because the integration time of GEO SAR is very long, $S_o(\bar{\tau},f)$ is similar to the delta function. So (4.72) can be approximated as

$$S_{oe}(\bar{\tau},f) = S_o(\bar{\tau},f) + S_{A^2}(\bar{\tau},f) + j2S_{\varphi}(\bar{\tau},f) - 2S_{\varphi^2}(\bar{\tau},f) \tag{4.73}$$

Thus, the output PSF $PSF(\bar{\tau},f)$ can be expressed as

$$\begin{aligned}
PSF(\bar{\tau},f) &= \left[S_{oe}(\bar{\tau},f)S_{oe}^*(\bar{\tau},f)\right]^{\frac{1}{2}} \\
&= \Big[\big(S_o(\bar{\tau},f) + S_{A^2}(\bar{\tau},f) + j2S_{\varphi}(\bar{\tau},f) - 2S_{\varphi^2}(\bar{\tau},f)\big) \\
&\quad \cdot\big(S_o(\bar{\tau},f) + S_{A^2}(\bar{\tau},f) + j2S_{\varphi}(\bar{\tau},f) - 2S_{\varphi^2}(\bar{\tau},f)\big)^*\Big]^{\frac{1}{2}} \\
&= \Big[\left|S_o(\bar{\tau},f)\right|^2 + \left|S_{A^2}(\bar{\tau},f)\right|^2 + \left|2S_{\varphi}(\bar{\tau},f)\right|^2 + \left|-2S_{\varphi^2}(\bar{\tau},f)\right|^2 \\
&\quad + 2\mathrm{Re}\big\{S_o(\bar{\tau},f)S_{A^2}^*(\bar{\tau},f) + j2S_o(\bar{\tau},f)S_{\varphi}^*(\bar{\tau},f) - 2S_o(\bar{\tau},f)S_{\varphi^2}^*(\bar{\tau},f) \\
&\quad + j2S_{A^2}(\bar{\tau},f)S_{\varphi}^*(\bar{\tau},f) - 2S_{A^2}(\bar{\tau},f)S_{\varphi^2}^*(\bar{\tau},f) - j4S_{\varphi}(\bar{\tau},f)S_{\varphi^2}^*(\bar{\tau},f)\big\}\Big]^{\frac{1}{2}}
\end{aligned} \tag{4.74}$$

where $\mathrm{Re}\{\cdot\}$ is the operation for extraction the real part of the expression.

The $S_o(\bar{\tau},f)$ can be regarded as a SINC function and its amplitude will decline to −30 dB beyond the cut-off frequencies of the ionospheric scintillation spectra. Moreover, the energy of $\varphi_{\Delta}^2(u)$ is less than 10% compared with the energy of $\varphi_{\Delta}(u)$, when the scintillation is weak. Thus, $PSF(\bar{\tau},f)$ is approximately re-written as

$$\begin{aligned}
PSF(\bar{\tau},f) &= \left[\left|S_o(\bar{\tau},f)\right|^2 + \left|S_{A^2}(\bar{\tau},f)\right|^2 + \left|2S_{\varphi}(\bar{\tau},f)\right|^2\right]^{\frac{1}{2}} \\
&= \left[\left|\mathrm{Sinc}(\bar{\tau}, T_a f)\right|^2 + S_I(\bar{\tau},f) + 4S_{\Phi}(\bar{\tau},f)\right]^{\frac{1}{2}}
\end{aligned} \tag{4.75}$$

where T_a is integration time, $S_I(\bar{\tau},f)$ is the intensity scintillation power spectrum.

4.4.3 Theoretical Analysis

According to (4.75), it can be shown that the ionospheric scintillation power spectra will disturb the ideal azimuth PSF, which will degrade the GEO SAR images. In the following part, we will analyze the detailed impacts of ionospheric scintillation on

GEO SAR based on the analysis of the classical imaging parameters (resolution, PSLR and ISLR).

4.4.3.1 Effects on Azimuth Resolution

As the Doppler frequency domain is the imaging domain after SPECAN imaging in azimuth, according to the definition of 3 dB resolution in a SAR image and considering the expression of PSF in the absence of ionospheric scintillation, the azimuth resolution satisfies

$$20\log_{10}\left(\frac{PSF(\overline{\tau},f_m)}{PSF(\overline{\tau},f_i)}\right) = 20\log_{10}\left(\frac{1}{|Sinc(\overline{\tau},T_a f_i)|}\right) = 3 \tag{4.76}$$

where f_i and f_m are the frequencies corresponding to azimuth resolution and the peak of the mainlobe in the absence of ionospheric scintillation, respectively.

Considering (4.75), the azimuth resolution in the presence of ionospheric scintillation is expressed as

$$\begin{aligned}20\log_{10}\left(\frac{PSF(\overline{\tau},f_{mp})}{PSF(\overline{\tau},f_r)}\right) &= 20\log_{10}\left\{\frac{\left[\left|\mathrm{Sinc}(\overline{\tau},T_a f_{mp})\right|^2 + S_I(\overline{\tau},f_{mp}) + 4S_\Phi(\overline{\tau},f_{mp})\right]^{\frac{1}{2}}}{\left[\left|\mathrm{Sinc}(\overline{\tau},T_a f_r)\right|^2 + S_I(\overline{\tau},f_r) + 4S_\Phi(\overline{\tau},f_r)\right]^{\frac{1}{2}}}\right\}\\ &= 3\end{aligned} \tag{4.77}$$

where f_r and f_{mp} are the frequencies corresponding to the azimuth resolution and the peak of the mainlobe in the presence of ionospheric scintillation, respectively.

Based on the description of the ionospheric scintillation power spectra in the previous section, ionospheric amplitude and phase scintillation power spectra do not have energy at zero frequency. Assuming f_m and f_{mp} are almost zero and combining (4.76) and (4.77), we have

$$\left|\mathrm{Sinc}(\overline{\tau},T_a f_r)\right|^2 + S_I(\overline{\tau},f_r) + 4S_\Phi(\overline{\tau},f_r) \approx \left|\mathrm{Sinc}(\overline{\tau},T_a f_i)\right|^2 \tag{4.78}$$

Because of the long integration time of GEO SAR, f_r (about 0.004 Hz in a case of 120 s integration time) is far lower than the cut-off frequencies of the ionospheric scintillation power spectra (about 0.1 Hz). $S_I(\overline{\tau},f_r)$ and $S_\Phi(\overline{\tau},f_r)$ are zero based on (4.60). Thus, (4.78) can be written as

$$\left|\mathrm{Sinc}(\overline{\tau},T_a f_r)\right|^2 \approx \left|\mathrm{Sinc}(\overline{\tau},T_a f_i)\right|^2 \tag{4.79}$$

Therefore, in GEO SAR cases, the azimuth resolution will almost not be affected ($f_r \approx f_i$).

4.4.3.2 Effects on Azimuth PSLR

In the presence of ionospheric scintillation, azimuth PSLR can be expressed as

$$\begin{aligned} PSLR &= 20\log_{10}\left(\frac{PSF(\bar{\tau},f_{sh})}{PSF(\bar{\tau},f_{mp})}\right) \\ &= 20\log_{10}\left\{\frac{\left[|\mathrm{Sinc}(\bar{\tau},T_a f_{sh})|^2 + S_I(\bar{\tau},f_1) + 4S_\Phi(\bar{\tau},f_{sh})\right]^{\frac{1}{2}}}{\left[|\mathrm{Sinc}(\bar{\tau},T_a f_{mp})|^2 + S_I(\bar{\tau},f_m) + 4S_\Phi(\bar{\tau},f_{mp})\right]^{\frac{1}{2}}}\right\} \end{aligned} \tag{4.80}$$

where f_{sh} is the frequency corresponding to the sidelobe with the highest level.

Since the ionospheric scintillation power spectra concentrates after the cutoff-frequencies, assuming $f_{mp} = 0$, (4.80) is expressed as

$$PSLR = 20\log_{10}\left\{\left[|\mathrm{Sinc}(\bar{\tau},T_a f_{sh})|^2 + S_I(\bar{\tau},f_{sh}) + 4S_\Phi(\bar{\tau},f_{sh})\right]^{\frac{1}{2}}\right\} \tag{4.81}$$

Extracting $|\sin c(T_a f_{sh})|$, we have

$$PSLR = \Theta_{PSLR,0} + 20\log_{10}(\Theta_A) \tag{4.82}$$

where $\Theta_{PSLR,0} = 20\log_{10}(|\sin c(T_a f_{sh})|)$ is the PSLR value in the absence of ionospheric scintillation and

$$\Theta_A = \left[1 + \frac{S_I(\bar{\tau},f_{sh})}{|\mathrm{Sinc}(\bar{\tau},T_a f_{sh})|^2} + \frac{4S_\Phi(\bar{\tau},f_{sh})}{|\mathrm{Sinc}(\bar{\tau},T_a f_{sh})|^2}\right]^{\frac{1}{2}} \tag{4.83}$$

The second term in (4.82) represents the increment of the PSLR in the presence of ionospheric scintillation. When the ionospheric scintillation is weak, the azimuth PSLR is usually dominated by the first sidelobe, which is the highest sidelobe. In this case, $f_{sh} \approx 1.43/T_a$. Because of the long integration time of GEO SAR, f_{sh} (about 0.01 Hz in the designed moderate azimuth resolution) is far lower than f_c (approximately 0.1 Hz). Thus, Θ_A is nearly 1 and the azimuth PSLR will almost not be affected ($PSLR \approx \Theta_{PSLR,0,}$). However, as for a moderate or a strong scintillation, the sidelobe with a frequency near the peak of the ionospheric power spectra will have a higher level than the first sidelobe. In this case, Θ_A is larger than 1 and the azimuth PSLR will deteriorate.

4.4.3.3 Effects on Azimuth ISLR

According to (4.71), the whole energy of the PSF is determined by

$$
\begin{aligned}
E\left[S_{oe}^2(\bar{\tau},f)\right] &= E\left[\left(S_o(\bar{\tau},f) + S_{A^2d}(\bar{\tau},f) + j2S_{\varphi d}(\bar{\tau},f) - 2S_{\varphi^2 d}(\bar{\tau},f)\right)\right. \\
&\quad \left.\left(S_o(\bar{\tau},f) + S_{A^2d}(\bar{\tau},f) + j2S_{\varphi d}(\bar{\tau},f) - 2S_{\varphi^2 d}(\bar{\tau},f)\right)^*\right] \\
&= E\left[S_o^2(\bar{\tau},f)\right] + E\left[S_{A^2d}^2(\bar{\tau},f)\right] + E\left\{\left[2S_{\varphi d}(\bar{\tau},f)\right]^2\right\} \\
&\quad + E\left\{\left[-2S_{\varphi^2 d}(\bar{\tau},f)\right]^2\right\} + 2\mathrm{Re}\left\{E\left[S_o(\bar{\tau},f)S_{A^2d}^*(\bar{\tau},f)\right]\right. \\
&\quad + E\left[j2S_o(\bar{\tau},f)S_{\varphi d}^*(\bar{\tau},f)\right] - E\left[2S_o(\bar{\tau},f)S_{\varphi^2 d}^*(\bar{\tau},f)\right] \\
&\quad + E\left[j2S_{A^2d}(\bar{\tau},f)S_{\varphi d}^*(\bar{\tau},f)\right] - E\left[2S_{A^2d}(\bar{\tau},f)S_{\varphi^2 d}^*(\bar{\tau},f)\right] \\
&\quad \left. - E\left[j4S_{\varphi d}(\bar{\tau},f)S_{\varphi^2 d}^*(\bar{\tau},f)\right]\right\}
\end{aligned}
\tag{4.84}
$$

where $E[\cdot]$ represents the mathematical expectation operation.

Firstly, taking $E\left\{\left[2S_{\varphi d}(\bar{\tau},f)\right]^2\right\}$ in (4.84) as an example, its mean square value is written as [32]

$$
\begin{aligned}
E\left\{\left[2S_{\varphi d}(\bar{\tau},f)\right]^2\right\} &= 4\int p(\varphi)\left[\int j\varphi_\Delta(u_1)\bar{W}_{r,a}(\bar{\tau},u_1)\exp(-j2\pi f u_1)du_1\right. \\
&\quad \left.\times \int -j\varphi_\Delta(u_2)\bar{W}_{r,a}(\bar{\tau},u_2)\exp(j2\pi f u_2)du_2\right]d\varphi \\
&= 4\iint R(u_1-u_2)\bar{W}_{r,a}(\bar{\tau},u_1)\bar{W}_{r,a}(\bar{\tau},u_2)\exp[-j2\pi f(u_1-u_2)]du_1du_2
\end{aligned}
\tag{4.85}
$$

where $p(\varphi)$ is the probability density function of phase scintillation. u_1 and u_2 represents two different moments in slow time. Here

$$
R(u_1-u_2) = \int p(\varphi)\varphi_\Delta(u_1)\varphi_\Delta(u_2)d\varphi = E[\varphi_\Delta(u_1)\varphi_\Delta(u_2)] \tag{4.86}
$$

Then, assuming

$$
\xi = u_1, \eta = u_1 - u_2 \tag{4.87}
$$

we have

$$
\begin{aligned}
E\left\{\left[2S_{\varphi d}(\bar{\tau},f)\right]^2\right\} &= 4\int R(\eta)\left[\int \bar{W}_{r,a}(\bar{\tau},\xi)\bar{W}_{r,a}(\bar{\tau},\xi-\eta)d\xi\right]\exp(-j2\pi f\eta)d\eta \\
&= 4\int R(\eta)A_W(\bar{\tau},\eta)\exp(-j2\pi f\eta)d\eta
\end{aligned}
\tag{4.88}
$$

where $A_W(\bar{\tau},\eta) = \int \bar{W}_{r,a}(\bar{\tau},\xi)\bar{W}_{r,a}(\bar{\tau},\xi-\eta)d\xi$.

Based on the Wiener-Khintchine theorem, we obtain

$$E\left\{\left[2S_{\varphi d}(\bar{\tau},f)\right]^2\right\} = 4\int R(\eta)\exp(-j2\pi f\eta)d\eta \otimes \int A_W(\bar{\tau},\eta)\exp(-j2\pi f\eta)d\eta \tag{4.89}$$

where

$$\begin{aligned}\int A_W(\bar{\tau},\eta)\exp(-j2\pi f\eta)d\eta &= \int\left[\int \bar{W}_{r,a}(\bar{\tau},\xi)\bar{W}_{r,a}(\bar{\tau},\xi-\eta)d\xi\right]\exp(-j2\pi f\eta)d\eta \\ &= \left[\int \bar{W}_{r,a}(\bar{\tau},\xi)\exp(-j2\pi f\xi)d\xi\right] \\ &\quad\cdot\left[\int \bar{W}_{r,a}(\bar{\tau},\xi)\exp(-j2\pi f\xi)d\xi\right] \\ &= S_0^2(\bar{\tau},f)\end{aligned} \tag{4.90}$$

Considering that $S_\Phi(f) = \int R(\eta)\exp(-j2\pi f\eta)d\eta$ is the power spectral density of the phase scintillation, we have

$$E\left\{\left[2S_{\varphi d}(\bar{\tau},f)\right]^2\right\} = 4\int S_\Phi(\bar{\tau},\varsigma)|S_o(\bar{\tau},f-\varsigma)|^2 d\varsigma \tag{4.91}$$

$S_o(\bar{\tau},f)$ is approximately a delta function and the power is concentrated to the mainlobe. Thus, (4.91) can be simplified as

$$E\left\{\left[2S_{\varphi d}(\bar{\tau},f)\right]^2\right\} \approx 4S_\Phi(\bar{\tau},\varsigma)\int|S_o(\bar{\tau},f-\varsigma)|^2 d\varsigma = 4S_\Phi(\bar{\tau},\varsigma)P_M \tag{4.92}$$

where P_M is the energy of main lobe of the PSF.

Thus, the increment of azimuth ISLR contributed by the phase scintillation is shown as

$$\Delta ISLR_\varphi = \frac{P_S}{P_M} = \frac{\int_{1/T_a}^{\infty} 4S_\Phi(\bar{\tau},\varsigma)P_M df}{P_M} = 4\int_{1/T_a}^{\infty} S_\Phi(\bar{\tau},\varsigma)df \tag{4.93}$$

where P_S is the energy of sidelobes of the PSF. Because of $1/T_a << f_c$ in GEO SAR case, the energy of the ionospheric scintillation is spread out of the mainlobe, thus $\Delta ISLR_\varphi = 4\sigma_\varphi^2$ is validated.

In GEO SAR, the other terms in (4.84) can be similarly deduced ignoring the small terms. The azimuth ISLR in GEO SAR in the present of ionospheric scintillation can be expressed as

$$ISLR \approx 10\log_{10}\left(\Theta_{ISLR,0} + S_4^2 + 4\sigma_\varphi^2\right) \tag{4.94}$$

where $\Theta_{ISLR,0}$ represents the ideal ISLR value in the absence of ionospheric scintillation and $10\log_{10}(\Theta_{ISLR,0})$ is $-9.7\,\mathrm{dB}$ without weighting operation in imaging.

According to (4.94), the relation between the azimuth ISLR and the measureable ionospheric scintillation indexes are directly established. The last two terms increase when the phase and amplitude scintillation turns more intense, which results in a higher azimuth ISLR. Since the coefficient of σ_φ^2 is four times of the coefficient of S_4^2, for the same numerical values of S_4 and σ_φ, the deterioration of azimuth ISLR is more sensitive to phase scintillation than amplitude scintillation.

4.4.3.4 Comparison with LEO SAR

The azimuth PSF highly depends on the ideal azimuth PSF (integration time T_a) and ionospheric scintillation power spectra (cut-off frequencies) in the presence of ionospheric scintillation. As a GEO SAR travels far higher than a LEO SAR, they have different travelling velocities, integration times, cut-off frequencies of the ionospheric scintillation power spectra and working geometries, which are shown in Table 4.9. Resultantly, the impacts of ionospheric scintillation on a GEO SAR and a LEO SAR show different patterns.

According to the previous analysis, the frequency zone of the contaminated PSF corresponding to the power-low regions of the scintillation power spectra (after the cut-off frequencies) will be severely impacted. Thus, the relationship of the width of mainlobe, the position of the first sidelobe and the cut-off frequencies of the

Table 4.9 Comparison between GEO SAR and LEO SAR in geometries, integration times, cut-off frequencies and effective size scope of the irregularities

Parameter	LEO SAR	GEO SAR
Orbit height (km)	600 (typically)	36,000
Velocity (km/s)	7 (typically)	1–3
IPPs velocity (km/s)	≈3.5	0.028 (median)
Incidence angle (°)	45 (typically)	45 (typically)
Wavelength (m)	0.24	0.24
Integration time (s)	1 (typically)	120 (typically)
Fresnel frequency (Hz)	≈7.2	≈0.1
Cut-off frequency of phase scintillation (Hz)	≈0.1	≈0.1
Effective size scope of the irregularities (km)	<35 (typically)	<1 (typically)

ionospheric scintillation power spectra is extremely important. AGEO SAR has a longer integration time than that in a LEO SAR case. Thus, compared with the broad mainlobe width of a LEO SAR (less than 2 Hz), GEO SAR has a narrower mainlobe width (less than 0.02 Hz). Likewise, the frequency position of the first sidelobe in a GEO SAR (about 0.01 Hz) is far smaller than that in a LEO SAR case (about 1 Hz). Meanwhile, because of the different orbit heights and geometries, the IPPs velocity of GEO SAR is only dozens of meters per second, while it is more than 3 km per second in a LEO SAR case. According to the previous analysis of f_F, in LEO SAR case, f_F is more than 7 Hz, which is far higher than that in GEO SAR (about 0.1 Hz). Moreover, as f_c is mainly determined by the correlation time, they are same in the both cases (about 0.1 Hz). Nevertheless, because of the different $|\mathbf{V_f}|$, the corresponding effective size scopes of the irregularities to produce phase scintillation are different in the two cases. The effective size scope of the irregularities can reach to dozens of kilometers in a LEO SAR case, while it is only small than 1 km in a GEO SAR case.

Figure 4.24 shows the predicted azimuth PSFs in the presence of ionospheric scintillation (weak scintillation) in a GEO SAR case (a) and a LEO SAR case (b).

As the cut-off frequencies of the ionospheric scintillation (f_F and f_c) are far higher than the frequency scope of the mainlobe and the first sidelobe, GEO SAR has the un-impacted azimuth resolution and azimuth PSLR. However, as the main power of the ionospheric scintillation falls into the sidelobes, the azimuth ISLR is obviously deteriorated. In contrast, as for a LEO SAR, although f_F is larger than the width of the mainlobe, f_c is small compared with the width of the mainlobe. Thus, (4.78) is validated only in the case of $f_r > f_i$ and the azimuth resolution of the LEO SAR will degrade. Likewise, in terms of the azimuth PSLR in LEO SAR case, f_{sh} is comparative with f_c. Resultantly, Θ_A is larger than 1 ($S_I(\overline{\tau},f_{sh})$ and $S_\Phi(\overline{\tau},f_{sh})$ are positive) and azimuth PSLR is deteriorated. Moreover, the azimuth ISLR of LEO SAR will be raised due to the increase of the sidelobe levels. However, because of the half width of the mainlobe is larger than f_c in LEO SAR case, the

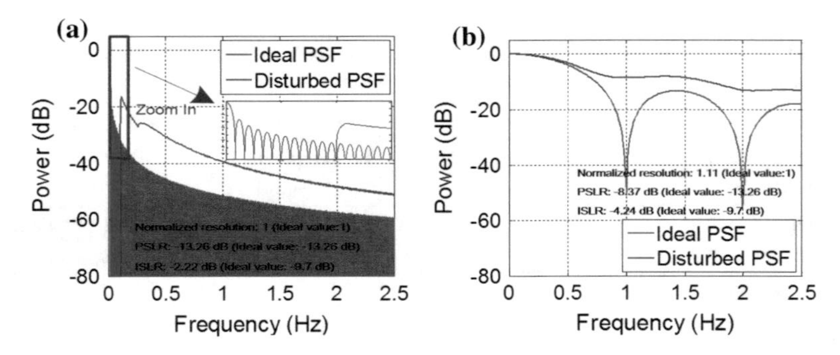

Fig. 4.24 Predicted azimuth PSFs in the presence of ionospheric scintillation in a GEO SAR case (**a**) and a LEO SAR case (**b**) (scintillation parameters: $S_4 = 0.3$, $\sigma_\varphi = 0.3$ rad, $p_A = 3.6$ and $p_\Phi = 2.8$)

power of the phase scintillation falls into both the mainlobe and the sidelobes. Thus, the deterioration of the azimuth ISLR in the LEO SAR case (−4.24 dB) is not as severe as that in the GEO SAR case (−2.22 dB). Summarily, the comparison of the impacts of the ionospheric scintillation on GEO SAR imaging and LEO SAR imaging is given in Table 4.10.

4.4.4 Simulations

4.4.4.1 Configuration and Parameter Settings

In order to testify the effectiveness of the previous analysis, simulation experiments are carried out and the flowchart is shown in Fig. 4.25. The experiment includes three steps: the scintillation indexes retrieval, the ionospheric scintillation signal generation, and the imaging and performance analysis. In the simulation, we set scintillations indexes with different intensities indexes to analyze the imaging performance of GEO SAR. Then, we use the scintillation indexes and ionospheric scintillation model to simulate ionospheric scintillation signal. Finally the simulated scintillation data will be superimposed on the accurate GEO SAR signal to generate the contaminated GEO SAR signal by ionospheric scintillation, and imaging and results evaluation are conducted. The GEO SAR system and imaging parameters in the experiment are listed in Table 4.11. The ionospheric scintillation power spectral indexes are assumed to be $p_A = 3.6, p_\Phi = 2.8$ in the following simulations.

4.4.4.2 Generation of Ionospheric Scintillation Signal

Scintillation sampling model [18] is used to generate the ionospheric scintillation signal with the designed scintillation intensity by only using S_4 index and σ_φ index. Since the two indexes are easy to be measured directly by GPS monitor, scintillation sampling model is a preferred ionospheric scintillation model to generate the ionospheric scintillation signal.

In the ionospheric scintillation signal generation part of Fig. 4.25, two correlated Gauss sequences are generated based on S_4 index and σ_φ index firstly. The correlation between the amplitude scintillation and the phase scintillation is considered,

Table 4.10 Comparison of the impacts of the ionospheric scintillation on GEO SAR imaging and LEO SAR imaging

Parameter	LEO SAR (short integration time)	GEO SAR (long integration time)
Azimuth resolution	Obviously degradation	Un-impacted
Azimuth PSLR	Obviously degradation	Un-impacted (weak scintillation)
Azimuth ISLR	Degradation	Seriously degradation

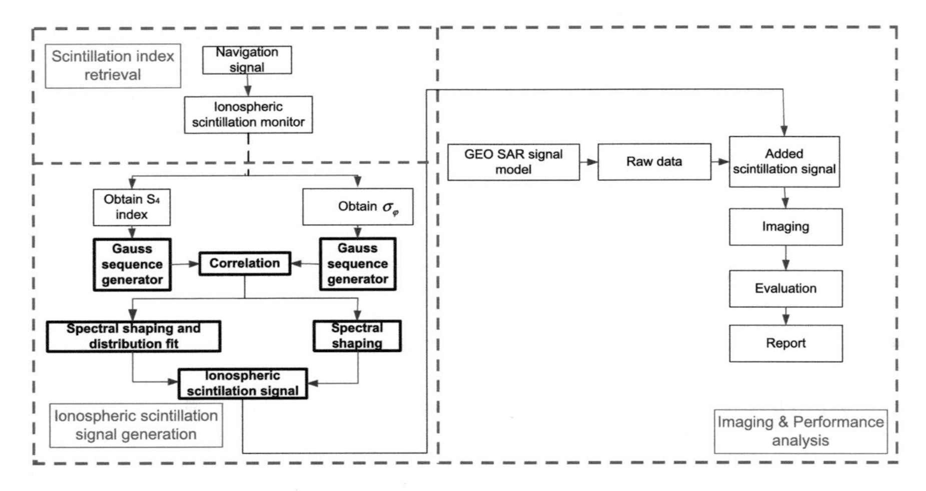

Fig. 4.25 Flowchart of the simulation experiments

Table 4.11 GEO SAR system and orbit elements

Parameter	Value
Carrier central frequency (GHz)	1.25
Pulse repeat frequency (Hz)	150
Semimajor axis (km)	42,164
Incidence angle (°)	30
Imaging position	Perigee
Imaging mode	Spotlight
Bandwidth (MHz)	18
Eccentricity	0.07
Inclination (°)	53
Sampling rate (MHz)	20

which satisfies that a greater amplitude variation is coupled with a lesser phase variation. In the scintillation sampling model, ionospheric amplitude and phase scintillation power spectra fit power-law, and amplitude scintillation and phase scintillation fit Nakagami-m distribution and normal distribution, respectively. Thus, the spectral shaping and distribution fitting are needed to make the produced signal fit the statistical behaviors of ionospheric scintillation signals. Finally, the ionospheric scintillation signal is generated by combining the generated amplitude scintillation sequence and the phase scintillation sequence.

4.4.4.3 Theoretical Validation Simulation

Based on the imaging method described in 4.4.3, we use the GEO SAR SPECAN algorithm for imaging in the following part. Point-target simulation experiments are conducted firstly. The integration time is 120 s, which can guarantee the 20 m azimuth resolution. We assume phase and amplitude scintillation occur simultaneously with the same intensity and the ionospheric scintillation indexes (S_4 and σ_φ) are 0.1, 0.2, 0.3 and 0.5 (rad) for imitating scintillations from weak to moderate. The imaging results are shown in Fig. 4.26. The azimuth PSLR, ISLR and normalized resolution of the imaged results (through 30 Monte Carlo experiments) and the corresponding predicted values based on the theoretical analysis are given in Fig. 4.27.

It can be seen from Figs. 4.26 and 4. 27, when the scintillation is weak, as the intensity of the ionospheric scintillation increases, the impacts of ionospheric scintillation on the normalized azimuth resolution and the azimuth PSLR is not obvious. This is accordance with the previous analysis of the normalized azimuth resolution and the azimuth PSLR. In GEO SAR case, the disturbance of the ionospheric scintillation on the PSF concentrates after the cut-off frequencies. Generally, the cut-off frequencies of the amplitude scintillation power spectrum (Fresnel frequency) and phase scintillation power spectrum are about 0.1 Hz (assuming a 350 km ionosphere, a 28 m/s IPPs velocity of the GEO SAR, a 75 m/s moving velocity of the ionospheric irregularities and 1–2 s correlation time). Thus, they are far larger than the width of the mainlobe (<0.02 Hz, an integration time of 120 s) and the frequency position of the first sidelobes (about 0.01 Hz). The related impacts on the normalized azimuth resolution and the azimuth PSLR are negligible. The variation of the normalized azimuth resolution and the azimuth PSLR derives

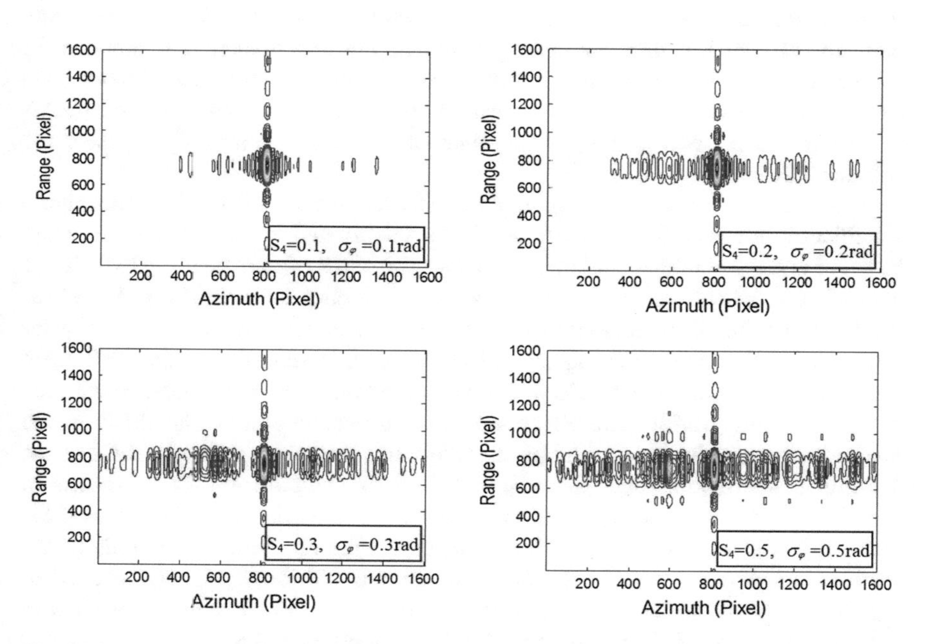

Fig. 4.26 Imaging results of GEO SAR in the presence of the ionospheric scintillations with different intensities

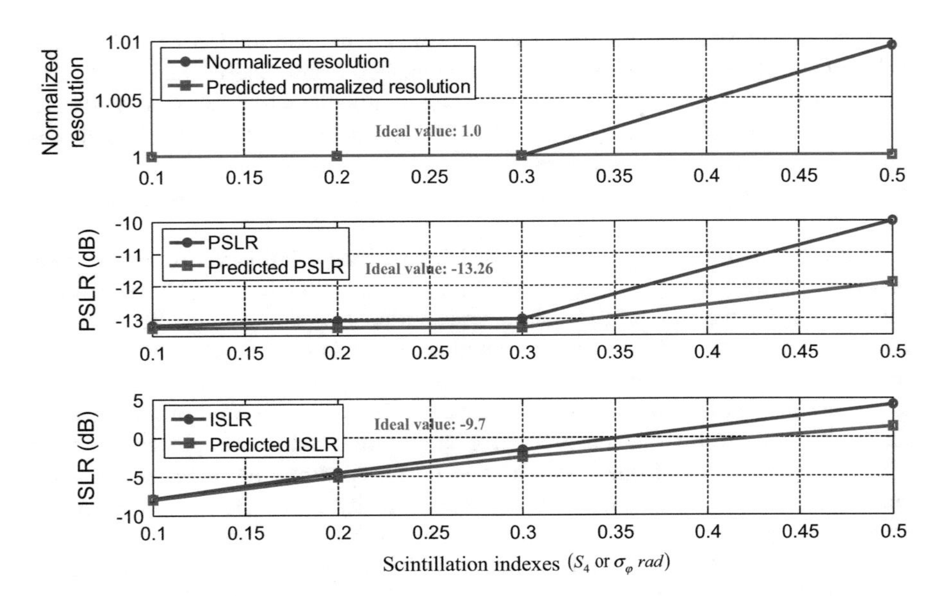

Fig. 4.27 Azimuth PSLRs, ISLRs and normalized resolutions and their predicted values in the presence of the ionospheric scintillations with different intensities

from the unideal detrending filtering in the spectral shaping (not absolute steep low pass filters), which cannot completely eliminate the low frequency component in the scintillation power spectra. However, as for the moderate scintillation, the sidelobe with a frequency near the peak of the ionospheric power spectra (about 0.1 Hz or larger) will dominate the value of PSLR and obviously increase the azimuth PSLR.

The azimuth ISLR is turning higher with the increasing of the ionospheric scintillation intensity. The azimuth ISLR has deteriorated to more than 4 dB when $S_4 = 0.5$ and $\sigma_\varphi = 0.5$ rad and there is an azimuth ISLR of only nearly -1 dB in the case of $S_4 = 0.3$ and $\sigma_\varphi = 0.3$ rad. Seriously deteriorated azimuth ISLR shows that the levels of the whole sidelobes have increased significantly. Thus, the deterioration of ISLR is the main impact of the ionospheric scintillation on GEO SAR imaging. In Fig. 4.27, the red line computed based on (4.94) is accordance with the blue line, which verifies the effectiveness of the expression of the ISLR in the presence of ionospheric scintillation. In addition, comparison of the impacts of the amplitude scintillation and phase scintillation on azimuth ISLRs is shown in Fig. 4.28. It can be concluded that under the same numerical levels of σ_φ and S_4, phase scintillation has a more serious impact to azimuth ISLR. When only phase scintillation with $\sigma_\varphi = 0.5$ rad exists, the azimuth ISLR has already reached to higher than 0 dB, while azimuth ISLR only increases to higher than -5 dB with only the existence of the amplitude scintillation with $S_4 = 0.5$.

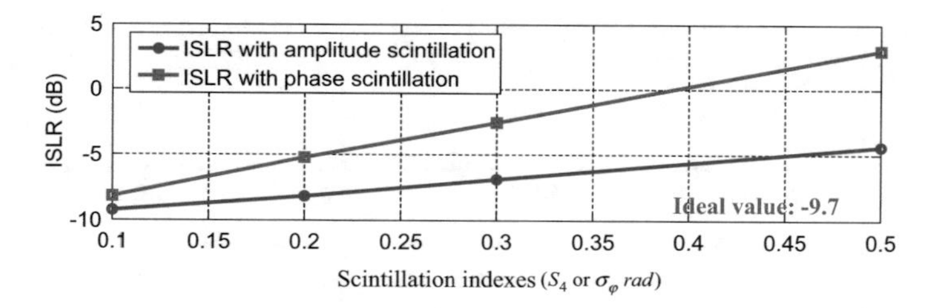

Fig. 4.28 Comparison of the impacts of the amplitude scintillation and phase scintillation on azimuth ISLR

Aside from the above discussions, the effects of ionospheric scintillation with various integration times in GEO SAR are analyzed. The simulated integration times are set from 20 to 250 s. The set ionospheric scintillation intensities are $S_4 = 0.3$ and $\sigma_\varphi = 0.3$ rad as an instance. The imaging results are shown in Fig. 4.29 and the imaging quality evaluation is summarized in Fig. 4.30.

The variation of the azimuth resolution, PSLR and ISLR is not obvious when the integration time of GEO SAR changes above 60 s. The frequency positions of the mainlobe and the first sidelobe of GEO SAR locate far away from the cut-off frequencies of the scintillation power spectra. Because there is no scintillation energy below the cut-off frequencies, the mainlobe and the first of GEO SAR are

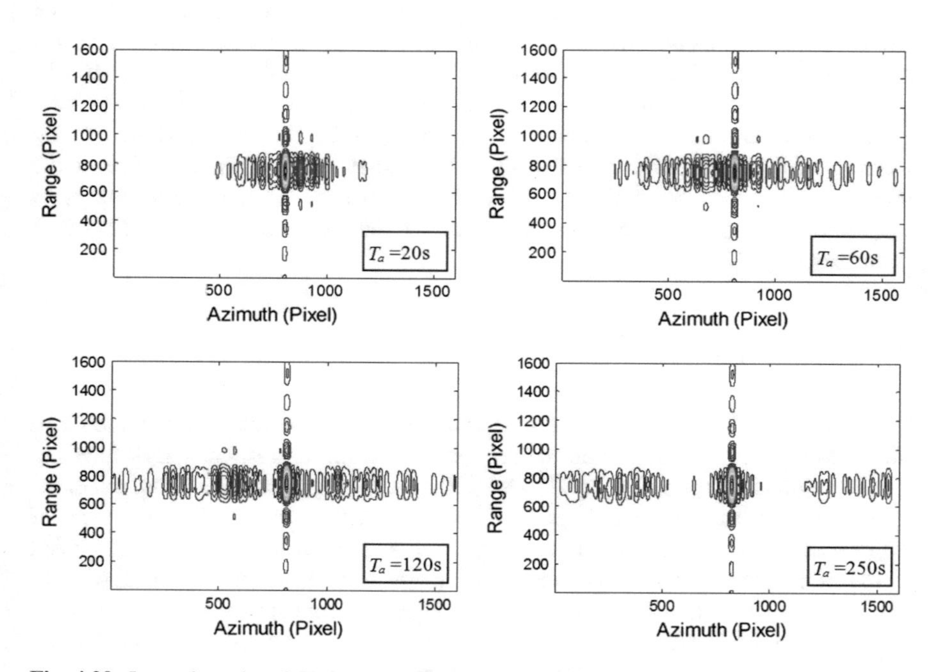

Fig. 4.29 Imaged results of GEO SAR with different integration times

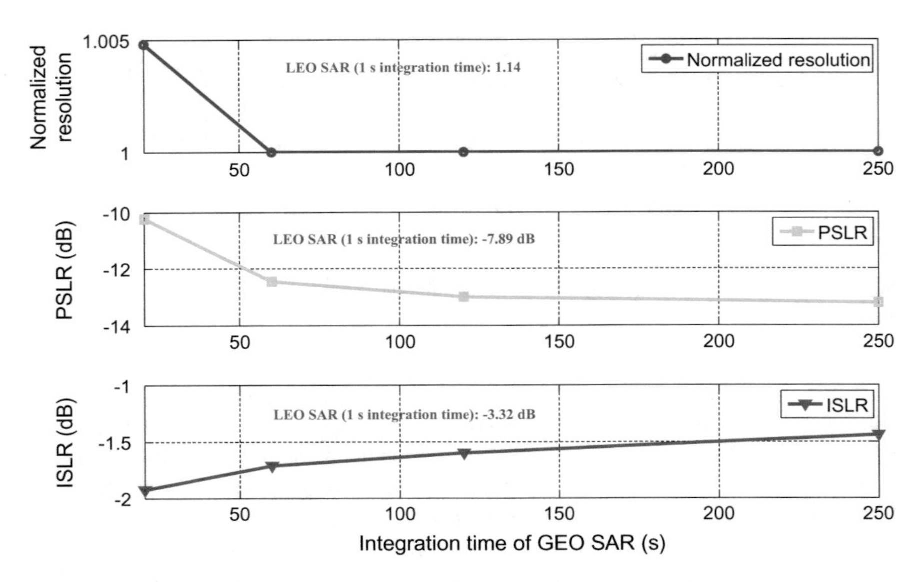

Fig. 4.30 Azimuth imaging quality evaluation in the present of ionospheric scintillation with different integration times in GEO SAR

not affected much and the values of the azimuth resolution, PSLR and ISLR of the imaged target are almost stable. However, when the integration time of GEO SAR decreases to 20 s, as the frequency position of the first sidelobe (nearly 0.1 Hz) is comparable to f_c (0.1 Hz), the azimuth PSLR will be raised to near −10 dB. Moreover, the wide of the mainlobe is increasing as the decreasing of the integration time. Because the residual energy brought by the non-ideal detrending filtering in the generation of the scintillation signal raises when the edge of the mainlobe close to f_c, the mainlobe is broadened. Meanwhile, since some energy of the scintillation power spectra is absorbed by the mainlobe, the azimuth ISLR decreases slightly.

4.5 Summary

Troposphere and ionosphere have temporal-spatial variant reflectivity which will induce time delays or phase errors to the signals traversing atmosphere. SAR focusing will be degraded. Especially for GEO SAR, the long integration time (from hundreds to thousands of seconds) and large imaging scene will deteriorate the influences of troposphere and ionosphere on focusing. In this chapter, the accurate GEO SAR signal model considering tropospheric and ionospheric influences are constructed. The two dimension frequency spectrum influenced by troposphere and ionosphere is obtained. The GEO SAR range and azimuth image shift and phase errors are derived analytically and their specific effects are analyzed.

- For the troposphere, the atmospheric reflectivity profile data from FY-3 satellite are employed to analyze the effects of troposphere on GEO SAR focusing and verify the theoretical analysis. Compared with the ionospheric influences, the influences are more dependent on the integration time and can be ignored for the 20-m GEO SAR.
- For the background ionosphere, it will cause the two-dimensional image shift and image defocusing which are derived analytically and verified by the Klobuchar model. Thresholds for the effects of the absolute and relative *TEC* on GEO SAR focusing under various resolutions are also given. The specific thresholds are dependent on the background ionosphere status, the integration time and the GEO SAR imaging geometry.
- For the ionospheric scintillation, it will induce the amplitude and phase fluctuations of signals and cause defocusing. The imaging model and azimuth PSF were determined considering the ionospheric irregularities. Influences on the azimuth resolution, azimuth PSLR and azimuth ISLR are quantitatively analyzed. The results show that the deterioration of azimuth ISLR should be specifically concerned in GEO SAR imaging.

Appendix 1: Derivation of the Two-Dimensional Spectrum of the GEO SAR Signal Under the Effects of Temporal-Spatial Variant Background Ionosphere

Through the azimuth Fourier transform of the echo signal in (4.33), it is able to derive the analytical expression of the two-dimensional GEO SAR signal spectrum considering the background ionosphere. Since both the GEO SAR accurate slant range and the TEC variations are expressed in the form of high-order polynomials, it will be difficult to use the principle of stationary phase (POSP) to obtain the stationary phase point and then derive the two-dimensional spectrum. Herein, the series inversion [33] method is used to derive the two-dimensional spectrum of the GEO SAR signal in the context of the background ionosphere.

Firstly, the echo signal in (4.33) can be rewritten as

$$\begin{aligned} S(f_r, t_a) = {} & A_r(f_r) \cdot A_a(t_a) \cdot \exp\left(-j\frac{\pi f_r^2}{k_r}\right) \\ & \cdot \exp\left[-j\frac{4\pi(f_r + f_0)}{c}\left(r(t_a) + \frac{40.3 \cdot TEC(t_a)}{(f_r + f_0)^2}\right)\right] \end{aligned} \tag{4.95}$$

Assuming $R(t_a) = r(t_a) + \Delta r_{iono}(t_a) = r(t_a) + 40.3 \cdot TEC(t_a) \Big/ (f_0 + f_r)^2$, wherein, $r(t_a)$ is the slant range history in the absence of ionosphere, and $\Delta r_{iono}(t_a) = 40.3 \cdot TEC(t_a) \Big/ (f_0 + f_r)^2$ is the slant range difference introduced by the ionosphere. The specific expressions of $r(t_a)$ and $\Delta r_{iono}(t_a)$ are written as follows

$$\begin{aligned} r(t_a) &== r_0 + q_1 \cdot t_a + q_2 \cdot t_a^2 + q_3 \cdot t_a^3 + \cdots \\ \Delta r_{iono}(t_a) &= \frac{40.3 \cdot TEC_0}{(f_0+f_r)^2} + \frac{40.3 \cdot k_1}{(f_0+f_r)^2} \cdot t_a + \frac{40.3 \cdot k_2}{(f_0+f_r)^2} \cdot t_a^2 + \frac{40.3 \cdot k_3}{(f_0+f_r)^2} \cdot t_a^3 + \cdots \end{aligned} \tag{4.96}$$

According to the properties of Taylor series expansion, then

$$\frac{1}{(f_r+f_0)^2} = \frac{1}{f_0^2} - \frac{2}{f_0^3} \cdot f_r + \frac{3}{f_0^4} \cdot f_r^2 - \frac{4}{f_0^5} \cdot f_r^3 + \cdots \tag{4.97}$$

As the first-order, second-order and third-order terms of $\Delta r_{iono}(t_a)$ are relatively small, the constant term in (4.97) for those three were approximated. Hence, $\Delta r_{iono}(t_a)$ was simplified as

$$\Delta r_{iono}(t_a) = \frac{40.3 \cdot TEC_0}{(f_0+f_r)^2} + \frac{40.3 \cdot k_1}{f_0^2} \cdot t_a + \frac{40.3 \cdot k_2}{f_0^2} \cdot t_a^2 + \frac{40.3 \cdot k_3}{f_0^2} \cdot t_a^3 + \cdots \tag{4.98}$$

$R(t_a)$ can be denoted in the form as follows

$$R(t_a) = R_0 + Q_1 \cdot t_a + Q_2 \cdot t_a^2 + Q_3 \cdot t_a^3 + \cdots \tag{4.99}$$

where,

$$\begin{aligned} R_0 &= r_0 + \frac{40.3 \cdot TEC_0}{(f_0+f_r)^2} \quad Q_1 = q_1 + \frac{40.3 \cdot k_1}{f_0^2} \\ Q_2 &= q_2 + \frac{40.3 \cdot k_2}{f_0^2} \quad Q_3 = q_3 + \frac{40.3 \cdot k_3}{f_0^2} \end{aligned} \tag{4.100}$$

Next, the series inversion theorem was utilized to derive the accurate two-dimensional spectrum of the GEO SAR signal in the context of the background ionosphere. In order to use the series inversion, $s(f_r, t_a)$ should be re-expressed as

$$s(f_r, t_a) = s_1(f_r, t_a) \cdot \exp\left(-j\frac{4 \cdot \pi \cdot (f_r + f_c)}{c} \cdot Q_1 \cdot t_a\right) \tag{4.101}$$

In the subsequent derivation, it needs to firstly derive the two-dimensional spectrum of $s_1(f_r, t_a)$, and then use the following relations to calculate the two-dimensional spectrum of $s(f_r, t_a)$

$$s_1(f_r, t_a) \Leftrightarrow S_1(f_r, f_a) \quad (4.102)$$

$$s_1(f_r, t_a) \cdot \exp\left(-j\frac{4 \cdot \pi \cdot (f_r + f_c)}{c} \cdot Q_1 \cdot t_a\right) \Leftrightarrow S_1\left(f_r, f_a + \frac{2 \cdot (f_r + f_c)}{c} \cdot Q_1\right) \quad (4.103)$$

$$s(f_r, t_a) \Leftrightarrow S(f_r, f_a) = S_1\left(f_r, f_a + \frac{2 \cdot (f_r + f_c)}{c} \cdot Q_1\right) \quad (4.104)$$

First of all, $s_1(f_r, t_a)$ was Fourier transformed, and its integral phase expression was

$$\psi = -\pi \cdot \frac{f_r^2}{\beta} - \frac{4 \cdot \pi \cdot (f_r + f_c)}{c} \cdot \left(R_0 + Q_2 \cdot t_a^2 + Q_3 \cdot t_a^3 + \cdots\right) - 2\pi \cdot f_a \cdot t_a \quad (4.105)$$

Assuming $\frac{\partial \psi}{\partial t_a} = 0$, then

$$\left(-\frac{c}{2(f_r + f_0)}\right) \cdot f_a = 2Q_2 \cdot t_a + 3Q_3 \cdot t_a^2 + \cdots \quad (4.106)$$

Through the series inversion, it can get

$$t_a(f_a) = A_1 \cdot \left(-\frac{c \cdot f_a}{2(f_r + f_0)}\right) + A_2 \cdot \left(-\frac{c \cdot f_a}{2(f_r + f_0)}\right)^2 + A_3 \cdot \left(-\frac{c \cdot f_a}{2(f_r + f_0)}\right)^3 + \cdots \quad (4.107)$$

where,

$$A_1 = \frac{1}{2Q_2} \quad A_2 = -\frac{3Q_3}{8Q_2^3} \quad A_3 = \frac{18Q_3^2 - 8Q_2Q_4}{32Q_2^5} \quad (4.108)$$

Substituting (4.108) into (4.107), and we can get the two-dimensional spectrum of $s_1(f_r, t_a)$

$$\begin{aligned}
S_1(f_a, f_r) = {} & \sigma A_a(f_a) A_r(f_r) \cdot \exp\left(-j \cdot \pi \frac{f_r^2}{\beta}\right) \\
& \times \exp\left(-j \cdot 2\pi \cdot \frac{2(f_r + f_0)}{c} \cdot \left(r_0 + \frac{40.3 \cdot TEC_0}{(f_0 + f_r)^2}\right)\right) \\
& \times \exp\left(j \cdot 2\pi \cdot \frac{1}{4 \cdot \left(q_2 + \frac{40.3 \cdot k_2}{f_0^2}\right)} \cdot \frac{c}{2(f_r + f_0)} \cdot f_a^2\right) \\
& \times \exp\left(j \cdot 2\pi \cdot \frac{q_3 + \frac{40.3 \cdot k_3}{f_0^2}}{8 \cdot \left(q_2 + \frac{40.3 \cdot k_2}{f_0^2}\right)^3} \cdot \left(\frac{c}{2(f_r + f_0)}\right)^2 \cdot f_a^3\right)
\end{aligned} \quad (4.109)$$

According to (4.104), it can derive the final two-dimensional spectrum expression of $s(f_r, t_a)$ as shown in (4.34).

Appendix 2: Klobuchar Model and TEC Calculation

In this chapter, Klobuchar Model [16] was used to calculate the ionospheric TEC values on the BD IGSO satellite propagation path. This model is particularly suitable for United States, China, Western Europe and other countries and regions at mid-latitudes. The eight ionospheric delay parameters and Klobuchar model were used to calculate the vertical ionospheric delay $I_z'(t)$ of the IGSO satellite signal in seconds, specifically expressed as follows

$$I_z'(t) = \begin{cases} 5 \times 10^{-9} + A_2 \cos\left[\frac{2\pi(t - 50400)}{A_4}\right], & |t - 50400| < A_4/4 \\ 5 \times 10^{-9}, \quad |t - 50400| \geq A_4/4 \end{cases} \tag{4.110}$$

where t is the local time of the ionosphere at the IPP and its value is in the range of 0–86,400 in seconds. A_2 is the magnitude of Klobuchar cosine curve in the day time, which is dependent on the $\alpha_1 \sim \alpha_4$ and the latitude of the IPP. A_4 is the period of cosine curve in seconds, which is dependent on the $\beta_1 \sim \beta_4$ and the latitude of the IPP. The geographic latitude of the IPP is also determined by the user geographic latitude, satellite azimuth and geocentric aperture angle.

As shown in Fig. 4.31, the IPP at the time of t_1 was located in Point A, and the satellite was moving forward at the time of t_2, while the IPP was located in Point B. Then, the segment AB was the length of the IPP of the IGSO satellite.

In the light of the equation of $I_{B1I}(t) = \frac{1}{\sqrt{1-\left(\frac{R}{R+h}\cos E\right)^2}} I_z'(t)$, it is able to convert the vertical ionospheric delay $I_z'(t)$ into the ionospheric delay $I_{B1I}(t)$ along the signal propagation path.

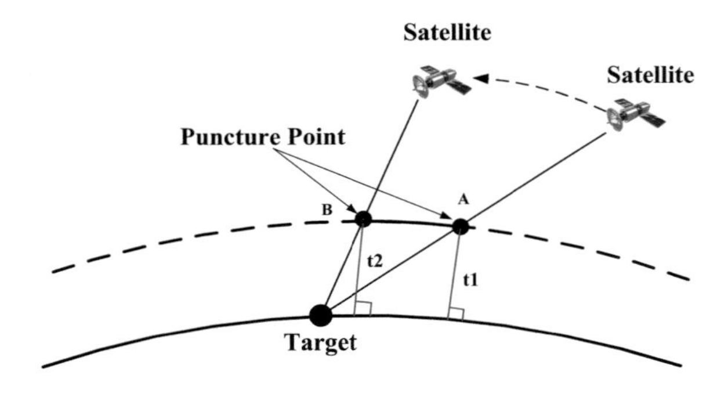

Fig. 4.31 Relations between the local time and the satellite motion within the Klobuchar model

In accordance with the relations between the ionospheric delay $I_{B1I}(t)$ along the signal propagation path and *TEC*, it is able to calculate the *TEC* values on the propagation path at the time of t by the following equation

$$TEC(t) = \frac{I_{B1I}(t) \cdot cf^2}{40.26} \tag{4.111}$$

Appendix 3: Derivation of (4.64)

Here we provide the derivation of (4.64). First, we assume A fits Nakagami-m distribution and its variance can be calculated as

$$\begin{aligned} \sigma_A^2 &= E(A^2) - [E(A)]^2 \\ &= \Omega - [E(A)]^2 \end{aligned} \tag{4.112}$$

where $\Omega = E(A^2)$.

According to the definition of mathematical expectation, $E(A)$ can be written as

$$\begin{aligned} E(A) &= \int_{-\infty}^{+\infty} xf(x, m, \Omega)dx \\ &= \int_{-\infty}^{+\infty} x\left[\frac{2m^m}{\Gamma(m)\Omega^m}x^{2m-1}\exp\left(-\frac{m}{\Omega}x^2\right)\right]dx \end{aligned} \tag{4.113}$$

where $f(x, m, \Omega)$ is the probability density function of Nakagami-m distribution, x is the variable and $m = 1/S_4^2$.

With a change of variable, (4.113) may be re-written as

$$E(A) = \sqrt{\frac{\Omega}{m}}\frac{1}{\Gamma(m)}2\int_{-\infty}^{+\infty}\left(\sqrt{\frac{m}{\Omega}}x\right)^{(2m+1)-1}\exp\left(-\left[\sqrt{\frac{m}{\Omega}}x\right]^2\right)d\left(\sqrt{\frac{m}{\Omega}}x\right) \tag{4.114}$$

Considering the definition of Gamma function, which is shown as

$$\Gamma(n) = 2\int_0^{+\infty} t^{2n-1}\exp\left(-t^2\right)dt \tag{4.115}$$

Based on (4.115) and $x > 0$, (4.114) can be written as

$$E(A) = \sqrt{\frac{\Omega}{m}} \frac{\Gamma\left(m + \frac{1}{2}\right)}{\Gamma(m)} \tag{4.116}$$

We consider (4.116) in (4.112) and then simplify it. Finally, we obtain (4.64).

References

1. Sun J, Bi Y, Wang Y, Hong W (2011) High resolution SAR performance limitation by the change of tropospheric refractivity. In: Proceedings of 2011 IEEE CIE International conference on radar, 24–27 Oct 2011. pp 1448–1451. https://doi.org/10.1109/cie-radar.2011.6159833
2. Muschinski A, Dickey FM, Doerry AW (2005) Possible effects of clear-air refractive-index perturbations on SAR images, pp 25–33
3. Ishimaru A, Kuga Y, Liu J, Kim Y, Freeman T (2016) Ionospheric effects on synthetic aperture radar at 100 MHz to 2 GHz. Radio Sci 34(1):257–268
4. Gao Y, Hu C, Dong X, Long T (2012) Accurate system parameter calculation and coverage analysis in GEO SAR. In: European conference on synthetic aperture radar, pp 607–610
5. Meyer F, Bamler R, Jakowski N, Fritz T (2006) The potential of low-frequency SAR systems for mapping ionospheric TEC distributions. IEEE Geosci Remote Sens Lett 3(4):560–564
6. Meyer FJ (2011) Performance requirements for ionospheric correction of low-frequency SAR data. IEEE Trans Geosci Remote Sens 49(10):3694–3702
7. Prati C, Rocca F, Giancola D, Guarnieri AM (1998) Passive geosynchronous SAR system reusing backscattered digital audio broadcasting signals. IEEE Trans Geosci Remote Sens 36 (6):1973–1976
8. Tian Y, Hu C, Dong X, Zeng T, Long T, Lin K, Zhang X (2015) Theoretical analysis and verification of time variation of background ionosphere on geosynchronous SAR imaging. IEEE Geosci Remote Sens Lett 12(4):721–725
9. Xu Z, Wu J, Wu Z (2004) A survey of ionospheric effects on space-based radar. Waves Random Media 14(2):S189–S273
10. Xu ZW, Wu J, Wu ZS (2008) Potential effects of the ionosphere on space-based SAR imaging. IEEE Trans Antennas Propag 56(7):1968–1975
11. Qi RY, Jin YQ (2007) Analysis of the effects of faraday rotation on spaceborne polarimetric SAR observations at P-Band. IEEE Trans Geosci Remote Sens 45(5):1115–1122
12. Liu J, Kuga Y, Ishimaru A, Pi X (2003) Ionospheric effects on SAR imaging: a numerical study. IEEE Trans Geosci Remote Sens 41(5):939–947
13. Aarons J (1995) 50 years of radio-scintillation observations. IEEE Antennas Propag Mag 39 (6):7–12
14. Aarons J, Whitney HE, Allen RS (2005) Global morphology of ionospheric scintillations. Proc IEEE 59(2):159–172
15. Bhattacharyya A, Basu S, Groves K, Valladares C, Sheehan R (2002) Space weather effects on the dynamics of equatorial f region irregularities. In: 34th COSPAR Scientific Assembly, pp SIA 20-21–SIA 20-27
16. Klobuchar JA (1987) Ionospheric time-delay algorithm for single-frequency GPS users. IEEE Trans Aerosp Electron Syst AES-23 (3):325–331. https://doi.org/10.1109/taes.1987.310829
17. Cheng H, Teng L, Tao Z, Feifeng L, Zhipeng L (2011) The accurate focusing and resolution analysis method in geosynchronous SAR. IEEE Trans Geosci Remote Sens 49(10):3548–3563. https://doi.org/10.1109/TGRS.2011.2160402

18. Pullen S, Opshaug G, Hansen A, Walter T, Enge P, Parkinson B (1998) A preliminary study of the effect of ionospheric scintillation on WAAS user availability in equatorial regions. In: ION GPS-98, pp 687–699
19. Li G, Ning B, Yuan H (2007) Analysis of ionospheric scintillation spectra and TEC in the Chinese low latitude region. Earth Planets Space 59(4):279–285
20. Pan L, Yin P (2014) Analysis of polar ionospheric scintillation characteristics based on GPS data. In: China Satellite Navigation Conference 2014, pp 11–18
21. Singleton DG (1974) Power spectra of ionospheric scintillations. J Atmos Terr Phys 36(1): 113–133
22. Fremouw EJ, Leadabrand RL, Livingston RC, Cousins MD, Rino CL, Fair BC, Long RA (2016) Early results from the DNA wideband satellite experiment—complex-signal scintillation. Radio Sci 13(1):167–187
23. Forte B, Radicella SM (2002) Problems in data treatment for ionospheric scintillation measurements
24. Ganguly S, Jovancevic A, Brown A, Kirchner M, Zigic S, Beach T, Groves KM (2002) Ionospheric scintillation monitoring and mitigation using a software GPS receiver. Radio Sci 39(1):241–262
25. Yannick B, Pierrick H (2011) A global ionosphere scintillation propagation model for equatorial regions. J Space Weather Space Clim 1(1):315–322
26. Humphreys T (2007) Modeling ionospheric scintillation and its effects on GPS carrier tracking loops and two other applications of modeling and estimation
27. Wernik AW, Alfonsi L, Materassi M (2004) Ionospheric irregularities, scintillation and its effect on systems. Acta Geophysica Polonica 52(2):237–249
28. Béniguel Y, Romano V, Alfonsi L, Aquino M (2009) Ionospheric scintillation monitoring and modelling. Ann Geophys 52(3–4):391–416
29. Yeh KC, Liu CH (1982) Radio wave scintillations in the ionosphere. Proc IEEE 70(4): 324–360
30. Humphreys TE, Psiaki ML, Kintner PM (2010) Modeling the effects of ionospheric scintillation on GPS carrier phase tracking. IEEE Trans Aerosp Electron Syst 46(4): 1624–1637
31. Rogers NC, Cannon PS (2009) The synthetic aperture radar transionospheric radio propagation simulator (SAR-TIRPS). In: The Institution of Engineering and Technology International conference on ionospheric radio systems and techniques, pp 1–5
32 Cumming IG, Wong HC (1980) Digital processing of synthetic aperture radar data: algorithms and implementation. In: International radar conference, pp 168–175
33 Neo YL, Wong F, Cumming IG (2007) A two-dimensional spectrum for bistatic SAR processing using series reversion. IEEE Geosci Remote Sens Lett 4(1):93–96. https://doi.org/10.1109/LGRS.2006.885862
34 Hu C, Li Y, Dong X, Wang R, Ao D (2018) Performance analysis of L-band geosynchronous SAR imaging in the presence of ionospheric scintillation. IEEE Transactions on Geoscience and Remote Sensing 55(1):159–172
35 Hu C, Tian Y, Yang X, Zeng T, Long T, Dong X (2016) Background ionosphere effects on geosynchronous SAR focusing: theoretical analysis and verification based on the BeiDou navigation satellite system (BDS). IEEE Journal of Selected Topics in Applied Earth Observations and Remote Sensing 9(3):1143–1162

Chapter 5
Ionospheric Experiment Validation and Compensation

Abstract The L-band Geosynchronous SAR (GEO SAR) is very susceptible to the temporal-spatial variant ionosphere as its long integration time and large coverage, leading to image drift and degradation. This chapter demonstrates an experimental study of ionospheric influences on GEO SAR, including both the background ionosphere and the ionospheric scintillation. In the experiment, we employ the Global Positioning Satellites (GPS), probe the ionosphere and collect the trans-ionosphere GPS signals. Then the recorded signals are used to create the data basis on which simulations and analysis are based. But GEO SAR has very different orbit trajectories from GPS. Thus in the real operation, the transformation of the temporal-spatial frame between GPS and GEO SAR should be first performed before the focusing and the evaluation are carried out. Then the influences of the background ionosphere and the ionospheric scintillation are analyzed based on the experimental data. Finally, the corresponding compensation or mitigation approaches are presented. For the temporal-spatial variant background ionosphere, the autofocus method can produce a well-focused image. In comparison, the random fluctuations of amplitude and phase induced by ionospheric scintillation are more difficult to deal with. Thus an orbit-optimization strategy is first proposed by utilizing the diurnal and geographical pattern of the ionospheric scintillation occurrence. It can avoid being interfered by ionospheric scintillation by tuning the orbit parameters. Alternatively, once interfered, an iterative algorithm based on entropy minimum is derived to jointly compensate the signal amplitude and phase fluctuations in GEO SAR.

T. Long et al., *Geosynchronous SAR: System and Signal Processing*,
https://doi.org/10.1007/978-981-10-7254-3_5

5.1 Introduction

Geosynchronous SAR (GEO SAR) has become a hot topic recently. There mainly exist two research groups. The groups from US and China mainland carried out lots of studies about the high inclined GEO SAR which works in L-band. Most researches are focused on the system design and performance analysis [1–4], imaging formation algorithms [5–8] and applications [9–12]. The other group is from Europe, including UK, Italy and Spain [13–17]. Their study is about a GEO SAR with a nearly stationary orbit which has a nearly zero inclination and the ultra-long integration time (about 8 h or above). It was designed to work in Ku- or X-band for the preferred weather applications. Compared with the high inclined GEO SAR, it has the lower cost and the higher feasibility in techniques. But it is easier to be impacted by the tropospheric disturbance during its ultra-long integration time (around several hours) which will seriously degrade the signal coherence and the image quality.

In comparison, for the L-band GEO SAR, the ionosphere is one of most important considerations and has been studied a lot in theory and experiment. Canadian Centre for Remote Sensing (CCRS) [18] analyzed the Radarsat data and pointed out that the fluctuations in ionosphere electron density would result in the azimuthal drifts of SAR images, and would be certain to affect the InSAR processing. Pi et al. [19] and Carrano et al. [20] modelled the ionospheric scintillation and simulated its effects based on the phase screen theory. They explained the physical mechanism of the azimuth fringes observed by ALOS PALSAR [21]. There are also several kinds of climatological or physical models, such as the WideBand MODel (WBMOD) [22–24] and Global Ionospheric Scintillation Propagation Model (GISM) [25, 26]. These climatological or physical models link the behavior of the ionosphere itself with the trans-ionosphere signals. Wu et al. [27] and Wang et al. [28, 29] deduced the impacts of the ionospheric scintillation on the long wavelength radar by virtue of the generalized ambiguity function theory. Meyer et al. [30, 31] utilized the L-band satellite data to extract TEC values through measuring the phase delay and the group delay, and then corrected the ionospheric impacts on imaging and InSAR processing. Rosen et al. [32] realized the separation of the ionospheric disturbance and the other disturbance introduced by the non-dispersion media by the split spectrum method, and accomplished the estimation of the ionosphere within the InSAR repeated tracks. Recently, Quegan et al. [33–35] carried out lots of studies on estimations and compensations of the ionospheric scintillation and the Faraday effects in the P-band spaceborne SAR systems.

However, the ionospheric impacts on GEO SAR are less concerned. The characteristics of GEO SAR make it suffer from the ionosphere differently from LEO SAR. The ultra-long synthetic aperture time and the extended wide swath in GEO SAR makes it more easily and seriously affected by the changing ionosphere temporally and spatially. Hobbs et al. [4] presented a general description of various influences on GEO SAR, and pointed out that the ionospheric impacts should be considered in GEO SAR. In other studies [36], a modified GEO SAR signal model considering ionospheric impacts is proposed and the theoretical analysis and verification of the background ionosphere on GEO SAR imaging are given.

Besides these theoretical analysis and simulations, the experimental study is a promising supplementary approach for demonstrating the ionospheric impacts on GEO SAR. Unfortunately, there is no operational GEO SAR and thus no real data are available. However, the Global Navigation Satellite System such as the Global Position System (GPS) currently operates on the medium or higher orbits and employs the similar frequency band as GEO SAR (L-band). The GEO SAR and GPS satellites are all above the top of ionosphere which is around 1000 km. So they have the similar propagation path covering the whole ionosphere layer. The similarity between their propagation paths makes it possible that the generation of the ionosphere-induced phase errors in GEO SAR based on the GPS data. Conclusively, it is feasible and promising to exploit GPS data for analyzing the ionospheric impacts on GEO SAR. We will employ the GPS satellites, probe the ionosphere and collect the trans-ionosphere GPS signals. Then the recorded signals will be used to create the data basis on which simulations and analysis are based.

5.2 Experiment Principle and Signal Model

5.2.1 Background Ionosphere

As the theoretical analysis in Chap. 4, GEO SAR signals are affected by the total electrons in the ionosphere as they can induce propagation time delays when signals traverse the ionosphere. The time delays can be expressed as

$$\Delta t = -40.28 \cdot TEC/c \cdot f^2 \tag{5.1}$$

where *TEC* is the number of electrons per unit area of the Earth's surface for a vertical column to the "top" of the atmosphere, c is the vacuum light velocity and f is the signal frequency.

The propagation time delays will surely induce phase errors and then affect the spaceborne SAR focusing. Considering the round-trip in SAR, phase errors can be modified as

$$\varphi_{iono}(t_a) = -2\pi \cdot 80.6 \cdot TEC(t_a)/cf \tag{5.2}$$

where $TEC(t_a)$ represents the variant *TEC* with respect to the slow time t_a within the long integration time in GEO SAR.

Inferred from (5.2), phase errors are related to the signal frequency and TEC along the propagation path. In LEO SAR, the integration time is relatively small ($\sim$1 s) and *TEC* is assumed to be frozen (i.e. to be constant). However, in GEO SAR, the integration time can achieve as high as the level of 100–1000 s. Thus, the *TEC* variance cannot be neglected and the *TEC* frozen model fails.

The temporal variant ionosphere will induce the linear, quadratic or higher orders of phase errors, and be sure to result in the image drifts and defocusing. Based on the Taylor expansion, $TEC(t_a)$ can be modelled as the all-order derivatives of the slow time t_a. The expanded $TEC(t_a)$ is

$$TEC(t_a) = TEC_0 + k_1 \cdot t_a + k_2 \cdot t_a^2 + k_3 \cdot t_a^3 + \cdots \tag{5.3}$$

where TEC_0 is the constant part which is not variant with the time, and k_i is the ith temporal derivatives. Thus, the phase errors induced by the temporal background ionosphere $TEC(t_a)$ can be modified as

$$\varphi_{iono}(t_a) = -\frac{2\pi \cdot 80.6}{cf} \cdot \left(TEC_0 + k_1 t_a + k_2 t_a^2 + k_3 t_a^3 + \cdots\right) \tag{5.4}$$

In nature, the ionosphere distributes along the altitudes between 50 and 1000 km and it can be represented by a layer at the height of 350 km. When the satellite transmits and receives signals, the line-of-sight will intercept this ionospheric layer and the intercept point is named as the puncture point. When the satellite runs along with its own trajectory, the puncture point also forms a trajectory called the puncture path. This gives an approximation about the description of the trans-ionosphere propagation and the relative movement between GEO SAR and the ionosphere.

The TEC values are also various at different puncture points where the propagation signals traverse the ionosphere. So considering the spatial variance, the phase errors induced by the background ionosphere can be modified as

$$\begin{aligned}\varphi_{iono}(t_a, P) = &-\frac{2\pi \cdot 80.6}{cf} \\ &\cdot \left(TEC_0(P) + k_1(P)t_a + k_2(P)t_a^2 + k_3(P)t_a^3 + \cdots\right)\end{aligned} \tag{5.5}$$

where P indicates the position of the puncture point.

5.2.2 *Ionospheric Scintillation*

In the real cases, besides the large-scale ionosphere as the background ionosphere, the small or medium irregularities in the ionosphere will result in the fluctuations of the phase and the amplitude of the trans-ionosphere signals, reducing the spatial and temporal correlation. The signals impacted by the ionospheric scintillation can be expressed as

$$S_i = S_0(\delta A \exp(j\delta\phi)) \tag{5.6}$$

where S_0 is the ideal signal, and δA and $\delta\phi$ represent the random fluctuations of the amplitude and phase. When there is no ionospheric scintillation, δA is 1 and $\delta\phi$ is 0.

These random fluctuations of the amplitude and phase can be presented by the ionospheric scintillation sampling model (SSM). It employs the amplitude scintillation index S_4 and the phase scintillation index σ_φ which can be used to describe and simulate the signals' characteristics directly. In the SSM model, the amplitude fluctuations fit the Nakagami-m distribution and the phase fluctuations fit the Gaussian distribution. Meantime, the power spectra of the amplitude and the phase follow the power law. The Nakagami-m distribution of the amplitude can be expressed as

$$f(A^2) = \frac{S_4^{-2S_4^{-2}}(A^2)^{S_4^{-2}-1}}{\Gamma(S_4^{-2})E[A^2]^{S_4^{-2}}}\exp(-S_4^{-2}A^2/E[A^2]) \tag{5.7}$$

where the amplitude scintillation index S_4 is defined by the normalized standard deviation of the amplitude A. The Gaussian distribution of the phase fluctuations can be expressed as

$$f(\varphi) = \frac{1}{\sigma_\varphi\sqrt{2\pi}}\exp\left(\frac{\varphi^2}{2\sigma_\varphi^2}\right) \tag{5.8}$$

where the phase scintillation index σ_φ is defined by the standard deviation of the phase φ.

After obtaining the parameters of S_4 and σ_φ which can be set manually in simulation or retrieved from the ionosphere monitor, the ionospheric scintillations can be produced through the steps shown in Fig. 5.1. The detailed steps are as follows. The parameters of S_4 and σ_φ are firstly used to generate a series of random numbers. The spacing of these numbers is the pulse repetition intervals (PRT) of GEO SAR. These numbers are correlated by the bi-linear transform and their spectra are then reshaped to follow the power law. At last, the generated ionospheric fluctuations of the amplitude and phase are then superimposed on the ideal GEO SAR signals.

Generally, in the SSM model, S_4 and σ_φ are negatively correlated and an empirical value of −0.6 is adopted in the step of the correlation. After spectral reshaping, the correlation is changed. The value lies around −0.4. In the operation, S_4 and σ_φ are estimated and output based on the recorded data. So they can be used to describe the ionosphere status during the recording period. The signal fluctuations at each azimuth sampling position within an aperture can be simulated by the generated random sequences.

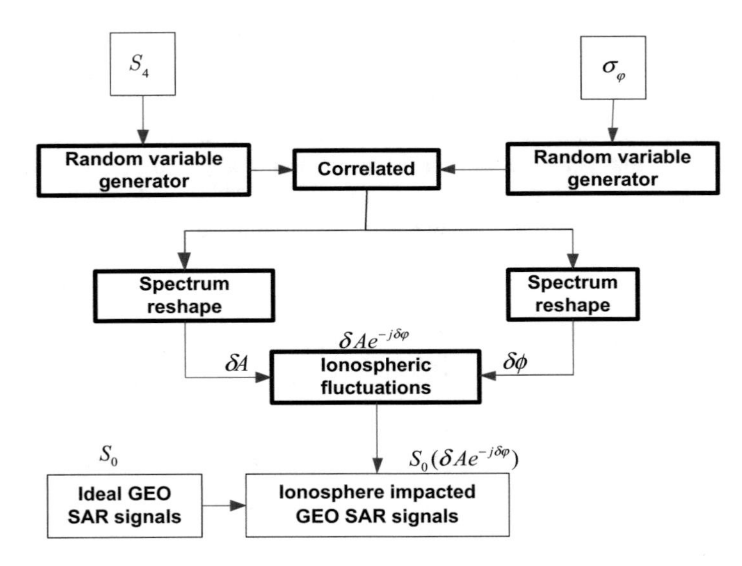

Fig. 5.1 Sketch map of generating the ionospheric scintillations using S_4 and σ_φ

5.2.3 GEO SAR Signal Models

Considering the phase errors are related to the signal frequency, the induced amplitude and phase errors can then be incorporated into the ideal GEO SAR echoes in the azimuth time domain and range frequency domain. Conclusively, the GEO SAR signals impacted by the background ionosphere and the ionospheric scintillation can be expresses as

$$S_i = S_0(\delta A \exp(-j\delta\phi - j\varphi_{iono}(t_a))) \tag{5.9}$$

5.2.4 Experiment Overview

In analyzing the ionospheric effects on GEO SAR, the temporal-spatial variance of the ionosphere should be considered. The temporal variance is the ionospheric variance within an aperture during the acquisition of the targets' echoes. The spatial variance is the ionospheric variance when recording the echoes of various targets located in the scene. In fact, the key problem in analyzing the ionospheric impacts on GEO SAR is how to generate or simulate the temporal and spatial variant ionosphere during the data acquisition. But it is difficult to record the variant ionospheric data in time and space simultaneously. So in the experiment [37], we employ the Global Positioning Satellites (GPS), probe the ionosphere and collect the trans-ionosphere GPS signals. These GPS data can indicate the status and the level of the ionosphere along its puncture path. The recorded GPS signals will be used to create the data basis on which the simulations are based. Then we generate the temporal and spatial variant ionosphere-impacted GEO SAR signals by combining the GPS data and the mathematic GEO SAR signal models. Unfortunately, GEO SAR has a 60° inclination and a high orbit altitude of 36,000 km. Its puncture path is quite different from that of GPS. The direct employment of the GPS data is not permitted. So our work is carried out to transfer data from the GPS to the GEO SAR puncture paths, using the temporal behavior of the ionosphere as measured by the GPS satellites along puncture paths selected to be similar to those of the GEO SAR.

In our experiment, we employ the ionosphere monitor to record the values of TEC, S_4 and σ_{φ}. The measurement only records the ionospheric status at the puncture point rather than the whole observation scene. When the satellite moves along its trajectory, the ionosphere along the trajectory of the puncture point is recorded. So the ionospheric variance during this period can be used to approximate the ionospheric variance within an aperture for one point target at the place of the monitor. This gives the temporal variance simulation. But for simulating the other targets in the scene, theoretical models should be adopted. Among these models, the simplest method is to consider the experiment measurement as the vertical TEC at the puncture point and expand it to the other targets by the transformation of the vertical TEC to the slant TEC based on a cosine law. This can provide a scene-dependent variance and give the spatial variance of the ionosphere.

In the experiment, the aims are to extract the TEC values and its variations from the GPS signals, and then simulate the ionospheric impacts on the amplitude and phase fluctuations. The experimental equipment is the GPS signal monitor (as shown in Fig. 5.2) produced by the China Research Institute of Radiowave Propagation (CRIRP). The monitor has functions of measuring, recording and outputting the TEC values and its variations, along with the scintillation parameters of S_4 and σ_{φ}. The main technical specifications of the GPS signal monitor are listed as in Table 5.1.

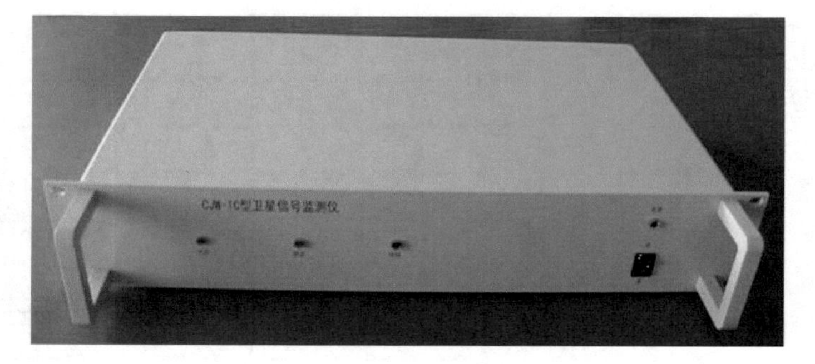

Fig. 5.2 GPS signal monitor (type: CJW-1C produced by China Research Institute of Radiowave Propagation)

Table 5.1 Main technical specifications of the GPS signal monitor

Operating band	Output	Sampling rate of the scintillation index	Sampling rate of TEC
L	TEC and its variations, scintillation index	1 per min	1 per 15 s

The experiment is carried out in Zhuhai and Beijing, China, respectively. The data were 24 h consecutively acquainted for 20 days. The first experiment was carried out in Zhuhai (113° 34′E, 22° 16′N) from 25th April 2014 to 5th May 2014. The second experiment was carried out in Beijing (116° 23′E, 39°54′N) from 19th to 27th June 2014. These two places are very representative of the ionosphere in China mainland. Zhuhai lies within the Tropic of Cancer with an active ionosphere and Beijing locates in the mid-latitude region whose ionosphere is much calmer.

5.3 Background Ionosphere Experiment and Compensation

5.3.1 GPS Data Recording

During the consecutive 20 days' experiment, the outputs of the ionosphere monitor could be used to record the daily TEC trends of the ionosphere. During the observation, a total of 30 GPS satellites flied overhead, including the GPS satellites with the pseudo random noise code (PRN) from 1 to 32 except PRN 6 and PRN 30. These data show that the daily TEC trends are similar. Consequently, the data recorded on 29th April 2014 are selected to analyze the ionospheric impacts. The sketch map of the TEC and its variations are presented in Fig. 5.3.

In comparison, the ionospheric scintillations were observed only once during the 20 days' experiment. It was captured in Zhuhai rather than in Beijing. Though lots

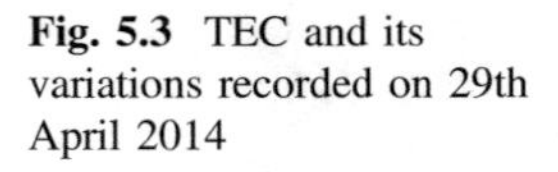
Fig. 5.3 TEC and its variations recorded on 29th April 2014

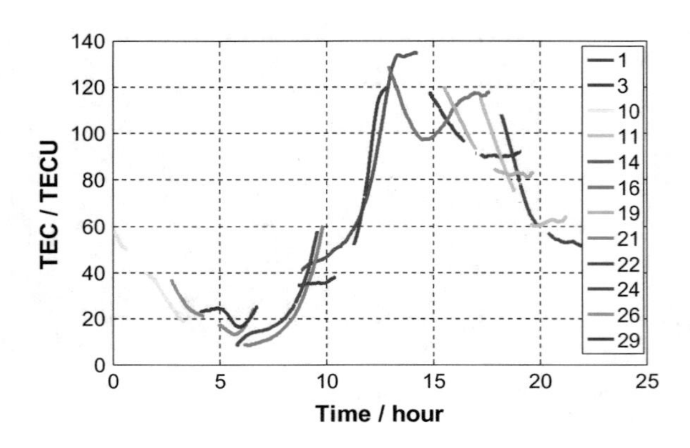

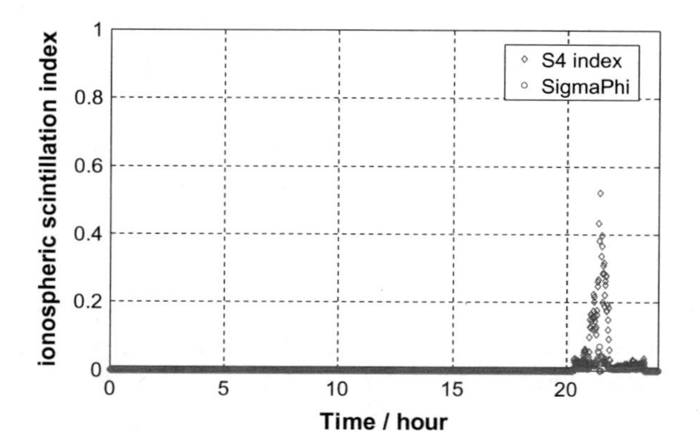

Fig. 5.4 Ionospheric scintillation data (GPS of PRN 20) recorded on 30th April 2014, in Zhuhai

of GPS satellites were in the scope of the ionosphere monitor, they had different puncture paths. Thus, not all of them were able to capture the occurrence of the scintillation. The captured scintillation lasted from 20:40 to 22:00 on 30th April 2014. A total of 5 GPS satellites (i.e. PRN 1, PRN 7, PRN 8, PRN 9 and PRN 20) recorded the scintillation data. The recorded scintillation data had 3 kinds of characteristics:

(1) the S_4 of these 5 GPS stayed above 0.1;
(2) the durations were greater than 10 min;
(3) the trends of the amplitude and the phase showed a good coincidence.

Consequently, the ionospheric scintillation could be judged to occur. Among these observations, the GPS of PRN 20 captured a strong scintillation which lasted over 1 h. Most of the recorded scintillations had the greater amplitude fluctuations than the phase fluctuations. The observation results are shown as in Fig. 5.4. It shows that S_4 and σ_φ stay almost zero when the ionosphere is silent. But when the ionospheric scintillation occurs, S_4 and σ_φ will arise rapidly. In the experiment, the raw signals are not the standard output. So the correlation coefficient between the amplitude and the phase cannot be estimated when it is not one of the standard outputs either. So in our experiment, an empirical value of −0.6 is adopted for the correlation coefficient.

5.3.2 Data Pre-processing

In the experiment, the ideal GEO SAR echoes can be generated through simulations directly [5]. The ionospheric errors can be extracted from GPS signals. However, compared with GEO SAR, the GPS satellites run on the orbit height of 22,000 km, and thus they have obviously different velocities and ionospheric puncture paths. The orbit parameters for GEO SAR and GPS satellites are listed in Table 5.2. RAAN stands for the right ascension of ascending node.

Table 5.2 Orbit parameters for GEO SAR and GPS satellites

	GEO SAR	GPS
Semi-major axis (km)	42,164	26,560
Eccentricity	0	0.0195
Inclination (°)	60	56.574
Argument of perigee (°)	–	261.9
RAAN (°)	115 E	25.218 E

Consequently, the difference in the data acquisition geometry makes it impossible to employ directly the TEC data extracted from GPS signals in analyzing ionospheric impacts on GEO SAR. The original GPS data should be first pre-processed and transformed into the temporal-spatial frame of GEO SAR. Two principles should be satisfied in order to ensure the close approximation to the real ionospheric errors in GEO SAR:

(1) Ionospheric puncture paths for GEO SAR and GPS should be identical or similar;
(2) Sampling time interval for GPS signals is identical to GEO SAR.

The detailed equivalent pre-processing procedures include the reproduction of GEO SAR and GPS orbits, the selection of the orbit trajectories and the ionospheric puncture paths, the interception and the interpolation of the GPS data and the following transformation to the temporal-spatial frame of GEO SAR, and the final generation of ionosphere-impacted GEO SAR signals and the focusing and evaluation. The data processing procedures are shown in Fig. 5.5. The procedures are summarized as follows:

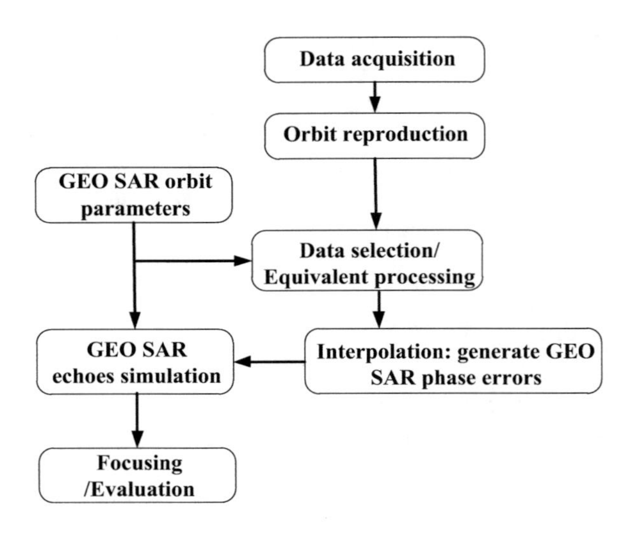

Fig. 5.5 Flowchart of the data processing procedures and the experiment

(1) record GPS signals;
(2) produce the GEO SAR and GPS orbits in System Tool Kits (STK) using GEO SAR orbit parameters and GPS data message;
(3) select and intercept GPS orbits whose nadir track and puncture path are similar to GEO SAR;
(4) extract the phase errors in the corresponding orbit section and generate the ones fit for GEO SAR using interpolation;
(5) simulate the GEO SAR signals considering ionosphere by incorporating the phase errors into GEO SAR signals;
(6) perform the image focusing and evaluation.

5.3.3 Experimental Data Processing

During the 20 days' observation, the trends of TEC variation each day are similar. Thus, the TEC data acquainted in Zhuhai on April 29th 2014 is selected for demonstration.

The recorded data are first intercepted and transformed according the equivalence of the acquisition geometry. The orbits of GEO SAR and GPS are both regenerated in STK according to the GEO SAR orbit parameters and the GPS data message (i.e. two-line orbital element data, TLE), respectively. Through comparing their tracks, the GPS of PRN 31 is selected as it has the partially parallel track as GEO SAR. Similarly, they both cover the whole ionospheric layer. Consequently, they will have similar nadir tracks and the corresponding ionospheric puncture paths which are shown in Fig. 5.6. Because GEO SAR and GPS have different orbit

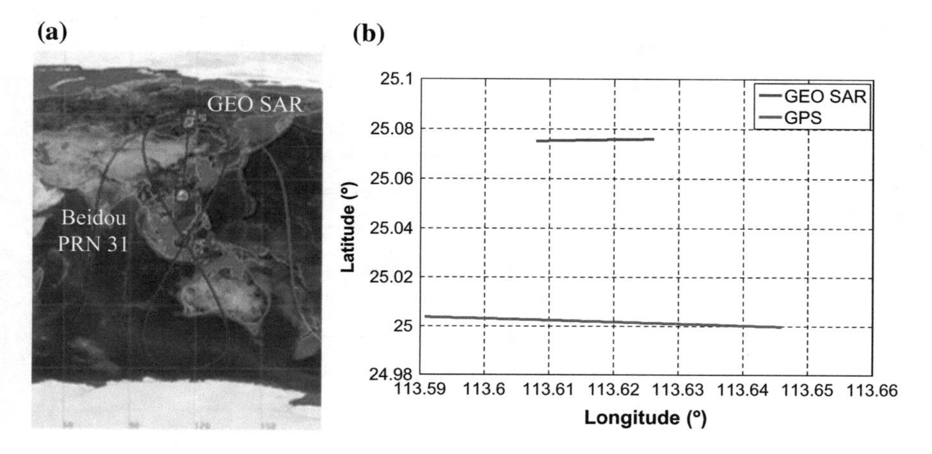

Fig. 5.6 Nadir tracks (left) and the corresponding ionospheric puncture paths (right) of GEO SAR and GPS, respectively

heights, their identical nadir track would not mean that they have the identical length or the duration of the orbit trajectories and the ionospheric puncture paths. So the data acquisition start time and the duration of GPS should be re-positioned in order to approximate to those of GEO SAR.

So the puncture path is selected following the rule that they have similar nadir tracks (in other words they have the same rotation angles seen from Earth center). But as they have different orbit heights, the corresponding puncture paths and the orbit trajectories are different. So the GPS puncture path should be intercepted or tailed according to that of GEO SAR which is identified by its satellite-Earth geometry and the resolution requirement. This means the selected GPS data have the same spatial length as GEO SAR. So the GPS puncture path will be intercepted in the assumption that they have the identical rotation angle as GEO SAR. The start time and the acquisition duration of GEO SAR data is first calculated according to the system configuration and the resolution requirement. The integration time of GEO SAR is 149 s, achieving an azimuth resolution of 20 m. Then the corresponding timing parameters of the PRN 31 GPS satellite are determined under the geometry limits.

The GPS satellites have a lower orbit height and greater velocities than GEO SAR. So in this case, only the 63 s' GPS data are used and thus their puncture paths are identical. The calculation is completed which meet with the criterion that the length of the puncture path of GPS is identical to that of GEO SAR during its integration time. The intercepted puncture paths for GEO SAR and GPS are shown in Fig. 5.7. After the data trimming, the next step is to interpolate the data to make them have the same sampling interval. The 63 s GPS data are intercepted and interpolated with the GEO SAR's PRT, leading to the equivalent data of GEO SAR's 149 s. However, the direct resampling from the 63 s of GPS observations to the 149 s of the ionospheric variations within a GEO SAR aperture is not accurate. In fact, the 63 s of GPS observations should be exploited as follows:

(1) the absolute TEC value at the center moment of GPS is employed as the absolute TEC value at the aperture center moment of GEO SAR;

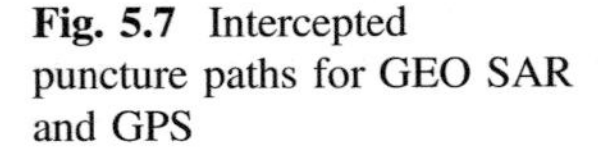

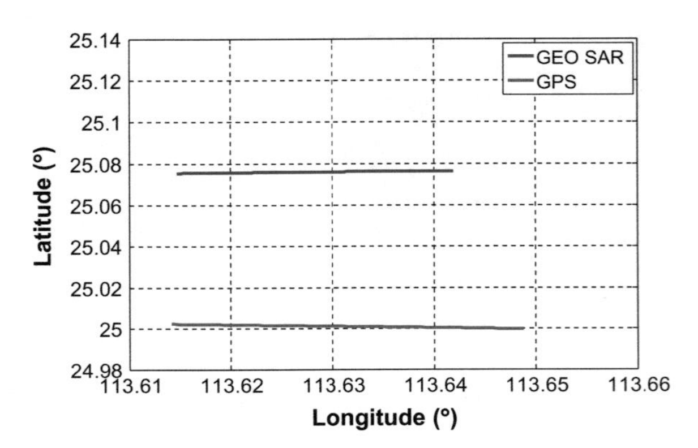

Fig. 5.7 Intercepted puncture paths for GEO SAR and GPS

(2) the temporal variation rates derived from the 63 s of GPS observations are used to depict the temporal variant characteristics of the 149 s' ionospheric variations in GEO SAR.

So the absolute TEC at the aperture center moment and the temporal variation rates of GPS will be both used to construct the temporal variant TEC values within the 149 s of GEO SAR aperture. Besides the above two items about TEC, the GEO SAR PRT is also used in the temporal variant TEC construction.

5.3.4 Result and Discussion

The GPS data are used for simulating the ionosphere impacted point target focusing. The plots in Figs. 5.8 and 5.9 are based on the recorded GPS data. The monitor records the variant TEC when the GPS satellites are visible. So each visible GPS can produce a curve of TEC variance and can be used to calculate the all orders of derivatives. These can be used to approximate the temporal TEC variance within an aperture in GEO SAR. After the data pre-processing, all orders coefficients of the TEC variations are extracted from the GPS data and then employed to construct the TEC variations in GEO SAR.

The GEO SAR system parameters are as shown in Table 5.3. The system parameters will affect the thresholds. For the L-band GEO SAR with the resolution of 20 m, the threshold of TEC values is around 900 TECU above which will lead to defocusing in range. But the maximum TEC value of the 20 days' observation is 192.7 TECU in Zhuhai in April 26th 2014 and 51.2 TECU in Beijing in June 20th 2014. The maximum TEC values of each day's measurement in the experiment are shown in Fig. 5.8. Generally speaking, the maximum of TEC values hardly achieves as high as 900 TECU in nature. Consequently, the range focusing would not be affected by the background ionosphere.

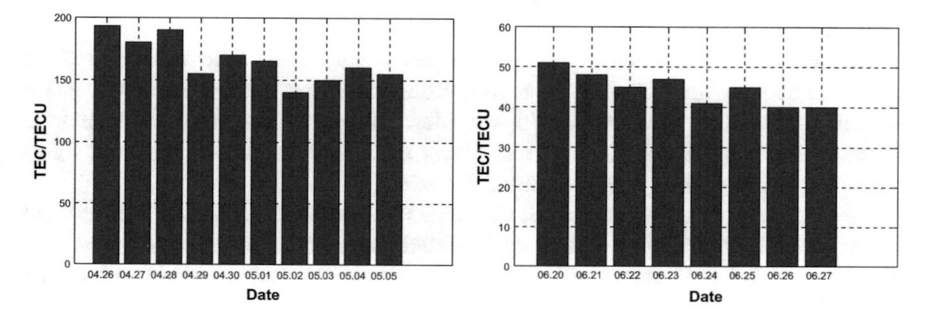

Fig. 5.8 Maximum TEC values of each day's measurement in the experiments in Zhuhai (left) and Beijing (right)

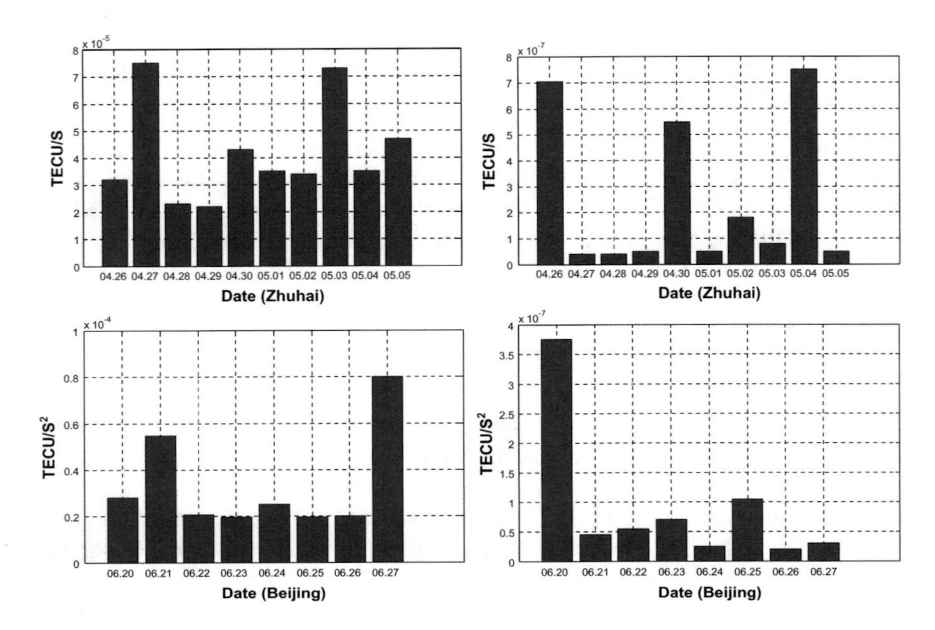

Fig. 5.9 Maximum (left) and minimum (right) second derivative of TEC values in Zhuhai and Beijing experiments

Table 5.3 GEO SAR system parameters

Resolution	Off-nadir angle	Orbit position	Bandwidth	Integration time
20 m	2°	Perigee	20 MHz	149 s

Here the data recorded in Zhuhai on April 29th 2014 will be exploited for analyzing the impacts on the azimuth focusing. The TEC data are Taylor expanded and the second derivative is 6.65e-8. The accumulated phase is below $\pi/4$. Thus the impacts on azimuth focusing can be neglected in this case.

In addition, the extracted phase errors are incorporated into the GEO SAR signals. The simulated GEO SAR echoes are focused and the azimuth profiles are shown in Fig. 5.10. Compared with the ideal imaging result, the first side lobe degrades around 0.2 dB. This also validates that these variant background ionosphere have little impacts on azimuthal focusing.

In the experiment, the image drifts in range induced by the constant TEC and the image drifts in azimuth induced by the first derivative of the variant ionosphere can be corrected by the image registration techniques. So the discussion about the image drifts is not the focus in the chapter.

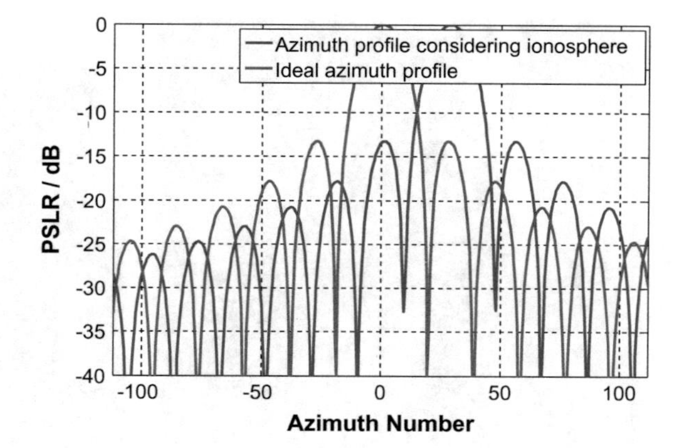

Fig. 5.10 Azimuth profiles considering ionospheric impacts

5.3.5 Compensation: Autofocus Methods

The phase error introduced by the background ionosphere was a high-order phase error both in azimuth and range, and the phase gradient autofocus (PGA) algorithm can precisely estimate such form of phase error [37]. Contemporarily, according to the theoretical analysis and simulation analysis results, the effects of background ionosphere on the GEO SAR range focusing can be ignored, therefore it is equivalent to that the range phase error is redundant substantially. In other words, the background ionosphere phase error is the function of azimuth within the entire scope of imaging, which is in consistence with the fundamental assumptions of the PGA algorithm. As a result, in theory, the PGA algorithm can be adopted to correct the effects of the background ionosphere.

The PGA algorithm is based on the range phase error redundancy assumptions, which is not only able to estimate the arbitrary order of phase error, but also has superior robustness. Firstly, the PGA algorithm was used to align the strong scattering points within each range cell through the circumferential shift process, in order to facilitate the subsequent phase error estimate. Secondly, the data after circumferential shift was conducted windowing process, in order to reduce interference from the neighboring targets and background clutter. Afterwards, the phase gradient estimation was carried out, which often adopted the linear unbiased minimum variance estimation (LUMV) method

$$\Delta\phi_{\varepsilon}(n) = \frac{\sum_{m} \mathrm{Im}\left[d(n)s_{w}^{*}(n)\right]}{\sum_{m} |s_{w}(n)|^{2}} \tag{5.10}$$

where, $s_w(n)$ is the range compression history domain signal, $d(n)$ is the first order difference of $s_w(n)$, and m is the number of range units. The accumulation of $\Delta\phi_{\varepsilon}(n)$ can be obtained the phase error $\phi_{\varepsilon}(n)$. The last step of the PGA algorithm

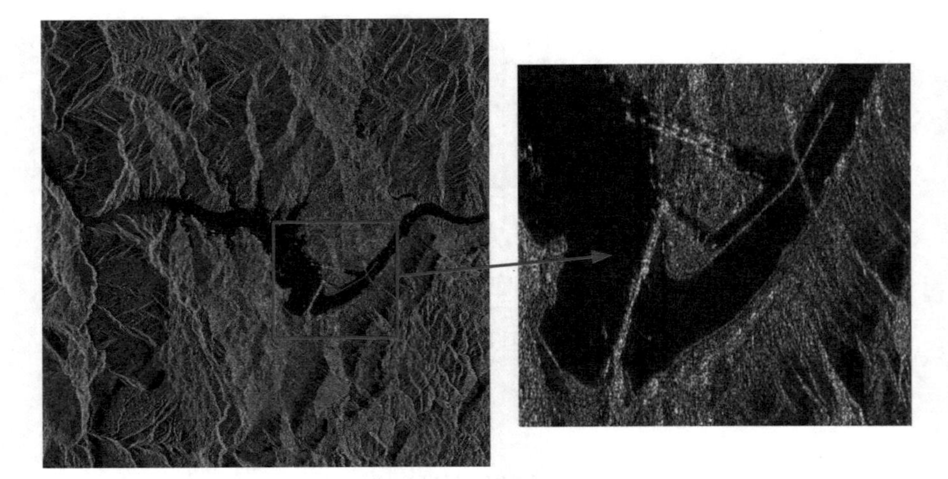

Fig. 5.11 Imaging results corrected by the PGA algorithm (the rectangular region is Three Gorges Dam)

was to use the estimated phase error $\phi_{\varepsilon}(n)$ to correct the entire image, so as to get a new image for the next phase error estimation, until the estimated phase error was less than a preset value. At this time, it was deemed that the image had been focused and the PGA algorithm was terminated.

The method described above was utilized to correct the GEO SAR image under the influence of the background ionosphere and the corrected imaging result is as shown in Fig. 5.11. The blurred Three Gorges Dam (as the rectangular region) under the influence of the ionosphere was clearly visible, and the double-line five-grade ship lock in the north side of the dam can be distinguished again.

5.4 Ionospheric Scintillation Monitoring Experiment

5.4.1 Experimental Data Processing

The ionospheric scintillation is caused by the small and medium scale ionospheric irregularities. In the natural environment, the ionospheric irregularities are kind of random media. The anisotropy is one important characteristic for these irregularities. So the resultant ionospheric scintillation is also anisotropic. But the study of the anisotropic ionosphere is a big challenge and goes beyond our research in this chapter. So we choose the suboptimal choice or the approximate assumption that they are isotropy for the principle validation. Actually, the spectrum of scintillation follows the power law. When the power law is adopted, the spectrum of scintillation lies in the inertial region where it can be considered to be approximately isotropic. On the other hand, it is limited by our experiment conditions. The output of the ionospheric monitor is once per minute. So for the simulation of a 100 s' synthetic aperture, we only have two S_4 index but we have to generate the variations at each azimuth sampling position. So we consider the ionospheric scintillation at each azimuth sample position follows the same power law and the parameters (i.e. S_4 and σ_φ) are same.

Assuming the ionospheric irregularities follow the isotropic distribution, the impacts induced by the ionospheric scintillation can be considered to be independent of the specified GEO SAR geometry. Consequently, during the observation, the outputs of the ionosphere monitor can be considered to be representative of the ionospheric scintillation changes within the GEO SAR integration time. Thus, differing from the mandatory requirement of the equivalent processing of the orbit and recorded data in the background ionosphere experiment, the scintillation data can be directly used. The only requirement is that the data duration is identical to the GEO SAR integration time.

In the experiment, the SSM model is employed. The scintillation index of S_4 and σ_φ from the monitor are used as the inputs of this model. The fluctuations are simulated and then incorporated in the ideal GEO SAR echoes for the generation of the ionospheric scintillation impacted GEO SAR signals. The imaging and the evaluation are then performed to analyze the ionospheric impacts.

The scintillation data, acquired on April 30th 2014 by the GPS of PRN 20, has the strong scintillation intensity and lasts a long time. The S_4 and σ_φ retrieved from these data are adopted in the model. However, the integration time for the 20 m resolution achieves around 149 s, but the output of the monitor is only one value per minute. Thus, during the analysis, two scintillation data are selected, i.e. the S_4 and σ_φ recorded at 21:24 and 21:25 on April 30th 2014. The specified values are S_4 = [0.3825 0.5222], $\sigma_\varphi = [0.05313\ 0.0703]$. The imaging results and the azimuthal profiles are shown in Fig. 5.12.

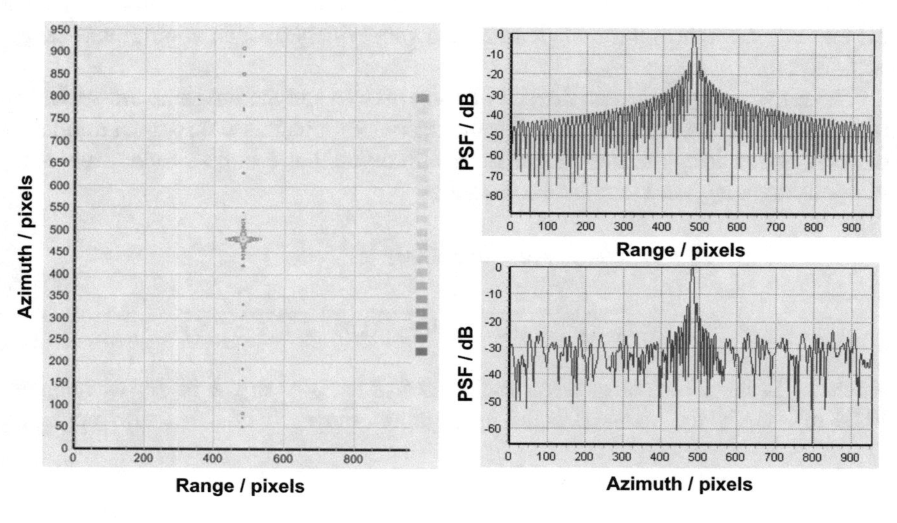

Fig. 5.12 Point target imaging result using the scintillation data acquired on April 30th by the PRN 20 GPS

From the imaging results, it can be seen that the ionospheric scintillation has few impacts on the range focusing. In comparison, the scintillation has significant impacts on the azimuth focusing. In cases of the scintillation captured in this experiment, the azimuth ISLR arise to the level −4.47 dB with a deterioration of over 4 dB, while the PSLR is −13.05 dB in which the impacts are small. Conclusively, the ionospheric scintillation mainly impacts the azimuth focusing and the ISLR is much more susceptible to be affected. But the China mainland mainly locates in the mid-latitude region where the ionosphere is calm. As in the experiment, scintillations are seldom captured. Thus, it is suggested to avoid the GEO SAR working during the occurrence of scintillations. Or if scintillations are encountered unfortunately, the approaches for estimations and compensations can be employed. The details are described in the following section.

5.4.2 Discussion

In our experiment, the scintillation indices are the outputs of our ionosphere monitor. So they are used to generate the ionospheric scintillation signals based on the SSM model. But for the representativeness of scintillation conditions, the scintillation indices are compared with the parameters of the climatological model such as the WBMOD.

For the WBMOD climatological model, the parameters of *CkL*, *p* and L_0 are used to represent the ionosphere status. *CkL* is the height-integrated electron-density irregularity strength and it is a measure of the total 'power' in the electron-density irregularities along a vertical path traversing the entire ionosphere. Generally, it ranges from 30 to 36 when it is transformed into dB values. *p* is the spectral index of the phase spectrum and it is used to describe the ionospheric irregularities. It generally ranges between 1 and 4 but is set with a mode between 2.5 and 2.8. L_0 is the outer scale lengths and vary widely over an approximate range of 5–20 km.

The relationships between the scintillation indices and the environment variables (*CkL* and *p*) can be found in [20]. For the one-way propagation through a thin layer of ionospheric irregularities in the Born approximation (single scatter), the scintillation intensity index can be expressed as

$$S_m^2 = r_e^2 \lambda^2 C_k L \sec\theta (2\pi/1000)^{2\nu+1} (\lambda d_R/4\pi)^{\nu-0.5} \cdot \frac{\Gamma[(2.5-\nu)/2]}{2\sqrt{\pi}\Gamma[(\nu+0.5)/2](\nu-0.5)} \cdot \frac{\Gamma(\nu)}{\sqrt{\pi}(\nu+0.5)} \tag{5.11}$$

where r_e is the classical electron radius (2.818×10^{-15} m), λ is the wavelength (0.24 m), d_R is the reduced slant propagation range, θ is the zenith angle of propagation (similar to the off-nadir angles which ranges from 2 to 7° in GEO SAR), Γ is Euler's gamma function. The orbit height of GEO SAR is 35,893 km and the ionospheric screen lies at the height of 350 km. So d_R approximate to 346.6 km (the ionosphere is assumed as a layer at the height of 350 km and the orbit height of GEO SAR is 35,893 km). The spectral slope v is defined such that $p = 2v$ is the phase spectral index.

The relationship between the scintillation intensity index for a two-way propagation path (S_4) and the scintillation intensity index for a one-way propagation path (S_m) is as follows

$$S_4^2 = 4S_m^2 + 2S_m^4/\left(S_m^2 + 1\right) \tag{5.12}$$

For the GEO SAR configuration, the relationship between the intensity index (S_4) and the *CkL* is shown in Fig. 5.13. The simulation parameters used are listed in Table 5.4. When the phase spectral index (p) is small and the integrated turbulence strength (*CkL*) is large, the S_4 will become greater.

For a typical value in WBMOD, the phase spectral index (p) is set as 2.7. The relationships of scintillation levels, S_4 and *CkL* are listed in Table 5.5. So when the medium scintillation occurs, the azimuth focusing will deteriorate seriously.

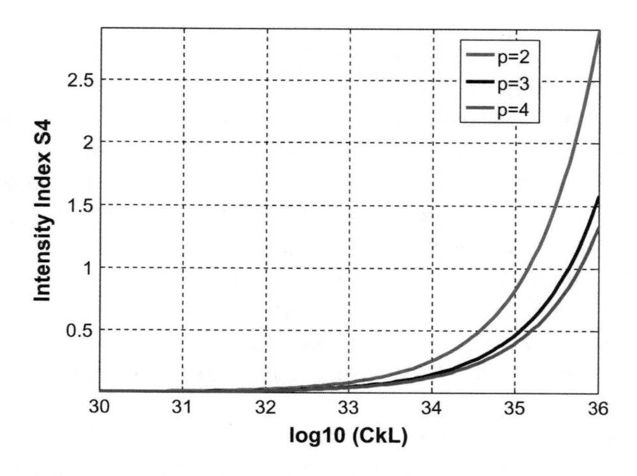

Fig. 5.13 Relationships between the intensity index (S_4) and the *CkL*

Table 5.4 Simulation parameters of the WBMOD models

r_e	Zenith angle	Ionosphere height	Orbit height	$\log_{10}$ (*CkL*)	p
2.818×10^{-15} m	2°	350 km	35,893 km	30–36	2–4

Table 5.5 Relationships of scintillation levels, S_4 and CkL

Scintillation levels	Weak	Medium	Strong
S_4	<0.3	0.3–0.5	>0.5
CkL	$<3.5 \times 10^{34}$	$(3.5\text{–}9.3) \times 10^{34}$	$>9.3 \times 10^{34}$

5.5 Ionospheric Scintillation Compensation

5.5.1 Avoidance Based on Orbit Design

5.5.1.1 Principles and Implementation

The occurrence of ionospheric scintillation shows a specified diurnal and geographical pattern. Because of the severe activities of Sun and Earth's magnetic field, the ionosphere in equatorial and polar region is more active than the mid-latitude region. In the equatorial region, massive temporary small-scale irregularities, which have lifetimes of several hours, are mainly generated after sunset. Thus, showing a dependency of scintillations on local time, ionospheric scintillation often happens from the early evening after sunset to midnight [38, 39]. Unfortunately, because the aurora is the source of generation of small-scale irregularities in the polar region, ionospheric scintillation depends on solar activity and can happen at all times in a day.

Therefore, an orbit optimization strategy [41] can be explored to make the GEO SAR avoid being interfered by ionospheric scintillation. The orbit optimization strategy will utilize the diurnal and geographical pattern of the ionospheric scintillation occurrence. Since the GEO SAR imaging is mainly aimed at middle and low latitude regions, we only consider the scintillation in the equatorial region and it has the time window for the occurrence of ionospheric scintillation (early evening after sunset to midnight). In fact, according to the various observation areas of GEO SAR, we consider a scintillation-sensitive region, which is equal to or smaller than the complete equatorial region. Based on the above description, we can optimize an orbit with the proper elements to avoid imaging for the area with the ionospheric puncture points (IPPs) within the scintillation-sensitive region during the time window.

Generally, satellite is determined by six orbit elements: inclination, eccentricity, semi-major axis, longitude of the ascending node, argument of perigee, and time past perigee. The first five orbit elements should be carefully designed to obtain the required performance of the coverage and the revisit toward the target area due to their relationships with orbit size, shape and scope, and position of the nadir trajectory. Nevertheless, time past perigee only represents the spacecraft's location within the orbit and can be optimized to make the GEO SAR have the capability of avoiding being impacted by ionospheric scintillation interference. The basic conception of the method is provided as follows.

The geocentric latitude ν and longitude ϑ of the nadir-point trajectory can be approximately described by

$$\begin{cases} \vartheta(t) = \vartheta_0 + \arctan(\tan(\omega+\zeta)\cos i) \\ -\left(\Omega_E - \frac{3\Omega_E J_2}{2}\left[\frac{a_e}{a(1-e^2)}\right]^2 \cos i\right)(t-t_p) \\ \nu(t) = \arcsin(\sin i \, \sin(\omega+\zeta)) \end{cases} \tag{5.13}$$

where t is the orbit time, ϑ_0 is the longitude of the ascending node, e is the eccentricity, Ω_E is the Earth's angular velocity, t_p is the time past perigee, i is the inclination, ω is the argument of the perigee, a is semi-major axis, a_e is the equatorial radius of the Earth, and ζ is the true anomaly, which has a relationship with t shown as

$$t = \frac{2\arctan\left(\sqrt{\frac{1-e}{1+e}}\tan\frac{\zeta}{2}\right) - e\,\sin\left(2\arctan\left(\sqrt{\frac{1-e}{1+e}}\tan\frac{\zeta}{2}\right)\right)}{\Omega_E} + t_p \tag{5.14}$$

The geometry between the nadir-point P and the IPP of the beam center P' is shown in Fig. 5.14. S represents the GEO SAR satellite. H is the altitude of the GEO SAR and θ is the look angle of the GEO SAR. The height of ionosphere h_{ion} is assumed to be 350 km. R is the radius of the Earth and O is the center of the Earth. r is the distance from S to P'.

The longitude ϑ_s and the latitude ν_s of the IPP of the beam center in GEO SAR can be approximately written as

$$\begin{aligned} \vartheta_s(t) &\approx \vartheta(t) \\ &- \frac{1}{\cos(\nu_s(t))}\cos\left(\arctan\left(\frac{\Delta(\vartheta_s(t))}{\Delta(\nu_s(t))}\right)\right)\left\{\pi - \left\{\theta + \arcsin\left[\frac{\sin\theta(a_e+H)}{a_e+h_{ion}}\right]\right\}\right\} \end{aligned} \tag{5.15}$$

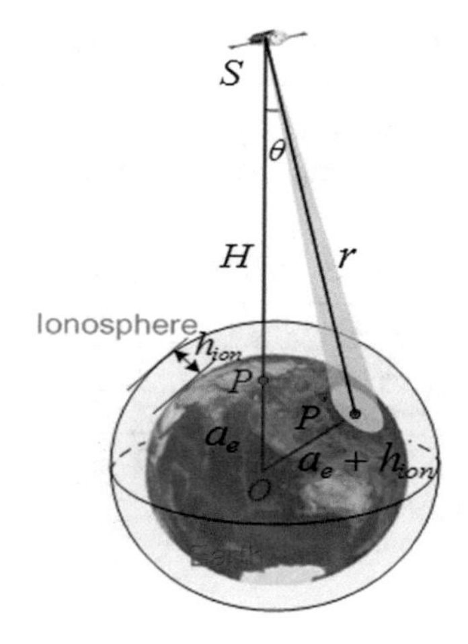

Fig. 5.14 Geometry between the nadir-point and the IPP of the beam center in GEO SAR

$$v_s(t) \approx v(t) - \sin\left(\arctan\left(\frac{\Delta(\vartheta_s(t))}{\Delta(v_s(t))}\right)\right)\left\{\pi - \left\{\theta + \arcsin\left[\frac{\sin\theta(a_e + H)}{a_e + h_{ion}}\right]\right\}\right\} \tag{5.16}$$

where $\Delta(\vartheta_s(t))$ and $\Delta(v_s(t))$ are the changing rate of ϑ_s and v_s, respectively. Ignoring the beam width and assuming that in the time interval $t \in [t_s, t_e]$, the IPPs of GEO SAR are within the scintillation-sensitive region, $v_s(t)$ satisfies

$$[v_s(t_s), v_s(t_e)] \subseteq \Phi_0 \tag{5.17}$$

where Φ_0 are the latitude scope of the scintillation-sensitive region.

To avoid suffering ionospheric scintillation, we should set the proper t_p to make $[t_s, t_e]$ satisfy

$$[t_s, t_e] \cap \prod = \varnothing \tag{5.18}$$

where $\prod$ is the time window in a local time form.

Therefore, based on (5.18), we could determine an effective t_p set to avoid visiting the scintillation-sensitive region within the time window. According to the above analysis, the effective t_p set relates to Φ_0, platform parameter (θ), and orbit elements (i, ω, and e). Different scintillation-sensitive regions, platform parameters, and orbits, determined by the practical applications, will correspond to various effective t_p sets, being relatively large sets or even null sets. Moreover, because the GEO SAR satellite will suffer the impacts of the perturbation forces inevitably, the orbit of GEO SAR will slowly drift away from the scheduled position. Although a proper t_p has been set initially, (5.18) will not validate after a period of time with the impacts of perturbation forces and the station-keeping is needed to compensate the drift of the orbit.

5.5.1.2 Simulation

In this part, we conduct the simulation to verify the previous analysis. Three classical orbits with different orbit elements are utilized. The corresponding orbit elements are given in Table 5.6 and their nadir trajectories are shown in Fig. 5.15. According to aforementioned analysis, we select the local time interval

Table 5.6 Orbit elements of 3 typical GEO SAR orbits

Parameters	'figure-8'	'figure-O'	Small 'figure-8'
Inclination (°)	53	7	16
Semimajor axis (km)	42,164	42,164	42,164
Argument of perigee (°)	270	270	180
Eccentricity	0.07	0.1	0.05
Lon. Ascn. Node (°)	113	109	93

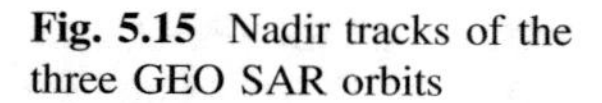

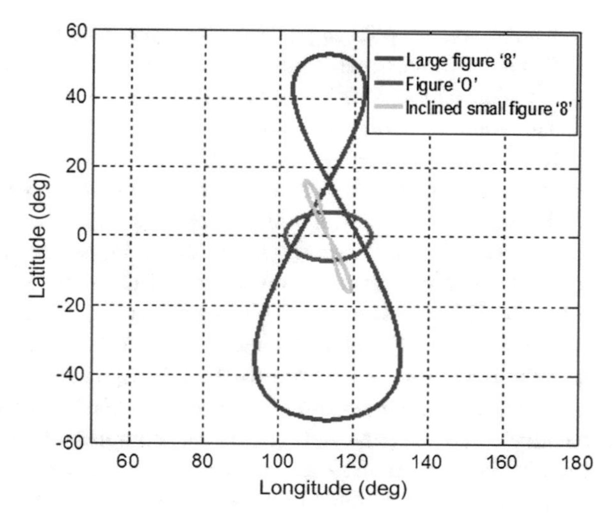

Fig. 5.15 Nadir tracks of the three GEO SAR orbits

[18:00, 24:00] as the time window and it can be adjusted based on the various applications practically.

We assume that the look angle of the designed GEO SAR is 2° (no ground squint angle). Based on (5.16), in Fig. 5.16, we can obtain the latitude changing of the IPPs of the beam center in GEO SAR (v_s) with respect to the local time in a case of t_p=0. We selected Φ_0 as [22.5°S, 22.5°N], marked as the pink rectangle. According to (5.13), the scope of v_s is determined by the inclination. Thus, because of the small inclinations, the IPPs of the beam center in the 'figure-O' nadir

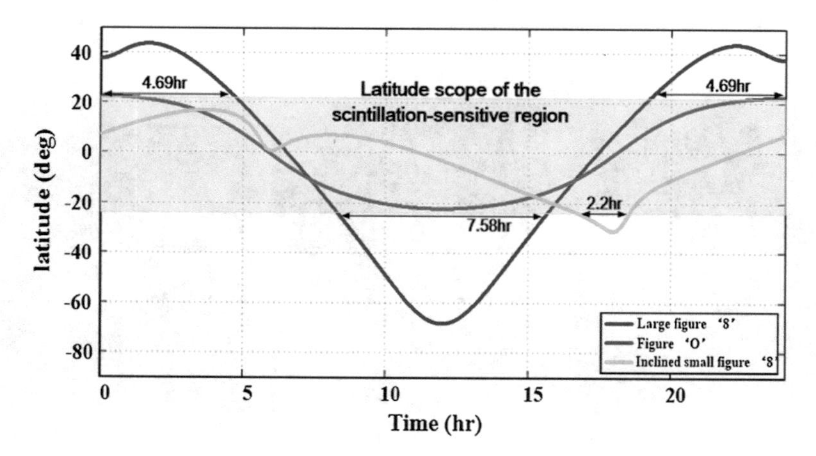

Fig. 5.16 Latitude changing of the IPPs of the beam center in GEO SAR (v_s) with respect to the local time ($t_p = 0$)

trajectory case keep falling into the scintillation-sensitive region and the IPPs of the beam center in the small 'figure-8' nadir trajectory case only has 2.2 h moving out of the scintillation-sensitive region [smaller than the duration of the time window (6 h)]. Thus, it is impossible to find the effective t_p for the two orbits to satisfy (5.18). When $t_p=0$, as the GEO SAR with the 'figure-8' nadir trajectory visits the scintillation-sensitive region at the time window, it does not have the capability to avoid ionospheric scintillation interference. Nevertheless, because the IPPs of the beam center in the large figure '8' nadir trajectory case have two separate time intervals of more than 6 h moving out of the scintillation-sensitive region, we can make the GEO SAR avoid the ionospheric scintillation interference by an optimal t_p. Based on (5.18) and assuming that t is in the local time form, we can determine its effective t_p sets, which are [0.19, 1.77 h] and [11.31, 14.68 h].

Practically, the interesting observation area of GEO SAR is a defined area (e.g. a country), which may not cover the complete latitude scope of the equator. Meanwhile, to improve the coverage and revisit performance, the GEO SAR platform is often designed to work in a wide range of look angle and ground squint angle. Moreover, complex impacts of orbit perturbation should be considered. Therefore, we should study the effective t_p sets in a more practical application.

We take the observation of China as our area of interest. Thus, given the geographical scope of China, we set the equatorial region with the geographical scope (0–22.5°N, 95–125°E) as the scintillation-sensitive region, which is marked as the red rectangles in Fig. 5.17. To form the beam-coverage area, the look angle of the designed GEO SAR ranges from 2° to 7° and the ground squint angle is limited to ±45°. STK software is utilized to conduct the simulation. The orbit start epoch is set at 4:00 Universal Time (UTC), October 20, 2015. High-Precision Orbit Propagator (HPOP) is used in the simulation, which concerns the impacts of the accurate perturbation of third-body gravity, atmospheric drag, and solar radiation pressure.

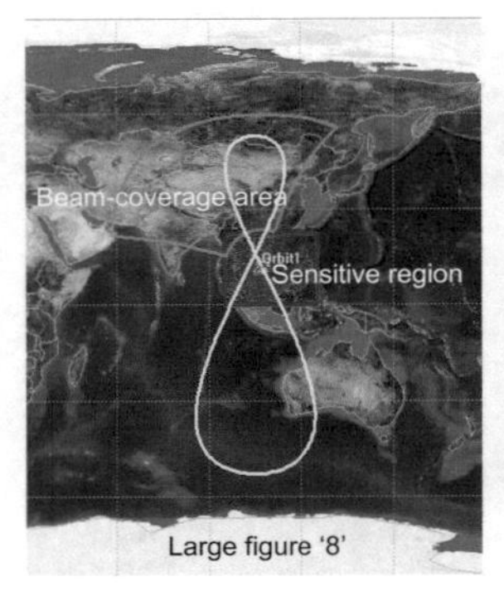

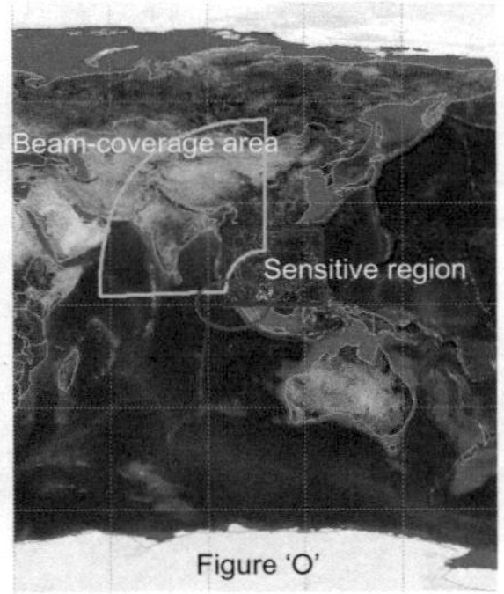

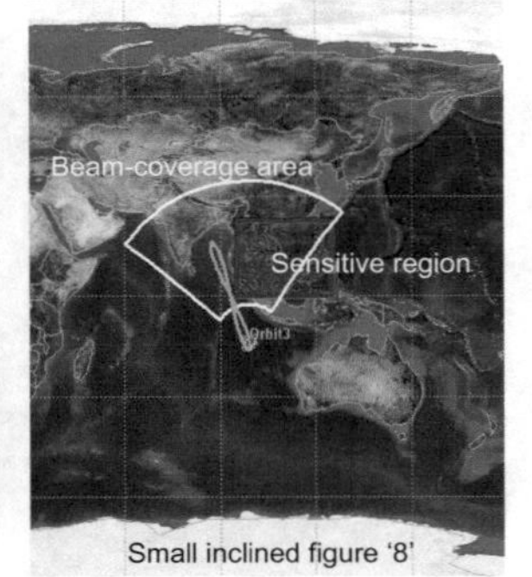

Fig. 5.17 Sketchmap of the three orbits, the scintillation-sensitive areas, and the beam-coverage areas

The visiting durations of the scintillation-sensitive region within the time window under different orbits in a year are given in Fig. 5.18. The boundaries of the initially effective t_p sets are marked as the light blue lines and the red lines after the GEO SAR gets into orbit. Thus, the initially effective t_p sets are [11.6, 15.7 h] ('figure-8'), [12.2, 18.7 h] ('figure-O'), and [9.5, 15.4 h] (small 'figure-8'). Selecting a t_p within the effective sets can make the GEO SAR satellite avoid visiting the scintillation-sensitive region within the time window after the satellite initially gets into orbit. In contrast, as the satellite spends more than 20,000 s visiting the scintillation-sensitive region within the time window, the designed GEO SAR orbit with the improper t_p (the dark blue line cases) will meet an ionospheric scintillation with a high probability after the satellite initially gets into orbit.

Moreover, because of the impacts of the perturbation, the optimized GEO SAR orbit cannot keep avoiding the impacts of the scintillation and will have the varied visiting duration of the scintillation-sensitive region within the time window under different orbits. Therefore, orbit station-keeping is needed to make the satellite have the stable capability to avoid the ionospheric scintillation after an initial orbit optimization. Comparing the red line with the light blue line, different t_p values

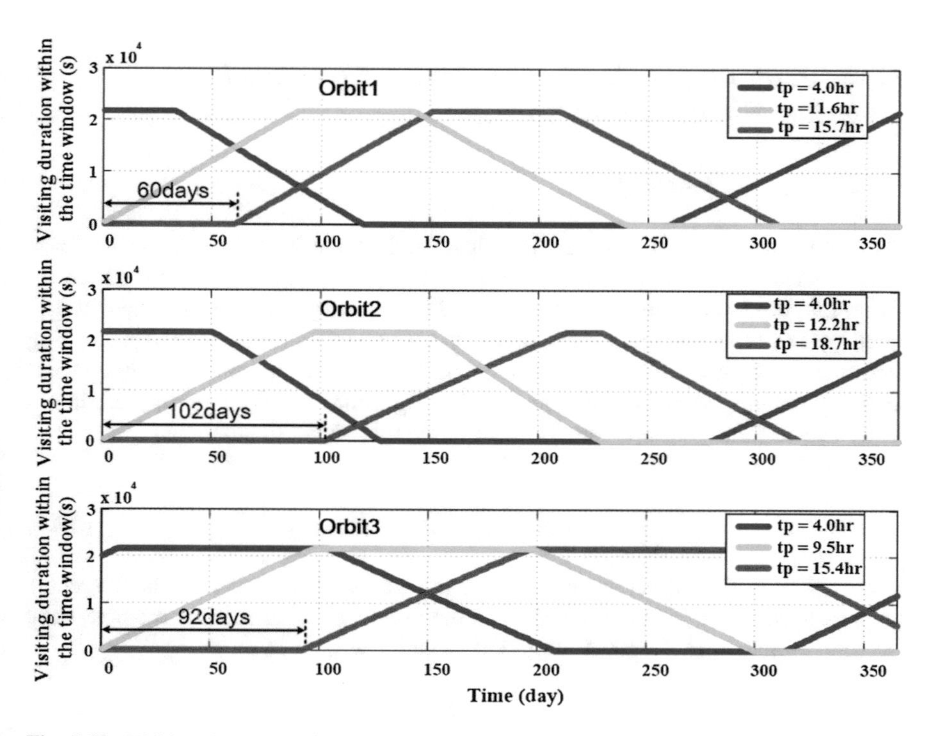

Fig. 5.18 Visiting durations of the scintillation-sensitive region within the time window under different orbits in a year (the light blue lines and the red lines are the boundary of the initially effective t_p sets after the GEO SAR gets into orbit)

correspond to different required time intervals for orbit station-keeping. Taking the orbit with the 'figure-O' nadir trajectory as an example, when the t_p is 18.7 h, orbit station-keeping is needed after 102 days, while orbit station-keeping should be conducted only after one day in the case of $t_p = 12.2$ h. Thus, we should select an optimal t_p with the longest time interval requirement for the orbit station-keeping to reduce the operation burden. By simulation, the designed optimal t_p values are 15.7 h for the orbit with the large figure '8' nadir trajectory, 18.7 h for the orbit with the 'figure-O' nadir trajectory, and 15.4 h for the orbit with the small 'figure-8' nadir trajectory. The orbit with the 'figure-O' nadir trajectory needs the orbit station-keeping after 102 days, which is the longest time interval for requiring the orbit station-keeping when compared with the other two cases (60 days in the 'figure-8' nadir trajectory case and 92 days in the small 'figure-8' nadir trajectory case). Thus, in view of fuel-saving, the orbit with the 'figure-O' nadir trajectory is more suitable for the orbit optimization in the three cases. Nevertheless, when compared with similar requirements in geostationary communication satellites (a few times a month), the orbit station-keeping requirements of the proposed three GEO SAR orbits are all feasible when they are operated in their orbit optimizations.

5.5.2 *Joint Amplitude-Phase Compensation Based on Minimum Entropy*

5.5.2.1 Signal Modelling

The ionospheric scintillation can principally affect the azimuthal pulse response function. This part will give the principles and implementation of compensation for signal amplitude and phase fluctuations [42]. The image is firstly segmented into small part to suppress the scintillation's space variance. Then, SPECAN imaging algorithm is adopted for each image segment because it is computationally efficient for small imaging scene. Further, an iterative algorithm based on entropy minimum is derived to jointly compensate the amplitude and phase fluctuations.

In GEO SAR, assume that the transmitted signal can be expressed as

$$s(\tau) = w(\tau)\exp\left\{j2\pi f_0\tau - j\pi K_r\tau^2\right\} \tag{5.19}$$

where $w(\tau)$ is the signal envelop, f_0 is the carrier frequency, K_r is the range frequency modulation (FM) rate and τ is the fast time. After the down conversion, the echo signal from Target A (Fig. 5.19) can be written as

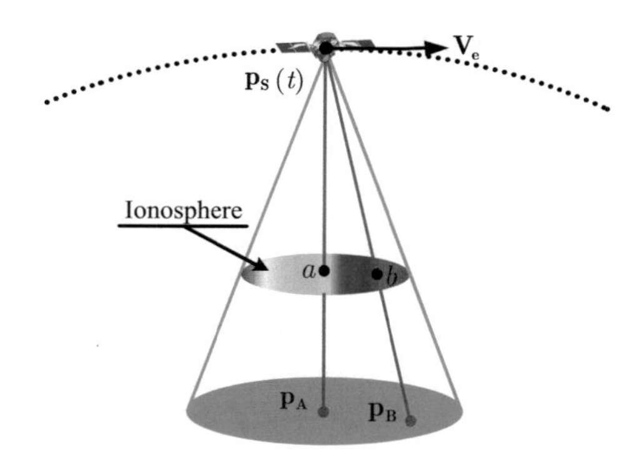

Fig. 5.19 Geometry of GEO SAR and ionospheric scintillation

$$\begin{aligned} s_R(t,\tau;A) &= \delta^A(t)e^{j\phi^A(t)}w\left(\tau - \frac{2|\mathbf{p_S}(t)-\mathbf{p_A}|}{c}\right) \\ &\times \exp\left\{-j\pi K_r\left(\tau - \frac{2|\mathbf{p_S}(t)-\mathbf{p_A}|}{c}\right)^2 - j\frac{4\pi}{\lambda}|\mathbf{p_S}(t)-\mathbf{p_A}|\right\} \end{aligned} \tag{5.20}$$

where t is the azimuth slow time and the beam center points at the imaging scene center while t is zero. The $|\mathbf{p_S}(t)-\mathbf{p_A}|$ is the slant range history of Target A, where $\mathbf{p_S}(t)$ and $\mathbf{p_A}$ are the position vectors of the satellite and Target A. In addition, $\delta^A(t)$ and $\phi^A(t)$ represent signal amplitude and phase fluctuations induced by scintillation, respectively. Then, after the range compression and range cell migration correction (RCMC), the following relationship can be obtained as

$$s_R(t,\tau;A) = \delta^A(t)e^{j\phi^A(t)}p\left(\tau - \frac{2R_A}{c}\right) \times \exp\left\{-j\frac{4\pi}{\lambda}|\mathbf{p_S}(t)-\mathbf{p_A}|\right\} \tag{5.21}$$

where $p(\cdot)$ is the pulse response function of range compression and R_A is the closest slant range from Target A to the satellite trajectory.

With respect to Target A and Target B, because the ionospheric puncture points (a and b in Fig. 5.19) of radar wave are different, the signal amplitude and phase fluctuations are also not identical. The coherence of echo signal between Target A and Target B can be measured by the generalized ambiguity function. The solution to the second moment of the generalize ambiguity function is the mutual correlation function, which can be expressed as

$$\Gamma(z,\rho,\zeta) = \left\langle \delta^A e^{j\phi^A} \cdot \delta^B e^{-j\phi^B} \right\rangle \tag{5.22}$$

where z and ζ represent the propagation path and propagation constant of radar wave, respectively. The denotation of $\langle\cdot\rangle$ represents the operator of mutual correlation function, which is the function of the distance ρ between the ionospheric puncture points a and b. The decorrelation distance ρ_0 is defined as

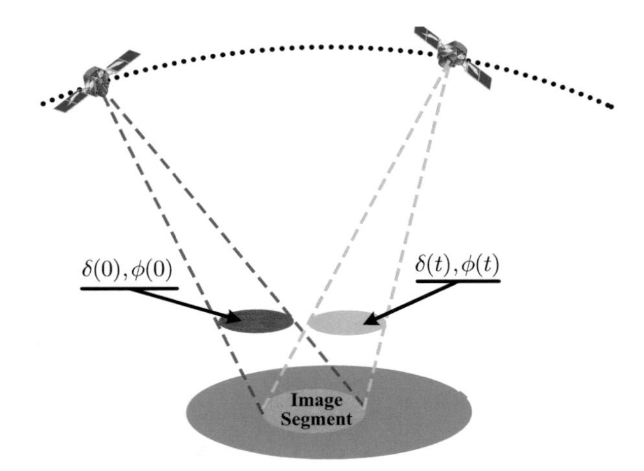

Fig. 5.20 Image segmentation

$$\frac{\Gamma\left(z,\rho_0,\zeta\right)}{\Gamma\left(z,\rho=0,\zeta\right)}=e^{-1} \tag{5.23}$$

In order to satisfy the autofocus assumption that all the scatterers in the imaging scene have the approximately identical signal fluctuations at each azimuth sampling time (as shown in Fig. 5.20), the SAR image segmentation must be performed. Thoroughly considering decorrelation distance analysis above and computational burden, the choice of several-kilometers is selected as an initial segmentation size in the SAR image correction. If the image quality after amplitude and phase compensations is not improved, the segmentation size will decrease further until the satisfied image is obtained.

As thus, the target echo in the same image segment can be written as

$$s_R(t,\tau)=\delta(t)e^{j\phi(t)}p\left(\tau-\frac{2R}{c}\right)\exp\left\{-j\frac{4\pi}{\lambda}|\mathbf{p}_\mathbf{S}(t)-\mathbf{p}|\right\} \tag{5.24}$$

5.5.2.2 Compensation Based on Minimum Entropy

In the real processing, the data is discrete and thus $s(t,\tau)$ is rewritten as $s(n,m)$ in the following statement, where the indices n and m refer to the azimuth and range sample respectively. Thus, (5.24) can be expressed as

$$\begin{aligned} s(k,m) &= \sum_{n=0}^{N-1}\delta_n e^{j\phi_n}s(n,m)\exp\left\{-\frac{2\pi}{N}kn\right\} \\ &= \sum_{n=0}^{N-1}s_{ap}(n,m)\exp\left\{-\frac{2\pi}{N}kn\right\} \end{aligned} \tag{5.25}$$

Based on (5.25), the compensation model can be formulated as

$$s_c(k,m) = \sum_{n=0}^{N-1} \alpha_n e^{j\psi_n} s_{ap}(n,m) \exp\left\{-\frac{2\pi}{N}kn\right\} \tag{5.26}$$

where α_n and ψ_n are used to compensate the signal amplitude and phase fluctuations induced by scintillation. The $s_c(k,m)$ represents the SAR image after amplitude and phase compensation while $s_{ap}(n,m)$ is the contaminated echo data in the azimuth time domain after range compression and azimuth SPECAN processing. Note that there is no any parametric model assumption about α_n and ψ_n.

In this chapter, image entropy is used to measure the SAR image quality. It is generally acknowledged that the better image quality is corresponding to the smaller image entropy. Consequently, the scintillation compensations can be considered to find a set of α_n and ψ_n by minimizing the image entropy, which can be expressed as

$$E = -\sum_{k=0}^{N-1}\sum_{m=0}^{M-1} \frac{|s_c(k,m)|^2}{\sum\limits_{k=0}^{N-1}\sum\limits_{m=0}^{M-1}|s_c(k,m)|^2} \ln \frac{|s_c(k,m)|^2}{\sum\limits_{k=0}^{N-1}\sum\limits_{m=0}^{M-1}|s_c(k,m)|^2} \tag{5.27}$$

Note that it is rather difficult to derive the analytical solution of the minimum of image entropy in (5.27). Next, an iterative method is proposed to find the minimum image entropy. Firstly, (5.27) is expanded to a Taylor series at $\alpha^i = \{\alpha_1^i, \alpha_2^i, \ldots, \alpha_{N-1}^i\}$ and $\psi^i = \{\psi_1^i, \psi_2^i, \ldots, \psi_{N-1}^i\}$, where the superscript refers to the ith iteration. The cubic and higher order items are ignored.

$$\begin{aligned} E &= E|_{\alpha^i,\psi^i} + E'_{\alpha_n}\Big|_{\alpha^i,\psi^i}(\alpha_n - \alpha_n^i) + \frac{1}{2}E''_{\alpha_n}\Big|_{\alpha^i,\psi^i}(\alpha_n - \alpha_n^i)^2 \\ E &= E|_{\alpha^i,\psi^i} + E'_{\psi_n}\Big|_{\alpha^i,\psi^i}(\psi_n - \psi_n^i) + \frac{1}{2}E''_{\psi_n}\Big|_{\alpha^i,\psi^i}(\psi_n - \psi_n^i)^2 \end{aligned} \tag{5.28}$$

The detailed derivations are quite cumbersome, so the derivatives in (5.28) are directly given as follows.

$$\begin{aligned} E'_{\alpha_n} = &\frac{2N\alpha_n \sum\limits_{m=0}^{M-1}\left|s_{ap}(n,m)\right|^2}{\left[\sum\limits_{k=0}^{N-1}\sum\limits_{m=0}^{M-1}|s_c(k,m)|^2\right]^2} \sum_{k=0}^{N-1}\sum_{m=0}^{M-1}|s_c(k,m)|^2 \ln|s_c(k,m)|^2 \\ &- \frac{2N\,\mathrm{Re}\left\{\sum\limits_{m=0}^{M-1} e^{j\psi_n} s_{ap}(n,m)\mathrm{IDFT}\left[s_c(k,m)\ln|s_c(k,m)|^2\right]_k^*\right\}}{\sum\limits_{k=0}^{N-1}\sum\limits_{m=0}^{M-1}|s_c(k,m)|^2} \end{aligned} \tag{5.29}$$

where IDFT[·] represents the inverse discrete Fourier transform.

Note that the computational burden can be significantly reduced by using FFT to replace discrete Fourier transform in (5.29)–(5.32).

$$
\begin{aligned}
E''_{\alpha_n} = & \left\{ \frac{2N\sum\limits_{m=0}^{M-1}\left|s_{ap}(n,m)\right|^2}{\left[\sum\limits_{k=0}^{N-1}\sum\limits_{m=0}^{M-1}|s_c(k,m)|^2\right]^2} - \frac{8\left[N\alpha_n\sum\limits_{m=0}^{M-1}\left|s_{ap}(n,m)\right|^2\right]^2}{\left[\sum\limits_{k=0}^{N-1}\sum\limits_{m=0}^{M-1}|s_c(k,m)|^2\right]^3} \right\} \\
& \times \sum_{k=0}^{N-1}\sum_{m=0}^{M-1}|s_c(k,m)|^2\ln|s_c(k,m)|^2 + \frac{8N^2\alpha_n\sum\limits_{m=0}^{M-1}\left|s_{ap}(n,m)\right|^2}{\left[\sum\limits_{k=0}^{N-1}\sum\limits_{m=0}^{M-1}|s_c(k,m)|^2\right]^2} \\
& \times \mathrm{Re}\left\{\sum_{m=0}^{M-1} e^{j\psi_n}s_{ap}(n,m)\cdot \mathrm{IDFT}\left[s_c(k,m)\ln|s_c(k,m)|^2\right]_k^*\right\} \\
& + \frac{\left[2N\alpha_n\sum\limits_{m=0}^{M-1}\left|s_{ap}(n,m)\right|^2\right]^2}{\left[\sum\limits_{k=0}^{N-1}\sum\limits_{m=0}^{M-1}|s_c(k,m)|^2\right]^2} - \frac{2\sum\limits_{m=0}^{M-1}\left|s_{ap}(n,m)\right|^2 + 2\sum\limits_{k=0}^{N-1}\sum\limits_{m=0}^{M-1}\left|s_{ap}(n,m)\right|^2\ln|s_c(k,m)|^2}{\sum\limits_{k=0}^{N-1}\sum\limits_{m=0}^{M-1}|s_c(k,m)|^2} \\
& - \frac{2\mathrm{Re}\left\{\sum\limits_{k=0}^{N-1}\sum\limits_{m=0}^{M-1} e^{j\psi_n}s_{ap}(n,m)^2\frac{2s_c(k,m)^*}{s_c(k,m)}\exp\left\{-j\frac{4\pi}{N}kn\right\}\right\}}{\sum\limits_{k=0}^{N-1}\sum\limits_{m=0}^{M-1}|s_c(k,m)|^2}
\end{aligned}
\tag{5.30}
$$

$$
\begin{aligned}
E'_{\psi_n} = & \frac{2N}{\sum\limits_{k=0}^{N-1}\sum\limits_{m=0}^{M-1}|s_c(k,m)|^2}\cdot \\
& \mathrm{Im}\left\{\sum_{m=0}^{M-1}\alpha_n e^{j\psi_n}s_{ap}(n,m)\cdot \mathrm{IDFT}\left[s_c(k,m)\ln|s_c(k,m)|^2\right]_k^*\right\}
\end{aligned}
\tag{5.31}
$$

$$
\begin{aligned}
E''_{\psi_n} = & \frac{2}{\sum\limits_{k=0}^{N-1}\sum\limits_{m=0}^{M-1}|s_c(k,m)|^2} \\
& \cdot \left\{ \begin{array}{l} N\mathrm{Re}\left\{\sum\limits_{m=0}^{M-1}\alpha_n s_{ap}(n,m)e^{j\psi_n}\cdot \mathrm{IDFT}\left[s_c(k,m)\ln|s_c(k,m)|^2\right]_k^*\right\} \\ +\mathrm{Re}\left\{\sum\limits_{m=0}^{M-1}\left[\alpha_n s_{ap}(n,m)e^{j\psi_n}\right]^2\cdot\sum\limits_{k=0}^{N-1}\frac{s_c^*(k,m)}{s_c(k,m)}\exp\left\{-j\frac{4\pi}{N}kn\right\}\right\} \\ -\sum\limits_{m=0}^{M-1}\left|\alpha_n s_{ap}(n,m)\right|^2\sum\limits_{k=0}^{N-1}\ln|s_c(k,m)|^2 - \sum\limits_{m=0}^{M-1}\left|\alpha_n s_{ap}(n,m)\right|^2 \end{array} \right\}
\end{aligned}
\tag{5.32}
$$

Subsequently, the minimums of (5.28) can be derived

$$\alpha_n = \alpha_n^i - \frac{E'_{\alpha_n}\big|_{\alpha^i,\psi^i}}{E''_{\alpha_n}\big|_{\alpha^i,\psi^i}}, \ \psi_n = \psi_n^i - \frac{E'_{\psi_n}\big|_{\alpha^i,\psi^i}}{E''_{\psi_n}\big|_{\alpha^i,\psi^i}} \tag{5.33}$$

The amplitude and phase corrections in $i + 1$th iteration will be updated using (5.33), namely

$$\begin{bmatrix} \alpha_1^{i+1} \\ \alpha_2^{i+1} \\ \vdots \\ \alpha_{N-1}^{i+1} \\ \psi_1^{i+1} \\ \psi_2^{i+1} \\ \vdots \\ \psi_{N-1}^{i+1} \end{bmatrix} = \begin{bmatrix} \alpha_1^{i} \\ \alpha_2^{i} \\ \vdots \\ \alpha_{N-1}^{i} \\ \psi_1^{i} \\ \psi_2^{i} \\ \vdots \\ \psi_{N-1}^{i} \end{bmatrix} - \begin{bmatrix} E'_{\alpha_1}/E''_{\alpha_1} \\ E'_{\alpha 2}/E''_{\alpha_2} \\ \vdots \\ E'_{\alpha_{N-1}}/E''_{\alpha_{N-1}} \\ E'_{\psi_1}/E''_{\psi_1} \\ E'_{\psi_2}/E''_{\psi_2} \\ \vdots \\ E'_{\psi_{N-1}}/E''_{\psi_{N-1}} \end{bmatrix}_{\alpha^i,\psi^i} \tag{5.34}$$

This iteration principle can be considered that a series of local quadratic curves are constructed to gradually approach the extreme of the objective function of (5.27). Note that the second derivative should be positive so as to make sure the iteration could converge toward the minimum.

5.5.2.3 Estimation Accuracy Analysis

Firstly, the simulation is used to illustrate the focusing performance under entropy minimum criterion. An ideal rectangular window without amplitude and phase fluctuations is adopted as the input of the proposed method. The amplitude and phase estimates are expected to the constant and zero, respectively. By applying our proposed autofocus method to the ideal rectangular window data, the amplitude and phase estimates are shown in Fig. 5.21. It can be seen that the phase estimate is zero, but the amplitude estimate does not keep constant. Namely, the optimal amplitude profile under entropy minimum criterion is some weighed window function W_n, rather than the rectangular window. Consequently, after compensation based on entropy minimum, the final amplitude profile will become this weighed window function W_n. If the amplitude error inducted by scintillation is denoted as δ_n and the amplitude compensation is denoted as α_n, the final amplitude profile can be denoted as $\delta_n \cdot \alpha_n$, which is equivalent to this optimal amplitude profile W_n under entropy minimum criterion. The impulse response of this weighed window is shown in Fig. 5.22. Compared with the sinc function, the side lobes is below −20 dB but the width of main lobe is spread slightly.

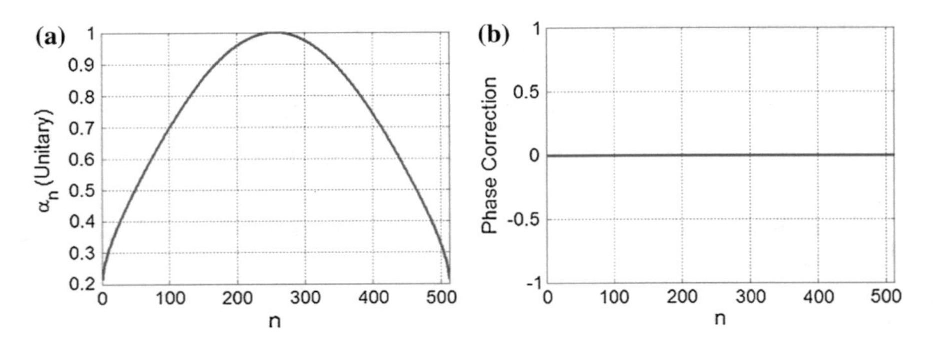

Fig. 5.21 Amplitude and phase corrections for the ideal rectangular window using the proposed method

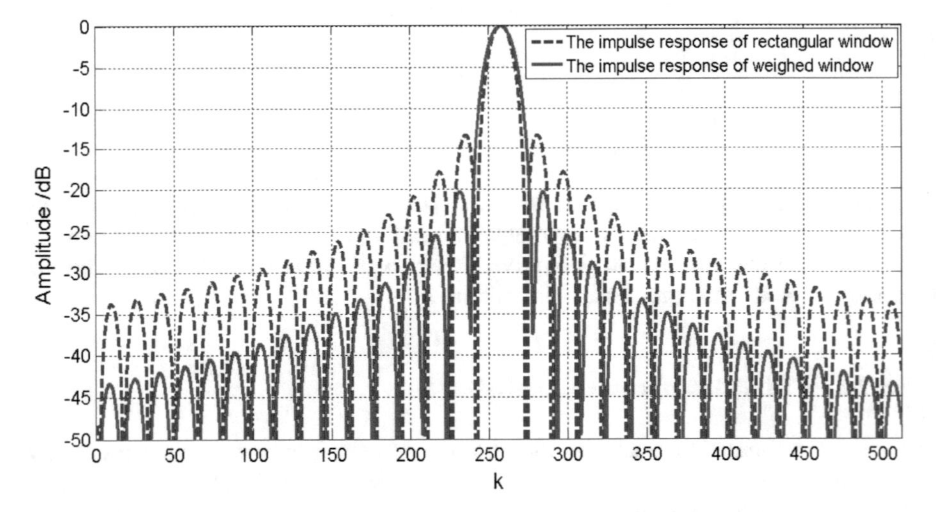

Fig. 5.22 Impulse response of the rectangular window and the weighed window under entropy minimum criterion

Next, the estimation accuracy of the proposed method will be analyzed. The echo data with different number of point scatterers and range cells are simulated. Without losing generality, the magnitudes of these point targets are of the Rayleigh distribution with the scale parameter of one while the phases are uniformly distributed between $-\pi$ and π. The amplitude and phase fluctuations due to scintillation are assumed to obey Nakagami and Gaussian distributions, respectively. S_4 and σ_ϕ are both set as 0.4. The total sample time is 100 s with the sample rate of 100 Hz. The complex white Gaussian noise is added to adjust the signal noise ratio (SNR). Monte Carlo simulations are performed so as to reveal more accurate curves. Figure 5.23 shows the rms estimate errors of amplitude and phase

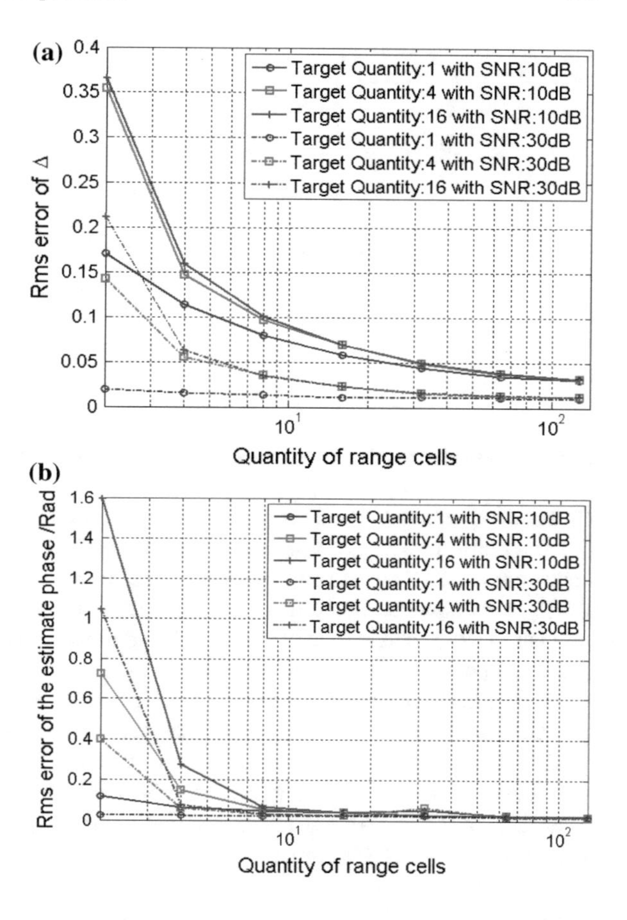

Fig. 5.23 Estimation accuracy versus the number of range cells and point scatterers in each range cell for **a** amplitude; **b** phase

fluctuations. The weighed window is removed in the rms calculation of the amplitude estimate (as shown in Fig. 5.24), which is done by an 8-order polynomial fitting as follows.

$$\Delta_n = \delta_n \alpha_n - \text{polyfit}\{\delta_n \alpha_n, 8\} \tag{5.35}$$

From Fig. 5.23, the estimation accuracies of amplitude and phase fluctuations are directly proportional to the number of range cell and inversely proportional to target quantity in each range cell. It implies that the isolated dominant scatterer is helpful to improve the estimation accuracy. However, the same accuracy can also be achieved by increasing the number of range cells in the estimation. Therefore, our proposed method is still valid in case of no isolated dominant scatterer. In addition, note that the amplitude estimation is more sensitive to SNR.

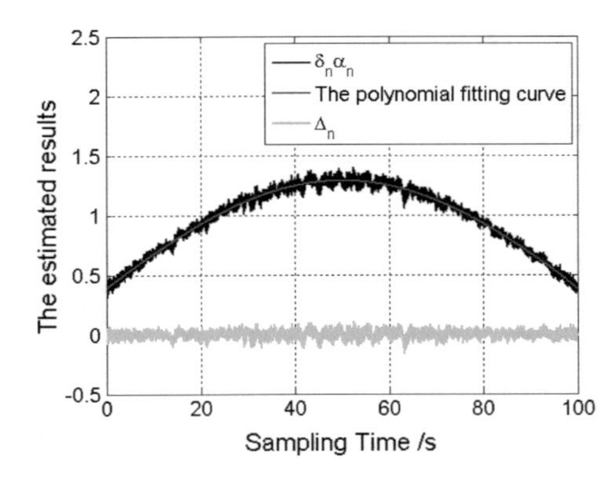

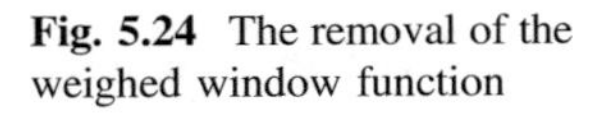
Fig. 5.24 The removal of the weighed window function

5.5.2.4 Space-Variance Compensation Validation

In the compensation validation, we use the phase screen technique to generate the 2D (azimuth and range) amplitude and phase fluctuations induced by ionosphere scintillation. The parameters of phase screen are set to be consistent with that in [20]. The magnetic heading of satellite motion is set to −6.5°. The turbulence intensity (*CkL*), phase spectral index (*p*) and outer scale (L_0) are 17.5×10^{33}, 9 and 5 km, respectively in our simulation.

A real SAR scene of 3.5 km (Ground Range) × 2.9 km (Azimuth) is used to generate SAR echo data using our developed simulation software regarding L-band GEO SAR parameters. The semi-major axis of satellite ellipsoid orbit is 42,164 km with the eccentricity of 0.07 while the orbit inclination is 60°. The local time of the ascending node (LTAN) is about 20:35:10 and the geographic coordinate of simulated position is 3.8°S and 135.2°E. The transmitted signal band is 18 MHz and the incidence angle is around 30°. For the central target in the scene, the frequency-modulation rate is about 0.62 Hz/s and the integration time is 100 s. Consequently, the azimuth signal bandwidth is about 62 Hz while PRF is 100 Hz. Because the satellite's equivalent velocity is 1.32 km/s, the azimuth resolution of the ideal SAR images is about 21 m.

Because the irregularities are aligned to the local geomagnetic field, it is found that space-variance in the range direction is severer and cannot be neglected. Consequently, the segmentation is further performed and the segmentation length is 350 m along the range direction. The imaging results are shown in Fig. 5.25. It can be clearly seen from Fig. 5.25d that the autofocus result using our proposed method has significant improvement. The result using our proposed method without range segmentation is also shown in Fig. 5.25c and the increase in ISLR can be still observed. Therefore, the segmentation is necessary and effective to deal with the space-variance problem of ionospheric scintillation.

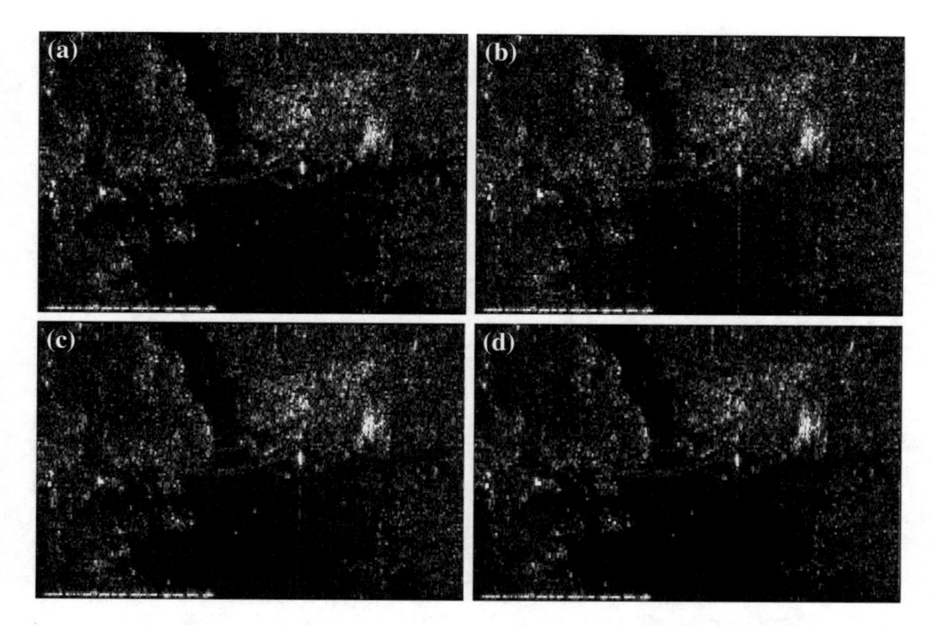

Fig. 5.25 GEO SAR imaging results: **a** the ideal SAR image; **b** the degraded image by amplitude and phase fluctuations; **c** the autofocus image using our proposed method without range segmentation; **d** the autofocus image using our proposed method with range segmentation

5.5.2.5 Experimental Validation Based on GPS-Derived Scintillation Signal

The simulations using this real GPS-derived scintillation signal are used to show the autofocus performance. A real SAR scene of 3.5 km (Ground Range) × 2.9 km (Azimuth) is firstly simulated and the derived scintillation signal is introduced into SAR image along the azimuth direction. The SAR echo data is generated according to L-band GEO SAR parameters [5]. The ideal SAR image is shown in Fig. 5.26a with the image entropy of 11.22. The degraded SAR images by amplitude and phase fluctuations are shown in Fig. 5.26b–d. It can be seen that either amplitude fluctuation or phase fluctuation could contribute to the increase in ISLR, but the effect of phase fluctuation is more severe relatively.

The autofocus image with only phase compensation using the method in [40] is shown in Fig. 5.26e, so as to illustrate the necessity of amplitude compensation. Its final image entropy is 11.46. In addition, PGA in [37] cannot be adopted because it is quite difficult to pick up the isolated dominant scatterers in Fig. 5.26d. The autofocus SAR image in Fig. 5.26f reveals the good performance of our proposed method and the image quality has a significant improvement with the image entropy from 11.96 to 11.18 (Fig. 5.27). To further evaluate the estimation accuracy of the proposed method, the estimates of signal amplitude and phase fluctuations are shown in Fig. 5.28. The rms estimate errors of residual amplitude and phase

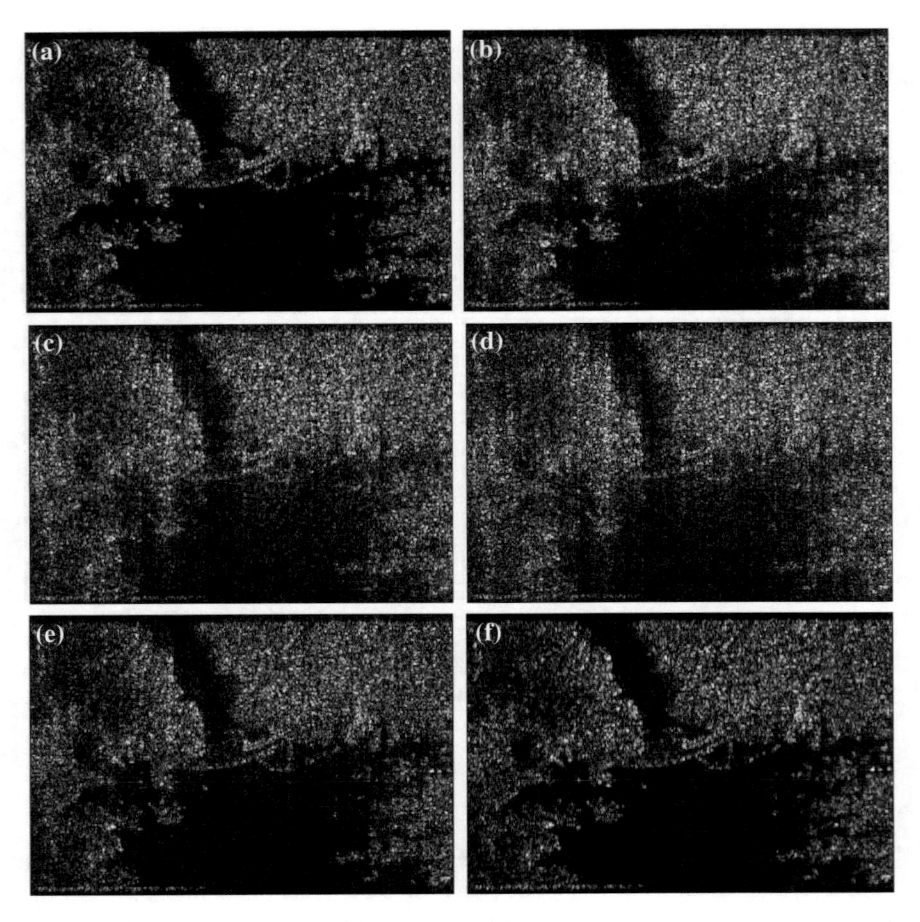

Fig. 5.26 GEO SAR imaging results: **a** the ideal SAR image; **b** the degraded image by amplitude fluctuation; **c** the degraded image by phase fluctuation; **d** the degraded image by amplitude and phase fluctuations; **e** the autofocus image with phase compensation using the method in [40]; **f** the autofocus image with amplitude and phase compensation using our method

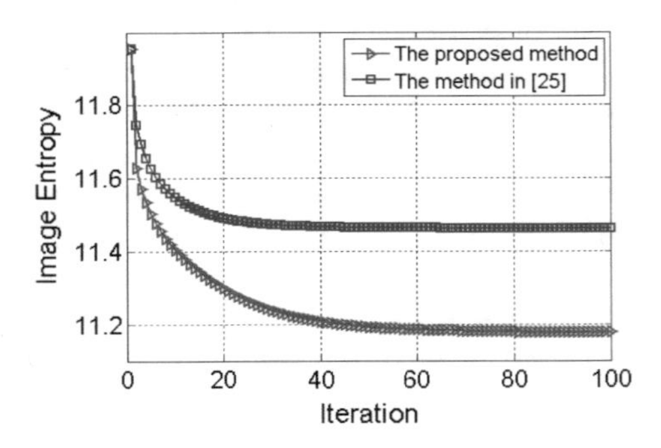

Fig. 5.27 Image entropy per iteration using the proposed jointly amplitude and phase compensation method

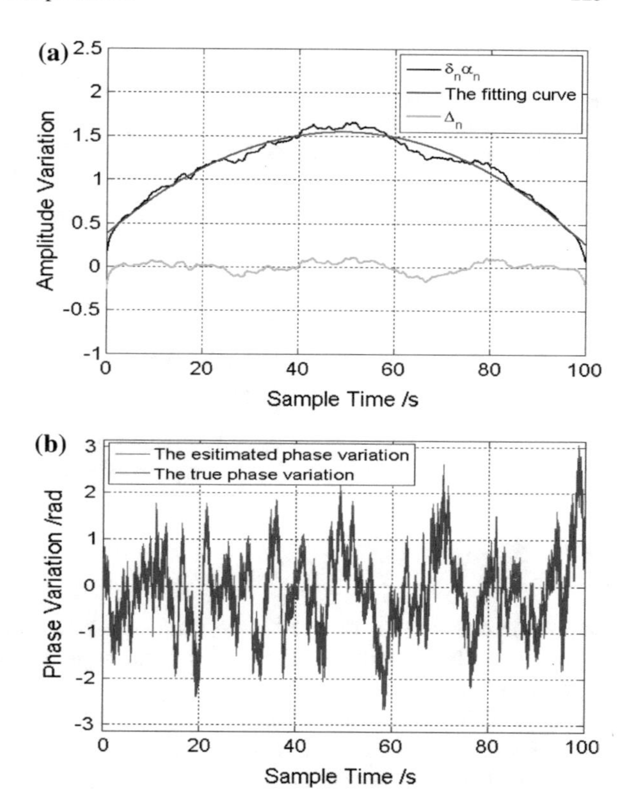

Fig. 5.28 The estimate results: **a** amplitude fluctuation; **b** phase fluctuation

fluctuations are about 0.05 and 0.08, respectively. Therefore, the proposed method has the satisfactory performance on estimation accuracy.

In addition, to show the effects of the number of range and azimuth cells on autofocus performance, the simulations are performed by applying different segmentations for SAR image scene. The original simulated scene size is 3.5 km 2.9 km. The segmentation along the range direction are done as 100, 150, 350, 500, 700, 1750, 3500 m, respectively, while it is done along the azimuth direction as 1000, 1450, 2900 m. By applying our proposed autofocus method, the rms of residual phase estimation errors at different segmentation sizes is shown in Fig. 5.29. It can be clearly seen that the phase estimate accuracy is improved with the increase in segmentation size. This is consistent with the simulation analysis in Fig. 5.23 that the better estimation accuracy could be achieved by applying more sample cells. The smallest rms of phase estimation error is about 0.08 rad regarding to maximal image segmentation of 3.5 km $\times$ 2.9 km. Therefore, the image segmentation should be as large as possible to achieve better estimation under the constraint of space-variance.

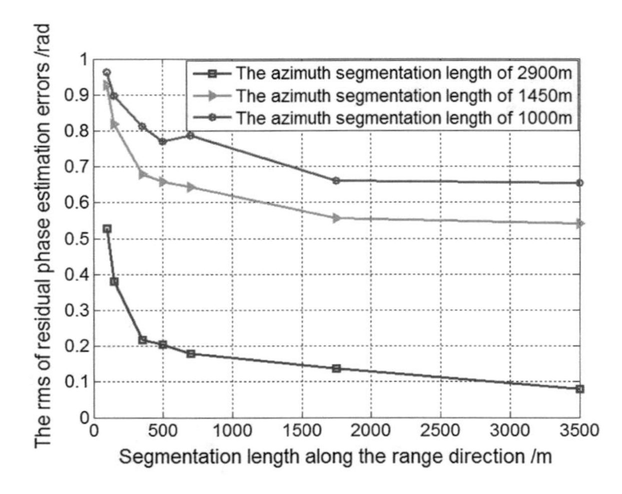

Fig. 5.29 The rms of residual phase estimation errors by applying our autofocus method on different segmentation sizes

5.6 Summary

Ionosphere is one of the most significant considerations in the L-band GEO SAR operation. The experiment study is a great supplementary and also a validation approach to the current theoretical analysis of the ionospheric impacts on GEO SAR. This chapter demonstrates an equivalent validation experiment based on the GPS signals. The experiment consists of the data acquisition of the GPS signals, the equivalent pre-processing, the GEO SAR signal modelling, the image focusing and evaluating. The key steps are the equivalent transformation of the data acquisition geometry and the sampling time to the same temporal-spatial frame of GEO SAR, in order to eliminate the existing difference in the orbit trajectories and IPPs between GEO SAR and GPS satellites. The phase errors induced by ionosphere in GEO SAR can be generated equivalently from the GPS data and then incorporated into the GEO SAR signals. The experiment principles and the design procedures can be adopted in other similar validation experiment of the satellite missions in the development stage.

In cases of the current GEO SAR system with the high inclination, the existence of background ionosphere will induce image drift, but it can be corrected through image registration techniques. The range defocusing is not considered in the current GEO SAR configuration with a relatively small bandwidth. But for the azimuthal focusing, the second and higher derivatives of TEC will result in defocusing and the thresholds can be determined by the specific GEO SAR geometry and the integration time. The classical PGA method can be employed to compensate the defocusing caused by the temporal-spatial variant background ionosphere. In comparison, ionospheric scintillation will induce amplitude and phase random fluctuations. These sorts of random signal errors can lead to the increase in ISLR. In most serious case the image will be defocused. The ionospheric scintillation can be avoided by orbit optimization on the basis of the diurnal and geographical pattern of the ionospheric scintillation occurrence. The proper time past perigee of GEO SAR can be tuned so that the imaging with the IPPs falling into the scintillation-sensitive region during the time window (early evening after sunset to midnight) will be avoided. If unfortunately interfered, the errors induced by ionospheric scintillation can also be compensated by an iterative method based on entropy minimum. In the compensation, the image should be first segmented based on the decorrelation distance, and then the signal amplitude and phase fluctuations can be jointly estimated.

References

1. Edelstein WN, Madsen SN, Moussessian A, Chen C (2005) Concepts and technologies for synthetic aperture radar from MEO and geosynchronous orbits. In 2005, pp 195–203
2. Moussessian A, Chen C, Edelstein W, Madsen S, Rosen P (2005) System concepts and technologies for high orbit SAR. In: IEEE MTT-S international microwave symposium digest, 2005, 12–17 June 2005. p 4. https://doi.org/10.1109/mwsym.2005.1517017
3. Teng L, Xichao D, Cheng H, Zeng T (2011) A new method of zero-doppler centroid control in GEO SAR. IEEE Geosci Remote Sens Lett 8(3):512–516. https://doi.org/10.1109/LGRS.2010.2089969
4. Hobbs S, Mitchell C, Forte B, Holley R, Snapir B, Whittaker P (2014) System design for geosynchronous synthetic aperture radar missions. IEEE Trans Geosci Remote Sens 52 (12):7750–7763. https://doi.org/10.1109/TGRS.2014.2318171
5. Cheng H, Teng L, Tao Z, Feifeng L, Zhipeng L (2011) The accurate focusing and resolution analysis method in geosynchronous SAR. IEEE Trans Geosci Remote Sens 49(10):3548–3563. https://doi.org/10.1109/TGRS.2011.2160402
6. Cheng H, Zhipeng L, Teng L (2012) An improved CS algorithm based on the curved trajectory in geosynchronous SAR. IEEE J Sel Top Appl Earth Observ Remote Sens 5 (3):795–808. https://doi.org/10.1109/JSTARS.2012.2188096
7. Cheng H, Teng L, Zhipeng L, Tao Z, Ye T (2014) An improved frequency domain focusing method in geosynchronous SAR. IEEE Trans Geosci Remote Sens 52(9):5514–5528. https://doi.org/10.1109/TGRS.2013.2290133
8. Hu C, Tian Y, Zeng T, Long T, Dong X (2016) Adaptive secondary range compression algorithm in geosynchronous SAR. IEEE J Sel Top Appl Earth Observ Remote Sens 9(4): 1397–1413. https://doi.org/10.1109/JSTARS.2015.2477317
9. Madsen SN, Chen C, Edelstein W (2002) Radar options for global earthquake monitoring. In: Geoscience and remote sensing symposium, 2002. IGARSS'02. 2002 IEEE International, 2002, vol 1483, pp 1483–1485. https://doi.org/10.1109/igarss.2002.1026156
10. Cheng H, Xiaorui L, Teng L, Yangte G (2013) GEO SAR interferometry: theory and feasibility study. In: Radar conference 2013, IET international, 14–16 April 2013, pp 1–5. https://doi.org/10.1049/cp.2013.0185
11. Ruiz Rodon J, Broquetas A, Monti Guarnieri A, Rocca F (2013) Geosynchronous SAR focusing with atmospheric phase screen retrieval and compensation. IEEE Trans Geosci Remote Sens 51(8):4397–4404. https://doi.org/10.1109/TGRS.2013.2242202
12. Tao Z, Wei Y, Zegang D, Teng L (2014) Geo-location error analysis in geosynchronous SAR. Electron Lett 50(23):1741–1743. https://doi.org/10.1049/el.2014.2542
13. Prati C, Rocca F, Giancola D, Guarnieri AM (1998) Passive geosynchronous SAR system reusing backscattered digital audio broadcasting signals. IEEE Trans Geosci Remote Sens 36(6):1973–1976. https://doi.org/10.1109/36.729370

14. Bruno D, Hobbs SE (2010) Radar imaging from geosynchronous orbit: temporal decorrelation aspects. IEEE Trans Geosci Remote Sens 48(7):2924–2929. https://doi.org/10.1109/TGRS.2010.2042062
15. Hobbs S, Guarnieri AM, Wadge G, Schulz D (2014) GeoSTARe initial mission design. In: Geoscience and remote sensing symposium (IGARSS), 2014 IEEE international, 13–18 July 2014, pp 92–95. https://doi.org/10.1109/igarss.2014.6946363
16. Ruiz-Rodon J, Broquetas A, Makhoul E, Monti Guarnieri A, Rocca F (2014) Nearly zero inclination geosynchronous SAR mission analysis with long integration time for earth observation. IEEE Trans Geosci Remote Sens 52(10):6379–6391. https://doi.org/10.1109/TGRS.2013.2296357
17. Monti Guarnieri A, Broquetas A, Recchia A, Rocca F, Ruiz-Rodon J (2015) Advanced radar geosynchronous observation system: ARGOS. IEEE Geosci Remote Sens Lett 12(7):1406–1410. https://doi.org/10.1109/LGRS.2015.2404214
18. Gray AL, Mattar KE, Sofko G (2000) Influence of ionospheric electron density fluctuations on satellite radar interferometry. Geophys Res Lett 27(10):1451–1454. https://doi.org/10.1029/2000GL000016
19. Pi X, Freeman A, Chapman B, Rosen P, Li Z (2011) Imaging ionospheric inhomogeneities using spaceborne synthetic aperture radar. J Geophys Res Space Phys 116(A04303):1–13. https://doi.org/10.1029/2010JA016267
20. Carrano CS, Groves KM, Caton RG (2012) Simulating the impacts of ionospheric scintillation on L band SAR image formation. Radio Sci 47(4):1–12. https://doi.org/10.1029/2011RS004956
21. Shimada M, Muraki Y, Otsuka Y (2008) Discovery of anoumoulous stripes over the Amazon by the PALSAR onboard ALOS satellite. In: Geoscience and remote sensing symposium, 2008. IGARSS 2008. IEEE International, 7–11 July 2008, pp II-387–II-390. https://doi.org/10.1109/igarss.2008.4779009
22. Fremouw EJ, Secan JA (1984) Modeling and scientific application of scintillation results. Radio Sci 19(3):687–694. https://doi.org/10.1029/RS019i003p00687
23. Secan JA, Bussey RM, Fremouw EJ, Basu S (1995) An improved model of equatorial scintillation. Radio Sci 30(3):607–617. https://doi.org/10.1029/94RS03172
24. Secan JA, Bussey RM, Fremouw EJ, Basu S (1997) High-latitude upgrade to the Wideband ionospheric scintillation model. Radio Sci 32(4):1567–1574. https://doi.org/10.1029/97RS00453
25. Béniguel Y (2002) Global ionospheric propagation model (GIM): a propagation model for scintillations of transmitted signals. Radio Sci 37(3):4-1–4-13. https://doi.org/10.1029/2000rs002393
26. Béniguel Y, Hamel P (2011) A global ionosphere scintillation propagation model for equatorial regions. J Space Weather Space Clim 1(1):A04
27. Xu ZW, Wu J, Wu ZS (2008) Potential effects of the ionosphere on space-based SAR imaging. IEEE Trans Antennas Propag 56(7):1968–1975. https://doi.org/10.1109/TAP.2008.924695
28. Wang C, Zhang M, Xu Z-W, Zhao H-S (2014) TEC retrieval from spaceborne SAR data and its applications. J Geophy Res Space Phys 119(10):8648–8659. https://doi.org/10.1002/2014JA020078
29. Wang C, Zhang M, Xu ZW, Chen C, Guo LX (2015) Cubic phase distortion and irregular degradation on SAR imaging due to the ionosphere. IEEE Trans Geosci Remote Sens 53(6):3442–3451. https://doi.org/10.1109/TGRS.2014.2376957
30. Meyer F, Bamler R, Jakowski N, Fritz T (2006) The potential of low-frequency SAR systems for mapping ionospheric TEC distributions. IEEE Geosci Remote Sens Lett 3(4):560–564. https://doi.org/10.1109/LGRS.2006.882148
31. Meyer FJ (2011) Performance requirements for ionospheric correction of low-frequency SAR data. IEEE Trans Geosci Remote Sens 49(10):3694–3702. https://doi.org/10.1109/TGRS.2011.2146786

32. Rosen PA, Hensley S, Zebker HA, Webb FH, Fielding EJ (1996) Surface deformation and coherence measurements of Kilauea Volcano, Hawaii, from SIR-C radar interferometry. J Geophys Res Planets 101(E10):23109–23125. https://doi.org/10.1029/96JE01459
33. Rogers NC, Quegan S, Kim JS, Papathanassiou KP (2014) Impacts of ionospheric scintillation on the BIOMASS P-band satellite SAR. IEEE Trans Geosci Remote Sens 52(3):1856–1868. https://doi.org/10.1109/TGRS.2013.2255880
34. Rogers NC, Shaun Q (2014) The accuracy of faraday rotation estimation in satellite synthetic aperture radar images. IEEE Trans Geosci Remote Sens 52(8):4799–4807. https://doi.org/10.1109/TGRS.2013.2284635
35. Quegan S, Lomas MR (2015) The interaction between Faraday rotation and system effects in synthetic aperture radar measurements of backscatter and biomass. IEEE Trans Geosci Remote Sens 53(8):4299–4312. https://doi.org/10.1109/TGRS.2015.2395138
36. Ye T, Cheng H, Xichao D, Tao Z, Teng L, Kuan L, Xinyu Z (2015) Theoretical analysis and verification of time variation of background ionosphere on geosynchronous SAR imaging. IEEE Geosci Remote Sens Lett 12(4):721–725. https://doi.org/10.1109/LGRS.2014.2360235
37. Dong X, Hu C, Tian Y, Tian W, Li Y, Long T (2016) Experimental study of ionospheric impacts on geosynchronous SAR using GPS signals. IEEE J Sel Top Appl Earth Obs Remote Sens 9(6):2171–2183
38. Wahl DE, Eichel PH, Ghiglia DC, Jakowatz CV (1994) Phase gradient autofocus-a robust tool for high resolution SAR phase correction. IEEE Trans Aerosp Electron Syst 30(3):827–835. https://doi.org/10.1109/7.303752
39. Chatterjee S, Chakraborty SK (2013) Variability of ionospheric scintillation near the equatorial anomaly crest of the Indian zone. Ann Geophys (ANGEO). https://doi.org/10.5194/angeo-31-697-2013
40. Priyadarshi S (2015) A review of ionospheric scintillation models. Surv Geophys 36(2):295–324. https://doi.org/10.1007/s10712-015-9319-1
41. Hu C, Li Y, Dong X, Ao D (2016) Avoiding the ionospheric scintillation interference on geosynchronous SAR by orbit optimization. IEEE Geosci Remote Sens Lett 13(11):1676–1680
42. Wang R, Hu C, Li Y, Hobbs SE, Tian W, Dong X, Chen L (2017) Joint Amplitude-phase compensation for ionospheric scintillation in GEO SAR imaging. IEEE Trans Geosci Remote Sens 55(6):3454–3465
43. Kragh TJ, Kharbouch AA (2006) Monotonic iterative algorithm for minimum-entropy autofocus. In: Adaptive sensor array processing (ASAP) workshop, (June 2006)

Chapter 6
Geosynchronous InSAR and D-InSAR

Abstract Geosynchronous SAR has almost the same trajectory during its orbit period which makes it suitable for interferometric SAR and differential interferometric SAR (InSAR and D-InSAR) processing. But the very large orbit height, which is about 60 times larger than that in a low Earth orbit SAR, will cause lots of special issues in the repeated-track InSAR and D-InSAR. In this chapter, we firstly analyze the effects of the un-parallel repeated tracks and the squint-looking working mode introduced by the orbital perturbations and the Earth's rotation in the repeat-track GEO InSAR system. Then, a novel data acquisition method is presented based on a criterion of optimal minimal rotational-induced decorrelation (OMRD). It can significantly improve the coherence of the InSAR pair. In the meantime, a modified GEO InSAR height retrieval model is proposed to mitigate the height and localization errors induced by the conventional model. Moreover, we also introduce the processing steps in GEO InSAR and D-InSAR. The retrieved height and deformation results are shown for validating the good performances of GEO InSAR system. At last, both the height retrieval accuracy and the deformation retrieval accuracy are analyzed detailedly with the consideration of the variations of the baseline lengths and the orbital configurations.

6.1 Interferometry Basics

Interferometric synthetic aperture radar (InSAR) can provide the surface height maps, which is similar to the interferometry techniques in applied physics and optics. In physical optics, two coherent light beams illuminate on one object to produce interferometric fringes. The phase difference of the two lights can be utilized to realize a high accurate distance measurement. Analogously, InSAR exploits radio signals to produce interferometric fringes and generates topographic maps.

T. Long et al., *Geosynchronous SAR: System and Signal Processing*,
https://doi.org/10.1007/978-981-10-7254-3_6

Generally, SAR images include both amplitude and phase information of the target. In this chapter we will focus on the phase information which is mainly utilized in InSAR systems to obtain height retrieval [1]. There are three types of work patterns in InSAR: Along-Track InSAR, Cross-Track InSAR and Repeat-Track InSAR [2, 3].

Figure 6.1 shows the InSAR configuration: two SAR antennae S_1 and S_2 are separated by a baseline B, and observe the complex response at range R_1 and the look angle θ. α is the angle between the baseline and the horizontal plane.

The height Z can be derived from the orbit geometry:

$$Z = \mathrm{H} - \mathrm{R}_1 \cos\theta \tag{6.1}$$

and according to the cosine law, the relationship between R_1 and R_2 is expressed by

$$\begin{aligned} R_2^2 &= R_1^2 + B^2 - 2R_1 B \cos\left(\frac{\pi}{2} - \theta + \alpha\right) \\ &= R_1^2 + B^2 + 2R_1 B \sin(\alpha - \theta) \end{aligned} \tag{6.2}$$

Using (6.2), we find

$$\sin(\alpha - \theta) = \frac{R_2^2 - R_1^2 - B^2}{2R_1 B} \tag{6.3}$$

Different work modes have different relationships between the actual phase difference and the range difference. For the repeat-track InSAR, the range difference can be written as:

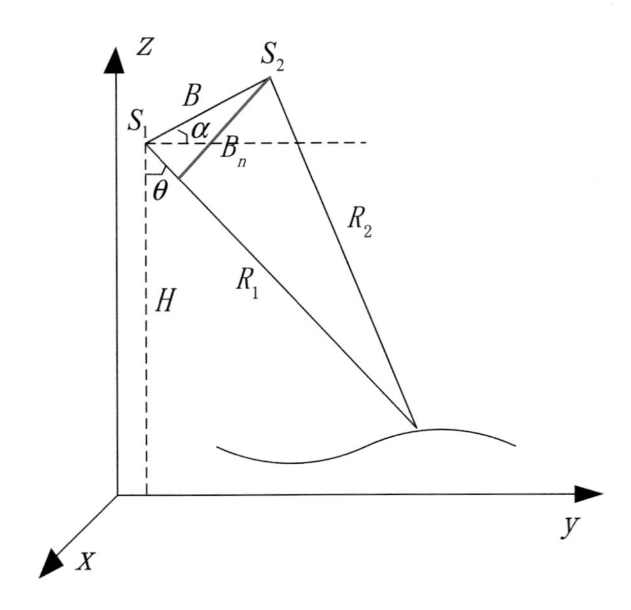

Fig. 6.1 Geometry of a spaceborne interferometric SAR system

$$\Delta R = R_2 - R_1 = -\frac{\lambda \Delta \phi}{4\pi} \tag{6.4}$$

For the cross-track InSAR, the range difference can be written as:

$$\Delta R = R_2 - R_1 = -\frac{\lambda \Delta \phi}{2\pi} \tag{6.5}$$

where λ is the operating wavelength of radar and $\Delta\phi$ is the actual phase difference between the two images for the one target. To conclude the height expression in the cross-track InSAR, we combine (6.1), (6.2) and (6.5) and obtain:

$$Z = H - \frac{\left(\frac{\lambda \Delta \phi}{2\pi}\right)^2 - B^2}{2B\sin(\alpha - \theta) + \frac{\lambda \Delta \phi}{\pi}} \cos\theta \tag{6.6}$$

In (6.6), it shows that the height Z is related to the baseline B, included angle α and the look angle θ.

It is worth noting that the baseline plays an essential role in the InSAR system. When the baseline is zero, the phase difference is absence and it is impossible to obtain the height of the target. Moreover, the length of the baseline also impacts the performance of InSAR. Too long baseline will result in the low correlation between the two SAR images, which means the correlation coefficient is small and the accuracy will be reduced. Also, small baseline will cause small phase difference and the low sensitivity on height.

With the continuous development of spaceborne SAR system and the requirement of rapid emergency response of geological disasters (earthquake, landslide, debris flow, volcanic eruption, etc.), geosynchronous Earth orbit SAR (GEO SAR) has more advantages than the low Earth orbit SAR (LEO SAR). The revisit time of LEO SAR is shown in Table 6.1 and it can be seen that traditional LEO SAR is hard to meet the demand as its long revisit time and small coverage area. The performance of GEO SAR can be seen directly in STK in Fig. 6.2 which shows that the short revisit time (from several to dozens of hours [4, 5]) and large coverage regions provide favorable condition for InSAR [5–7]. Thus, it is necessary to develop the research from LEO SAR to GEO SAR.

Table 6.1 Revisit time of LEO SAR

Satellite/constellation	Revisit time (days)
SENTINEL-1 A/B	6
COSMO-SkyMed	5
TerraSAR-X	11
RADARSAT-2	24
ALOS-2	14

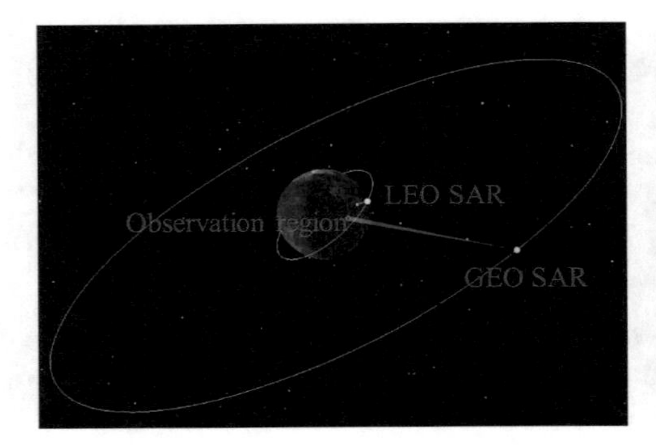

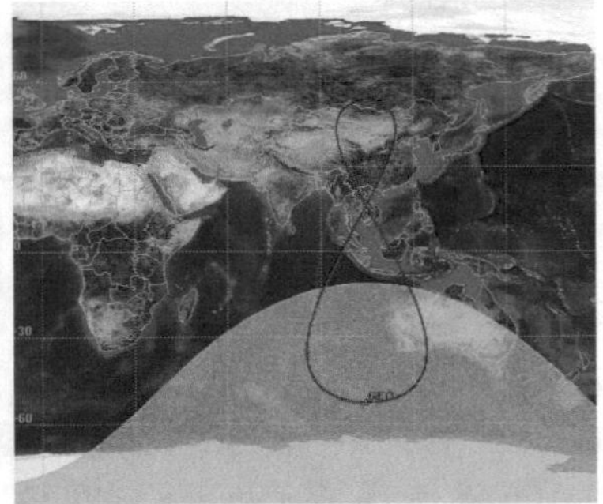

Fig. 6.2 Orbit characteristics of GEO SAR

6.2 GEO InSAR Special Issues

6.2.1 Un-parallel Repeated Tracks of GEO InSAR

Because of the high orbit, GEO SAR is more seriously impacted by the perturbation forces, such as the radiation pressure, the third-body and aspheric gravity and so on [5]. Due to these forces, satellites will have perturbation accelerations [8], which will force the orbital elements to vary with time. Using the High-Precision Orbit Propagator (HPOP) model in the Systems Tool Kit (STK) [9], the impacts of perturbations on GEO SAR orbital elements in a short period are given in Table 6.2. The HPOP is the numerical integration of the differential equations of motions to generate the high accurate satellite orbit in the presence of perturbations, including the accurate impacts of the third-body gravity, the atmospheric drag and the solar radiation pressure. In Table 6.2, it is concluded that there are the change of 2×10^{-3} degrees per day in the inclination and the drift of more than 1.2×10^{-2} degrees per day in the right ascension of the ascending node (RAAN). Thus, impacts of perturbations on the GEO SAR orbit cannot be ignored.

As the orbital elements of the GEO SAR vary with time, its orbit and the corresponding nadir-point trajectory will drift away from the scheduled ones. Taking the perigee as an example, the analysis of the drifts of the perigee nadir-point in GEO SAR is shown in Table 6.3. In one day, a drift of nearly 20 km distance and the geocentric angle difference of 0.2° for the perigee nadir-point are generated in the presence of perturbation. Furthermore, the distance drift and the geocentric angle difference become larger as the increase of the time interval between repeated tracks.

Table 6.2 Impacts of perturbations on some GEO SAR orbital elements in a short period

Orbital element	Perturbation value (degrees per day)
Inclination	2×10^{-3}
RAAN	1.2×10^{-2}

Table 6.3 Analysis of the perigee nadir-point drifts in GEO SAR

Time interval (day)	Distance drift (km)	Geocentric angle difference (°)
1	18.05	0.162
5	27.06	0.244
10	46.11	0.415

In addition, a GEO SAR has curved trajectories due to its high orbit and the effects of Earth rotation [10–12]. Figure 6.3a gives out the nadir-point trajectories of LEO SAR (ERS-2) and GEO SAR. It can be shown that GEO SAR has the curved trajectory for imaging rather than the straighter trajectory in LEO SAR. In the presence of impacts of perturbation and curved trajectories, the repeated trajectories of GEO InSAR are un-parallel (the corresponding nadir-point trajectories are given in Fig. 6.3b). Especially at the perigee and the apogee, the trajectories have large curvatures, thus the un-parallelism of the repeated trajectories is more serious.

The 3D sketch map of GEO InSAR repeated trajectories is shown in Fig. 6.3c. A yellow solid line and a blue solid line present the un-parallel repeated tracks of GEO SAR. In this case, if the GEO InSAR data is obtained under the zero-Doppler centroids (ZDC) data acquisition geometry, the obtained sub-apertures for the imaging and InSAR processing are marked in red color. M and S represent the data acquisition positions of the master and slave images based on ZDC, respectively. B is the baseline, B_a is the along-track baseline, B_I is the interferometric baseline and B_p is the perpendicular baseline. θ is the look angle. Because of the un-parallel GEO InSAR tracks, the obtained InSAR pair based on ZDC has an azimuth wave vector difference $\Delta\boldsymbol{\kappa}$, and the obtained InSAR pair has a rotation angle in azimuth ϑ shown as

$$\vartheta = \arccos(\langle \Delta\boldsymbol{\kappa}, \hat{\mathbf{r}}_{\mathbf{A}} \rangle) \tag{6.7}$$

where $\mathbf{r}_{\mathbf{A}}$ is the ground projection vector of azimuth direction, ^ stands for unit vector.

This bias makes the InSAR pair have an obvious spatial spectral shift in azimuth Λ_a, which is shown in Fig. 6.4. The shift Λ_a causes a rotation-induced decorrelation in the interferogram. The rotation-induced decorrelation γ_r is given as

$$\gamma_r = \frac{\Delta k_a(\vartheta)}{W_{ga}} \tag{6.8}$$

where Δk_a is the spatial spectral shift in azimuth and W_{ga} is the spatial spectral bandwidth in azimuth.

Table 6.4 shows the correlation coefficient due to the rotation-induced decorrelation of the InSAR pair obtained by different satellites with different time intervals based on the ZDC data acquisition. For GEO SAR, the correlation coefficient decreases to approximately 0.85 when the time interval is one-day. With the

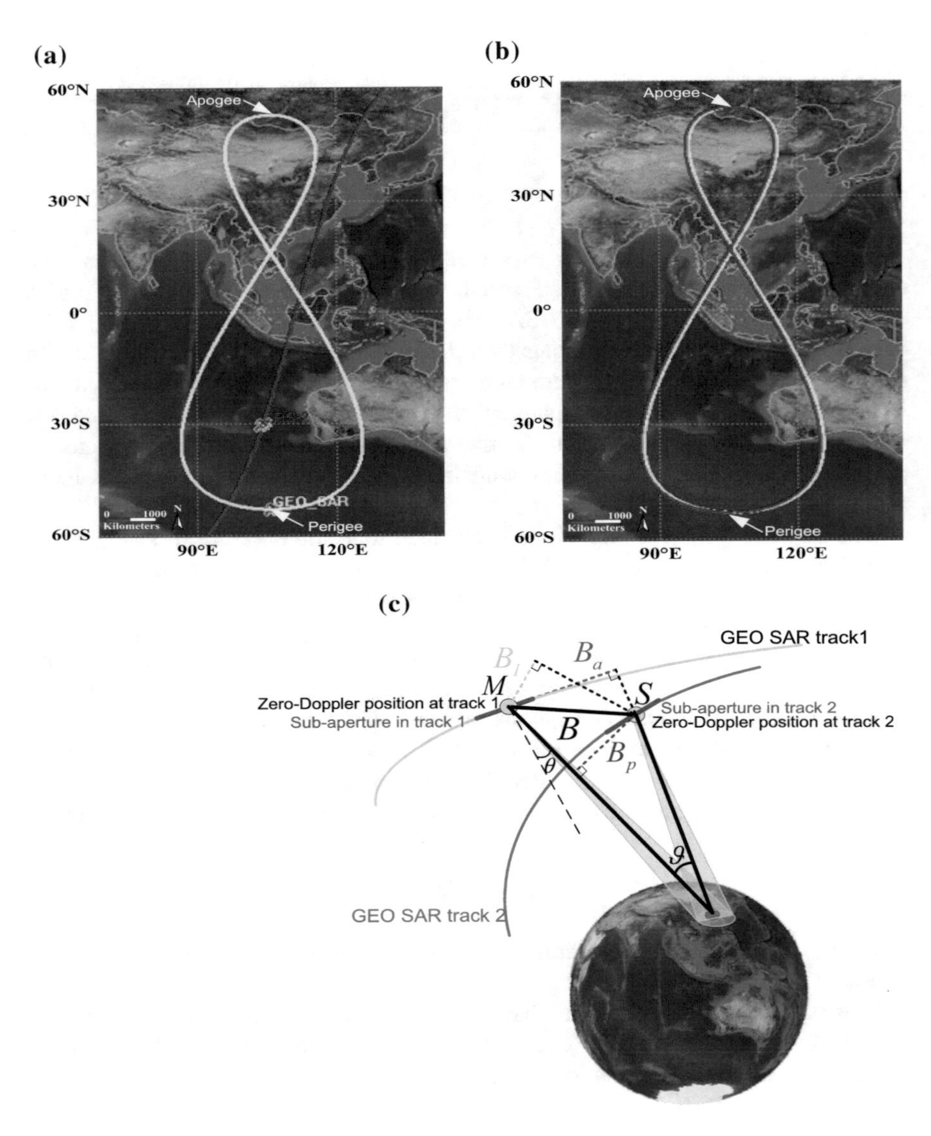

Fig. 6.3 **a** Nadir-point trajectories of ERS-2 (red) and GEO SAR (yellow); **b** nadir-point trajectories of GEO InSAR repeated trajectories (red line is the first track and the yellow one is the second track after ten days); **c** three-dimensional (3-D) sketch map of GEO InSAR repeated trajectories

increasing of the time interval, the correlation coefficient will be worse. It will be less than 0.6 when the time interval is five-day. Comparatively, the rotation-induced decorrelation is negligible in a LEO SAR case such as the TerraSAR-X, because their repeat tracks are nearly parallel.

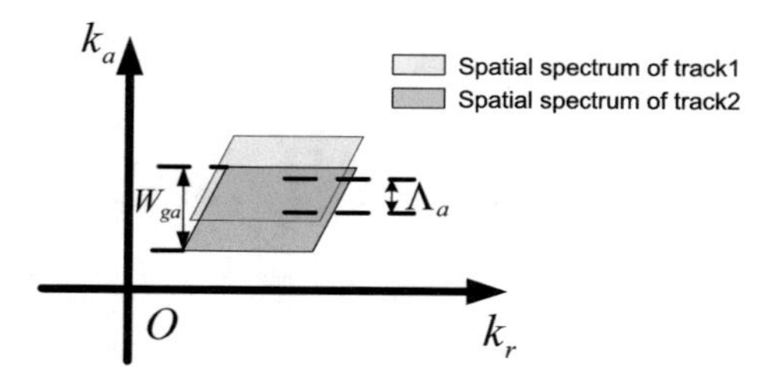

Fig. 6.4 Two-dimensional (2-D) spatial spectra of GEO InSAR pair. k_a is the spatial frequency in azimuth direction, k_r is the spatial frequency in range direction

Table 6.4 Correlation coefficient due to the rotation-induced decorrelation and the baseline analysis of GEO SAR (perigee) and TerraSAR-X (track 57, frame 9242 and track 57, frame 9409) under the ZDC data acquisition

Satellite	Correlation coefficient due to the rotation-induced decorrelation	Baseline length (m)	Along-track baseline length (m)
GEO SAR	0.849 (1 day interval)	13,600	13,500
	0.523 (5 day interval)	61,600	61,500
TerraSAR-X	0.973 (11 day interval)	314	62

Moreover, the un-parallel tracks make the GEO InSAR pair have the large along-track baseline (Table 6.4) which means that the length of the interferometric baseline cannot be approximated with the length of the perpendicular baseline in the height retrieval. The perpendicular baseline is dependent on both the look angle θ and the rotation angle in azimuth ϑ. Thus, the perpendicular baseline length is sure to have a bias with respect to the true interferometric baseline length.

6.2.2 Squint Looking in GEO InSAR

Generally, spaceborne SAR systems need the attitude control to compensate the effects of Earth rotation so that the squint angles can be eliminated, which can make the imaging easier and of a high quality. As impacted seriously by the Earth rotation, GEO SAR works in a significant squint-looking mode, resulting in the large Doppler centroid shifts (see the blue line in Fig. 6.5) and the degraded azimuth resolution performance. Under the optimal resolution steering, the GEO InSAR system also works in a slightly squint-looking mode and has Doppler centroid shifts (see the green line in Fig. 6.5). Only at the orbit positions near the perigee and apogee does the zero-Doppler centroid exist.

As the InSAR height retrieval depends on the geometric relationship, the GEO InSAR height retrieval is impacted by the serious squint looking. The LEO InSAR height retrieval model includes the range-Doppler localization equations in broadside imaging and a phase equation [13, 14] which are given by

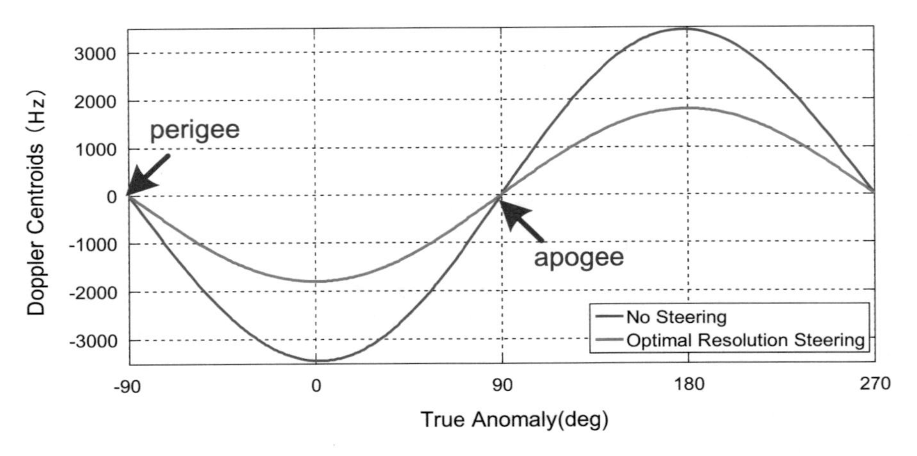

Fig. 6.5 Doppler centroid shifts along the GEO SAR track

$$\begin{cases} \left. \begin{array}{l} \left\|\mathbf{R}_{\widehat{\mathbf{M}},\mathbf{P}}\right\|_2 = \left\|\mathbf{R}_{\widehat{\mathbf{M}},\mathbf{Q}}\right\|_2 \\ f_{dc} = 0 \end{array} \right\} & \textit{Localization equations in broadside imaging} \\ \Delta\varphi_t = -\frac{4\pi}{\lambda\left\|\mathbf{R}_{\widehat{\mathbf{M}},\mathbf{P}}\right\|_2} \cdot \underbrace{\left[\left\|\mathbf{R}_{\widehat{\mathbf{M}},\mathbf{P}} - \mathbf{R}_{\widehat{\mathbf{S}},\mathbf{P}}\right\|_2\right]_P}_{\textit{Perpendicular baseline}} \cdot \frac{\Delta h}{\sin\theta_0} & \textit{Phase equation} \end{cases} \tag{6.9}$$

where $\mathbf{R}_{\widehat{\mathbf{M}},\mathbf{P}}$ and $\mathbf{R}_{\widehat{\mathbf{M}},\mathbf{P}}$ are the range vector from the zero-Doppler position to the target $\mathbf{P}$ and its corresponding ground projection with respect to the first track $\mathbf{Q}$, respectively, $f_{dc} = 2\left\langle \mathbf{v}_{\widehat{\mathbf{M}},\mathbf{P}}, \hat{\mathbf{R}}_{\widehat{\mathbf{M}},\mathbf{P}} \right\rangle / \lambda$ is the Doppler centroid, $\mathbf{v}_{\widehat{\mathbf{M}},\mathbf{P}}$ is the velocity vector at the zero-Doppler position in the first track, λ is the wavelength, $\Delta\varphi_t$ is the interferometric phase only related to the terrain, θ_0 is the local incidence angle, Δh is the height of the target with respect to the reference surface. The operator $\|\cdot\|_2$ and $\langle\cdot\rangle$ stand for the norm and inner product, respectively. $[\cdot]_P$ represents the operator of perpendicular baseline.

For GEO SAR orbital positions where large Doppler centroids residuals exist (such as when the GEO SAR satellite is above the equator), searching for the positions of zero-Doppler centroids fails, thus the employment of the localization equations in the broadside imaging geometry will make a target $\mathbf{P}$ in the generated DEMs have the localization error $\boldsymbol{\varepsilon}$, which is shown as

$$\boldsymbol{\varepsilon} = \mathbf{P}\left(\left\|\mathbf{R}_{\mathbf{M}_0,\mathbf{P}}\right\|_2, f_{dc0}\right) - \mathbf{P}\left(\left\|\mathbf{R}_{\widehat{\mathbf{M}},\mathbf{P}}\right\|_2, 0\right) \tag{6.10}$$

where $\mathbf{R}_{\mathbf{M}_0,\mathbf{P}}$ is the range between $\mathbf{M}$ and $\mathbf{P}$ at the aperture center moments, $f_{dc0} = 2\left\langle \mathbf{v}_{\mathbf{M}_0}, \widehat{\mathbf{R}}_{\mathbf{M}_0,\mathbf{P}} \right\rangle / \lambda$ is the Doppler centroid at the aperture center moment and $\mathbf{v}_{\mathbf{M}_0}$ is the velocity vector of the GEO SAR at the aperture center moment in the first track. Therefore, the squint-looking mode has to be specially considered in the GEO InSAR height retrieval.

As there exists non-zero-Doppler centroids along the full aperture in most parts of the track (see the green line in Fig. 6.5), the ZDC data acquisition is often impossible in the GEO InSAR system.

6.3 Optimal Data Acquisition and Height Retrieval

Considering the special issues of un-parallel repeated tracks and the squint-looking mode in GEO InSAR, a novel optimal minimal rotational-induced decorrelation (OMRD) data acquisition method and a GEO InSAR height retrieval model are proposed in this section to address the issues. In order to demonstrate the issues conveniently, GEO InSAR geometry is introduced firstly in Fig. 6.6. **M** and **S** represent the satellite positions of GEO InSAR tracks, and $\mathbf{R}_{\mathbf{M,S}}$ is their distance vector. **P** represents a target which has a height with respect to the global ellipsoid, and **Q** is its geometric projection on the global ellipsoid with respect to the first pass of the InSAR tracks. $\mathbf{R}_{\mathbf{M,P}}$ is the distance vector between **M** and **P**, $\mathbf{R}_{\mathbf{M,Q}}$ is the distance vector between **M** and **Q**, $\mathbf{R}_{\mathbf{S,P}}$ is the distance vector between **S** and **P**, $\mathbf{R}_{\mathbf{S,Q}}$ is the distance vector between **S** and **Q**. **l** is the distance vector from **P** to **Q**. $\mathbf{v}_{\mathbf{M}}$ and $\mathbf{v}_{\mathbf{S}}$ are the velocity vectors of the GEO SAR in the first track and the second track, respectively. $\widehat{\mathbf{r}}_{\mathbf{R}}$ is the project unit vector corresponding to the range direction unit vector $\widehat{\varsigma}$ with respect to the tangent plane of the global ellipsoid and $\widehat{\mathbf{r}}_{\mathbf{P}}$ represent the perpendicular direction of $\widehat{\mathbf{r}}_{\mathbf{R}}$. $\widehat{\mathbf{x}}$ is the azimuth direction vector. Δh is the height of the target with respect to the global ellipsoid, and $\widehat{\mathbf{H}}$ is the height direction unit vector.

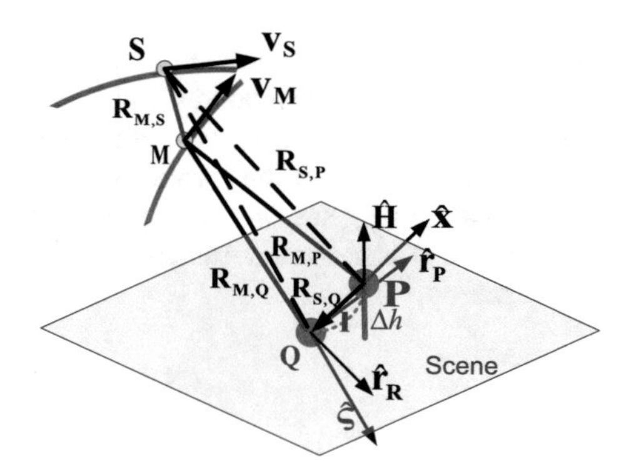

Fig. 6.6 Sketch map of the GEO InSAR geometry

6.3.1 OMRD Data Acquisition Method

The data acquisition of the repeat-track GEO InSAR is a process of the sub-apertures selection, and it is realized by determining of sub-aperture center moments. The sub-aperture center moment in the first track is determined by the observation moment of the scene of interest and the corresponding sub-aperture is obtained according to the length of the integration time. Nevertheless, the selection of the sub-aperture center moment and the related sub-aperture in the second orbit is vital because it determines the quality of the produced InSAR pair. Because the obtained GEO InSAR pair based on the ZDCs has serious rotation-induced decorrelation, the OMRD data acquisition method is induced to deal with the issue in the following part.

In InSAR, correlation coefficient is mainly affected by the following factors [15] as

$$\gamma = \gamma_{the} \cdot \gamma_g \cdot \gamma_t \tag{6.11}$$

where γ_{the} is the thermal noise decorrelation, γ_g is the geometric decorrelation and γ_t is the temporal decorrelation.

Because γ_g is mainly impacted by data acquisition methods, it is studied in details. Based on the basic concept of wavenumber domain analysis in [13], γ_g is expressed as

$$\gamma_g = \frac{\dfrac{\exp(-j\psi)\int\limits_V [\sigma_0(\mathbf{r})\exp(-2j\langle(\mathbf{\kappa_M} - \mathbf{\kappa_S}), \mathbf{r}\rangle)}{\sqrt{\int\limits_V \sigma_0(\mathbf{r})\left|h_\Theta\left(u(u_{mc}), t\left(\left\|\mathbf{R_{M_0,P}}\right\|_2\right)\right)\right|^2 dV''}}}{\dfrac{\cdot h_\Theta\left(u(u_{mc}), t\left(\left\|\mathbf{R_{M_0,P}}\right\|_2\right)\right)h_\Xi^*\left(u(u_{sc}), t\left(\left\|\mathbf{R_{S_0,P}}\right\|_2\right)\right)\Big]dV'}{\cdot\sqrt{\int\limits_V \sigma_0(\mathbf{r})\left|h_\Xi\left(u(u_{sc}), t\left(\left\|\mathbf{R_{S_0,P}}\right\|_2\right)\right)\right|^2 dV}}} \tag{6.12}$$

where $\mathbf{r}$ represents the surface scatter unit, $\sigma_0(\mathbf{r})$ is the surface backscatter coefficient, V represents the integral unit, $\mathbf{R_{S_0,P}}$ is the range between $\mathbf{S}$ and $\mathbf{P}$ at the aperture center moment, u and t are the azimuth time and the range time, respectively, u_{mc} and u_{sc} are the positions of targets at aperture center moments of the InSAR pair, $h_\Xi(\cdot)$ and $h_\Theta(\cdot)$ are the impulse response functions of the InSAR system, and ψ is the interferometric phase. $\mathbf{\kappa_M}$ and $\mathbf{\kappa_S}$ are wave vectors which are expressed by

$$\begin{cases} \mathbf{\kappa_M} = \kappa_0 \dfrac{\mathbf{R_{M_0,P}}}{\left\|\mathbf{R_{M_0,P}}\right\|_2} \\ \mathbf{\kappa_S} = \kappa_0 \dfrac{\mathbf{R_{S_0,P}}}{\left\|\mathbf{R_{S_0,P}}\right\|_2} \end{cases}. \tag{6.13}$$

$\kappa_0 = 2\pi/\lambda$ is the wavenumber. For simplicity, from (6.12) and (6.13) (see [16]) we have

$$\gamma_g = \frac{\exp(-j\psi)\int_V \left[\sigma_0(\mathbf{r})\exp\left(-2j\left(\frac{\kappa_0}{\left\|\mathbf{R}_{\mathbf{M_0},\mathbf{P}}\right\|_2}\left(\Delta\kappa_{\hat{\mathbf{x}}}x + \Delta\kappa_{\hat{\varsigma}}\varsigma\right)\right)\right)\right.}{\sqrt{\int_V \sigma_0(\mathbf{r})\left|h_\Theta\left(u(u_{mc}), t\left(\left\|\mathbf{R}_{\mathbf{M_0},\mathbf{P}}\right\|_2\right)\right)\right|^2 dV''}} \frac{\left.\cdot h_\Theta\left(u(u_{mc}), t\left(\left\|\mathbf{R}_{\mathbf{M_0},\mathbf{P}}\right\|_2\right)\right) h_\Xi^*\left(u(u_{sc}), t\left(\left\|\mathbf{R}_{\mathbf{S_0},\mathbf{P}}\right\|_2\right)\right)\right] dV'}{\cdot\sqrt{\int_V \sigma_0(\mathbf{r})\left|h_\Xi\left(u(u_{sc}), t\left(\left\|\mathbf{R}_{\mathbf{S_0},\mathbf{P}}\right\|_2\right)\right)\right|^2 dV}} \tag{6.14}$$

where $\Delta\kappa_{\hat{\mathbf{x}}}$ and $\Delta\kappa_{\hat{\varsigma}}$ are spectral shifts in range and azimuth respectively, which can be expressed as

$$\Delta\kappa_{\hat{\xi}} = \frac{1}{D_1}\left\{\underbrace{-A_1\left\langle \mathbf{R}_{\mathbf{M_0},\mathbf{S_0}}, \hat{\mathbf{l}}\right\rangle}_{interferometry-baseline\ term} \underbrace{+ B_1\left\langle \mathbf{R}_{\mathbf{M_0},\mathbf{S_0}}, \hat{\boldsymbol{\psi}}_\otimes\right\rangle + C_1\left\langle \mathbf{R}_{\mathbf{M_0},\mathbf{S_0}}, \hat{\mathbf{v}}_{\mathbf{M_0}}\right\rangle}_{squint-looking\ mode\ term}\right\} \tag{6.15}$$

$$\Delta\kappa_{\hat{\mathbf{x}}} = \frac{1}{E_2}\left\{\underbrace{A_2\left\langle \mathbf{R}_{\mathbf{M_0},\mathbf{S_0}}, \hat{\boldsymbol{\psi}}_\otimes\right\rangle\left\langle \hat{\mathbf{R}}_{\mathbf{S_0},\mathbf{P}}, \Delta\mathbf{v}\right\rangle}_{unparallel-track\ term}\right. \underbrace{- B_2\left\langle \mathbf{R}_{\mathbf{M_0},\mathbf{S_0}}, \hat{\mathbf{l}}\right\rangle + C_2\left\langle \mathbf{R}_{\mathbf{M_0},\mathbf{S_0}}, \hat{\boldsymbol{\psi}}_\otimes\right\rangle}_{Squint-looking\ mode\ term} + \underbrace{D_2\left\langle \mathbf{R}_{\mathbf{M_0},\mathbf{S_0}}, \hat{\mathbf{v}}_{\mathbf{M_0}}\right\rangle}_{Coupling\ term}. \tag{6.16}$$

Variables A_1 to D_1 and A_2 to E_2 only depends on the GEO InSAR geometry and can be expressed as

$$\begin{cases} A_1 = \left\langle \hat{\mathbf{r}}_\mathbf{R}, \hat{\mathbf{l}}\right\rangle \\ B_1 = \left\langle \hat{\mathbf{R}}_{\mathbf{S_0},\mathbf{P}}, \hat{\mathbf{v}}_{\mathbf{M_0}}\right\rangle\left\langle \hat{\mathbf{r}}_\mathbf{R}, \hat{\boldsymbol{\psi}}_\otimes\right\rangle\left\langle \hat{\mathbf{r}}_\mathbf{R}, \hat{\mathbf{v}}_{\mathbf{M_0}}\right\rangle \\ C_1 = \left(\left\langle \hat{\mathbf{R}}_{\mathbf{M_0},\mathbf{P}}, \hat{\mathbf{v}}_{\mathbf{M_0}}\right\rangle\left\langle \hat{\mathbf{R}}_{\mathbf{S_0},\mathbf{P}}, \hat{\boldsymbol{\psi}}_\otimes\right\rangle\left\langle \hat{\mathbf{r}}_\mathbf{R}, \hat{\boldsymbol{\psi}}_\otimes\right\rangle + \left\langle \hat{\mathbf{r}}_\mathbf{R}, \hat{\mathbf{v}}_{\mathbf{M_0}}\right\rangle\left(\left\langle \hat{\mathbf{r}}_\mathbf{R}, \hat{\mathbf{v}}_{\mathbf{M_0}}\right\rangle\left\langle \hat{\mathbf{R}}_{\mathbf{S_0},\mathbf{P}}, \hat{\mathbf{v}}_{\mathbf{M0}}\right\rangle - 1\right)\right) \\ D_1 = \left\langle \hat{\mathbf{R}}_{\mathbf{M_0},\mathbf{P}}, \hat{\mathbf{r}}_\mathbf{R}\right\rangle \end{cases} \tag{6.17}$$

$$\begin{cases} A_2 = \left\langle \widehat{\mathbf{R}}_{\mathbf{M_0,P}}, \widehat{\psi}_{\otimes} \right\rangle \left\langle \widehat{\mathbf{r}}_{\mathbf{P}}, \widehat{\mathbf{v}}_{\mathbf{M_0}} \right\rangle \\ B_2 = \left\langle \widehat{\mathbf{r}}_{\mathbf{P}}, \widehat{\mathbf{l}} \right\rangle \\ C_2 = \left(\left\langle \widehat{\mathbf{r}}_{\mathbf{P}}, \widehat{\psi}_{\otimes} \right\rangle \left(\left\langle \widehat{\mathbf{R}}_{\mathbf{M_0,P}}, \widehat{\psi}_{\otimes} \right\rangle \left\langle \widehat{\mathbf{R}}_{\mathbf{S_0,P}}, \widehat{\psi}_{\otimes} \right\rangle - 1 \right) + \left\langle \widehat{\mathbf{R}}_{\mathbf{M_0,P}}, \widehat{\psi}_{\otimes} \right\rangle \left\langle \widehat{\mathbf{r}}_{\mathbf{P}}, \widehat{\mathbf{v}}_{\mathbf{M_0}} \right\rangle \left\langle \widehat{\mathbf{R}}_{\mathbf{S_0,P}}, \widehat{\mathbf{v}}_{\mathbf{S_0}} \right\rangle \right) \\ D_2 = \left(\left\langle \widehat{\mathbf{r}}_{\mathbf{P}}, \widehat{\mathbf{v}}_{\mathbf{M_0}} \right\rangle \left(\left\langle \widehat{\mathbf{R}}_{\mathbf{M_0,P}}, \widehat{\mathbf{v}}_{\mathbf{M_0}} \right\rangle \left\langle \widehat{\mathbf{R}}_{\mathbf{S_0,P}}, \widehat{\mathbf{v}}_{\mathbf{M_0}} \right\rangle - 1 \right) + \left\langle \widehat{\mathbf{r}}_{\mathbf{P}}, \widehat{\psi}_{\otimes} \right\rangle \left\langle \widehat{\mathbf{R}}_{\mathbf{S_0,P}}, \widehat{\psi}_{\otimes} \right\rangle \left\langle \widehat{\mathbf{R}}_{\mathbf{M_0,P}}, \widehat{\mathbf{v}}_{\mathbf{M_0}} \right\rangle \right) \\ E_2 = \left\langle \widehat{\mathbf{r}}_{\mathbf{P}}, \widehat{\mathbf{v}}_{\mathbf{M_0}} \right\rangle \end{cases} \tag{6.18}$$

where $\widehat{\mathbf{v}}_{\mathbf{S_0}}$ is the velocity at the aperture center moment in the second GEO SAR track, $\Delta\mathbf{v}$ is the vector difference of $\mathbf{v}_{\mathbf{M_0}}$ and $\widehat{\mathbf{v}}_{\mathbf{S_0}}$, $\mathbf{R}_{\mathbf{M_0,S_0}}$ is the vector difference of $\mathbf{R}_{\mathbf{M_0,P}}$ and $\mathbf{R}_{\mathbf{S_0,P}}$, $\widehat{\psi}_{\otimes}$ is determined by the cross product of the orthonormal basis $\widehat{\mathbf{l}}$ and $\widehat{\mathbf{v}}_{\mathbf{M_0}}$.

According to (6.15), $A_1\left\langle \mathbf{R}_{\mathbf{M_0,S_0}}, \widehat{\mathbf{l}} \right\rangle$ relates to the interferometric baseline and $B_1\left\langle \mathbf{R}_{\mathbf{M_0,S_0}}, \widehat{\psi}_{\otimes} \right\rangle + C_1\left\langle \mathbf{R}_{\mathbf{M_0,S_0}}, \widehat{\mathbf{v}}_{\mathbf{M_0}} \right\rangle$ depends on the squint-looking mode. As for (6.16), $\Delta\kappa_{\hat{\mathbf{x}}}$ is impacted by lots of components, including the effects of the un-parallel repeated tracks $A_2\left\langle \mathbf{R}_{\mathbf{M_0,S_0}}, \widehat{\psi}_{\otimes} \right\rangle \left\langle \widehat{\mathbf{R}}_{\mathbf{S_0,P}}, \Delta\mathbf{v} \right\rangle$, the squint-looking mode $B_2\left\langle \mathbf{R}_{\mathbf{M_0,S_0}}, \widehat{\mathbf{l}} \right\rangle + C_2\left\langle \mathbf{R}_{\mathbf{M_0,S_0}}, \widehat{\psi}_{\otimes} \right\rangle$ and their coupling term $D_2\left\langle \mathbf{R}_{\mathbf{M_0,S_0}}, \widehat{\mathbf{v}}_{\mathbf{M_0}} \right\rangle$.

According to (6.15) and (6.16), the mismatch of the 2-D spatial spectra of the GEO InSAR pair exists if $\Delta\kappa_{\hat{\mathbf{x}}}$ and $\Delta\kappa_{\hat{\varsigma}}$ are not zero, which makes the GEO InSAR pair decorrelate. Thus, γ_g can be expressed as

$$\gamma_g = \gamma_B \cdot \gamma_r \tag{6.19}$$

where γ_B (related to $\Delta\kappa_{\hat{\varsigma}}$) is mainly induced by the interferometric baseline decorrelation and γ_r (related to $\Delta\kappa_{\hat{\mathbf{x}}}$) is induced by the rotation-induced decorrelation.

γ_B should not be eliminated because the interferometric baseline should not equal to zero in the height retrieval. On the basis of the range spatial spectral bandwidth determined by the range bandwidth W_r, the critical baseline B_c which is the upper limitation of interferometric baseline is expressed as

$$B_c = \left\langle \mathbf{R}_{\mathbf{M_0,S_0}} \Big|_{\Delta\kappa_{\hat{\varsigma}} = \frac{2\pi \left\| \mathbf{R}_{\mathbf{M_0,P}} \right\|_2}{c\kappa_0} \cdot W_r}, \hat{\mathbf{l}} \right\rangle. \tag{6.20}$$

Especially, in the case of the slightly squint angle, (6.20) can be simplified as

$$B_c = \frac{2\pi \left\langle \widehat{\mathbf{R}}_{\mathbf{M_0,P}}, \widehat{\mathbf{r}}_{\mathbf{R}} \right\rangle \left\| \mathbf{R}_{\mathbf{M_0,P}} \right\|_2}{c\kappa_0 \left\langle \widehat{\mathbf{r}}_{\mathbf{R}}, \widehat{\mathbf{l}} \right\rangle} \cdot W_r. \tag{6.21}$$

$\Delta\kappa_{\hat{\mathbf{x}}}$ is useless for the height retrieval and γ_r should be 1 for obtaining high coherence. Conventionally, when the ZDCs are employed, (6.16) is expressed as

$$\Delta\kappa_{\hat{\mathbf{x}}} = \left\langle \widehat{\mathbf{R}}_{\mathbf{S_0},\mathbf{P}}, \Delta\mathbf{v} \right\rangle \left(\left\langle \mathbf{R}_{\mathbf{M_0},\mathbf{S_0}}, \widehat{\boldsymbol{\psi}}_{\otimes} \right\rangle + 1 \right). \tag{6.22}$$

In LEO SAR, the repeated tracks are nearly perfectly parallel. Thus, $\Delta\mathbf{v}$ is zero and $\Delta\kappa_{\hat{\mathbf{x}}}$ is zero. The spatial spectra of the InSAR pair in azimuth are coherent and the minimum rotation-induced decorrelation can be obtained (i.e. $\gamma_r = 1$). However in GEO SAR, the repeated tracks are un-parallel and the geometry is squint. Thus, $\Delta\mathbf{v}$ is non-zero and $\Delta\kappa_{\hat{\mathbf{x}}}$ is non-zero if using ZDC data acquisition. In this case, the rotation-induced decorrelation exists (i.e. $\gamma_r \neq 1$) as the overlapped azimuth spatial spectra of GEO InSAR pair decrease. By employing the azimuth bandwidth W_a, γ_r is expressed as

$$\gamma_r = 1 - \frac{\kappa_0 \|\mathbf{v}_{\mathbf{M_0}}\|_2 |\Delta\kappa_{\hat{\mathbf{x}}}|}{\pi \left\|\mathbf{R}_{\mathbf{M_0},\mathbf{P}}\right\|_2 W_a}. \tag{6.23}$$

In order to maximize the overlaps of the spatial spectra of the GEO InSAR pair in azimuth and minimize the rotation-induced decorrelation, the criterion is proposed to realize the optimal selection of the aperture center moment and the corresponding sub-aperture in the second track. It is given as

$$\mathbf{S}_0 = \left\{ \mathbf{S}(\tau_0) \big|_{\tau_0 = \max \|\gamma_r[\Delta\kappa_{\hat{\mathbf{x}}} \mathbf{S}((\tau))]\|_2} \right\} \tag{6.24}$$

where $\mathbf{S}(\tau_0)$ is the proper satellite position at the second track, τ is the azimuth moment of the second aperture and τ_0 is the aperture center moment of the second sub-aperture.

The operation is summarized as follows:

Step 1: Based on (6.24), search the proper satellite position $\mathbf{S}(\tau_0)$ by the step corresponding to the pulse repeated time along the full aperture of the second track $\mathbf{S}(\tau)$;
Step 2: Use the determined moment τ_0 as the aperture center moment of the second sub-aperture;
Step 3: According to the integration time T_a, the corresponding sub-aperture of the second track is determined as $\left[\mathbf{S}\left(\tau_0 - \frac{T_a}{2}\right), \mathbf{S}\left(\tau_0 + \frac{T_a}{2}\right)\right]$.

Hereby, the GEO InSAR pair is obtained based on (6.24). Though the Doppler centroid shifts exist, the maximal coherence of the GEO InSAR pair in azimuth is achieved as the rotation-induced decorrelation is minimized and the azimuth spatial spectra are coherent.

Because of the long full aperture (nearly one thousand seconds for one full aperture) and the relatively limited un-parallelism of the repeated trajectories of GEO SAR, the proposed OMRD data acquisition method can avoid complete azimuth decorrelation in the designing phase. Assuming the observation in the first track is at perigee, it is one of places with the most serious un-parallel repeated trajectories. The GEO SAR coverage time analysis of a target (78.36°S, 105.58°E) is shown in Table 6.5 by STK simulations. Since the revisit time of GEO SAR is not exactly 24 h (3 min 56.4 s bias), the access time in Table 6.5 is corrected by the bias. It can be shown that the full apertures time of the InSAR pair are overlapped about 739 s (99.6% of the full aperture) in a one-day interval case and about 728 s (98.1% of the full aperture) in a five-day interval case. The full apertures of the InSAR pair will have no common part (giving raise to the complete azimuth decorrelation) after nearly 2 months. Since the OMRD can work when the full apertures of the InSAR pair have the overlapped part, the GEO InSAR pair with the complete azimuth coherence (the coherent azimuth spatial spectra) can be obtained by the OMRD method when the time interval of the access time is within 2 month. As for the longer time, station-keeping is needed for GEO SAR.

After the rotation-induced decorrelation is removed by the OMRD method, the obtained better azimuth coherence will result in the lower phase noise and the higher height accuracy. If the total coherence raises from γ_1 to γ_2 (γ_2 is not equal to 1 because of γ_{the} and γ_B), the phase variance in the single-look image pair will decrease by [13, 17]

$$\begin{aligned}\Delta\sigma_\varphi^2 &= -\pi[\arcsin(|\gamma_1|) - \arcsin(|\gamma_2|)] \\ &\quad + \left[\arcsin^2(|\gamma_1|) - \arcsin^2(|\gamma_2|)\right] - \left[\frac{Li_2\left(|\gamma_1|^2\right)}{2} - \frac{Li_2\left(|\gamma_2|^2\right)}{2}\right]\end{aligned} \tag{6.25}$$

where $\Delta\sigma_\varphi^2$ is the decrease of the phase variance, Li_2 is Euler's dilogarithm, defined as

$$Li_2\left(|\gamma|^2\right) = \sum_{k=1}^{\infty} \frac{|\gamma|^{2k}}{k^2}. \tag{6.26}$$

Table 6.5 GEO SAR coverage time analysis of a target by STK simulations (78.36°S, 105.58°E) (30 m diameter antenna)

Access time (day)	Access start (UTC)	Access end (UTC)	Overlapped time (s)	Full aperture (s)
1	1 Sept. 2015 03:50:12.247	1 Sept. 2015 04:02:34.008	–	741.761
2	2 Sept. 2015 03:50:14.769	2 Sept. 2015 04:02:36.536	739	741.768
6	6 Sept. 2015 03:50:28.494	6 Sept. 2015 04:02:50.326	728	741.832
31	1 Oct. 2015 03:54:49.051	1 Oct. 2015 04:07:10.908	466	741.856
46	16 Oct. 2015 03:58:49.970	16 Oct. 2015 04:11:11.867	224	741.897
61	31 Oct. 2015 04:03:49.464	31 Oct. 2015 04:16:11.608	0	742.144

6.3.2 GEO InSAR Height Retrieval Model

Considering the un-parallel repeated tracks and the squint-looking mode, the height retrieval model in LEO InSAR cannot be adopted in the GEO InSAR height retrieval directly. The localization equations and phase equations can be modified as follows.

Localization equations: Since GEO InSAR system works in the squint-looking mode, the conventional broadside imaging expression of geometric equations in the height retrieval model should be modified according to the squint-looking geometry with respect to the aperture center moment. It is important for avoiding localization errors in the final generated digital elevation model (DEM). Based on the Range-Doppler localization [18], the range equation and Doppler equation in the squint-looking mode are expressed as

$$\begin{cases} \left\|\mathbf{R}_{\mathbf{M_0},\mathbf{P}}\right\|_2 - \left\|\mathbf{R}_{\mathbf{M_0},\mathbf{Q}}\right\|_2 = 0 \\ f_{dc} = \frac{2\left\langle \widehat{\mathbf{v}_{\mathbf{M_0}}}, \widehat{\mathbf{R}_{\mathbf{M_0},\mathbf{Q}}} \right\rangle}{\lambda} \end{cases}. \tag{6.27}$$

Phase equation: In the case of the un-parallel tracks, the perpendicular baseline not only depends on the look angle but also is impacted by the rotation angle in azimuth ϑ. Thus, the perpendicular baseline in the phase equation of the conventional height retrieval model needs to be replaced by an accurate expression of the interferometric baseline. After the reference phase removal, the terrain phase $\Delta\varphi_T$ is expressed as

$$\Delta\varphi_t = \frac{4\pi}{\lambda\left\|\mathbf{R}_{\mathbf{M_0},\mathbf{P}}\right\|_2} \cdot \left\langle \mathbf{R}_{\mathbf{M_0},\mathbf{S_0}}, \widehat{\mathbf{l}} \right\rangle \cdot \frac{\Delta h}{\left\langle \widehat{\mathbf{l}}, \widehat{\mathrm{H}} \right\rangle}. \tag{6.28}$$

Differentiating (6.28) with respect to Δh, we have

$$\frac{\partial \Delta\varphi_t}{\partial \Delta h} = \frac{4\pi}{\lambda\left\langle \widehat{\mathbf{l}}, \widehat{\mathrm{H}} \right\rangle \left\|\mathbf{R}_{\mathbf{M_0},\mathbf{P}}\right\|_2} \cdot \left\langle \mathbf{R}_{\mathbf{M_0},\mathbf{S_0}}, \widehat{\mathbf{l}} \right\rangle. \tag{6.29}$$

On the basis of (6.29), the variation of the interferometric phase is directly related to the baseline component $\left\langle \mathbf{R}_{\mathbf{M_0},\mathbf{S_0}}, \widehat{\mathbf{l}} \right\rangle$. Thus, $\left\langle \mathbf{R}_{\mathbf{M_0},\mathbf{S_0}}, \widehat{\mathbf{l}} \right\rangle$ is the accurate expression of the interferometric baseline in GEO InSAR and it should be used to replace the perpendicular baseline in the phase equation of the GEO InSAR height retrieval model. The GEO InSAR height retrieval model is given as

$$\begin{cases} \left\|\mathbf{R}_{\mathbf{M_0},\mathbf{P}}\right\|_2 = \left\|\mathbf{R}_{\mathbf{M_0},\mathbf{Q}}\right\|_2 \\ f_{dc} = \frac{2\left\langle \hat{\mathbf{v}}_{\mathbf{M_0}}, \hat{\mathbf{R}}_{\mathbf{M_0},\mathbf{Q}} \right\rangle}{\lambda} \end{cases} \textit{Localization equations} \\ \Delta\varphi_t = -\frac{4\pi}{\lambda\left\|\mathbf{R}_{\mathbf{M_0},\mathbf{P}}\right\|_2} \cdot \underbrace{\left\langle \mathbf{R}_{\mathbf{M_0},\mathbf{S_0}}, \hat{\mathbf{l}} \right\rangle}_{\textit{Interferometric baseline}} \cdot \frac{\Delta h}{\left\langle \hat{\mathbf{l}}, \hat{\mathbf{H}} \right\rangle} \quad \textit{Phase equation} \tag{6.30}$$

When the InSAR system works in the broadside imaging and has parallel repeated tracks, $\left\langle \mathbf{R}_{\mathbf{M_0},\mathbf{S_0}}, \hat{\mathbf{l}} \right\rangle$ is equivalent to the perpendicular baseline.

6.3.3 Simulation Verifications

In this part, simulations are conducted to verify the proposed OMRD data acquisition method and the GEO InSAR height retrieval model. The inclined curved 'figure-8' GEO SAR orbit is utilized in the simulations [10, 19]. GEO SAR system parameters are listed in Table 6.6. The experimental scene size is 3.5 km × 3.5 km (a relative small scene scope only for the algorithm verification and a higher calculation efficiency), including a 260 m pyramid-like terrain variation. The bandwidth and the integration time are 18 MHz and 120 s to obtain about 20 m resolution both in range and azimuth directions. The signal to noise ratio (SNR) is set as 10 dB in the simulations.

Back-projection algorithm is adopted for the GEO SAR imaging. Perigee is used as an exemplary position in the data acquisition method verification due to the seriously un-parallel there, and the satellite position above the equator is added for the height retrieval model verification because of its obvious squint-looking status of the GEO SAR. Regarding to the InSAR height retrieval, we need to ensure a proper interferometric baseline to improve the height accuracy and a short temporal baseline to obtain a good coherence in the interferogram. In Fig. 6.7, the GEO InSAR baseline with respect to the time interval is shown. The baseline increases as the time interval increase. The variation of the interferometric baseline has the same regular pattern. As for a GEO InSAR pair with a small temporal baseline (e.g. 1 day time interval at perigee), the interferometric baseline is negligible, the height accuracy reduces seriously. Thus, we take the GEO InSAR pairs with a 5 day time interval as an example to conduct the simulations.

Table 6.6 GEO SAR system parameters

Parameters	Value	Parameters	Value
Wavelength (m)	0.24	Bandwidth (MHz)	18
Semimajor axis (km)	42,164	Eccentricity	0.07
Pulse repeated frequency (Hz)	150	Revisit time interval (day)	5

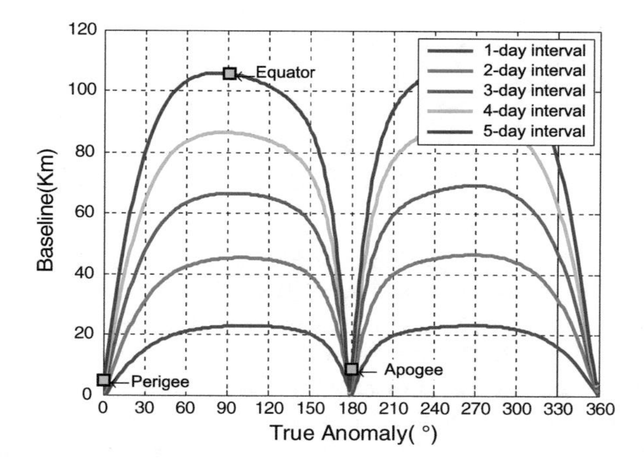

Fig. 6.7 GEO InSAR baseline with respect to various time intervals

6.3.3.1 Verification of OMRD Data Acquisition Method

In the simulations, with respect to the same repeated tracks of GEO InSAR, the ZDC data acquisition and the OMRD data acquisition are conducted to obtain the InSAR pair at perigee, respectively. The center point of the scene is used as the target point in the OMRD data acquisition method. Then, the obtained InSAR pairs are used for imaging and co-registering. After co-registrations, coherence maps are generated after a 3×3 looks averaging. In Fig. 6.8, coherence maps of the GEO InSAR pairs corresponding to two data acquisition methods are shown. Figure 6.8a is obtained using the ZDC data acquisition method with an averaged coherence coefficient of 0.475. Figure 6.8b is obtained based on the OMRD data acquisition with an averaged coherence coefficient of nearly 0.9. The sources of decorrelation of the GEO InSAR pairs corresponding to two data acquisition methods are given in Table 6.7.

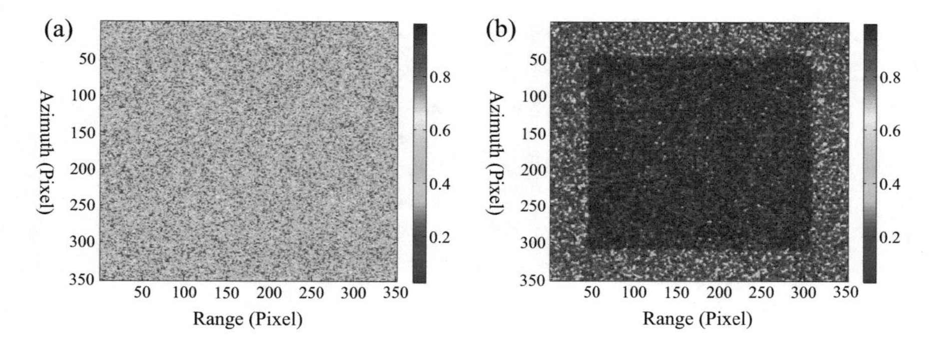

Fig. 6.8 Coherence maps of the GEO InSAR pairs corresponding to two data acquisition methods: **a** ZDC data acquisition method; **b** OMRD data acquisition method

Table 6.7 Sources of decorrelation of the GEO InSAR pairs corresponding to two data acquisition methods

Correlation coefficient	ZDC	OMRD
Averaged total correlation coefficient	0.475	0.891
Interferometric baseline decorrelation	0.986	0.985
Thermal noise decorrelation	0.909	0.909
Rotation-induced decorrelation	0.523	0.996

As for the ZDC data acquisition method and the OMRD data acquisition method, the critical baselines are about 320 km in two cases. According to the analysis of sources of decorrelation, the interferometric baseline decorrelation are obtained as 0.986 and 0.985 (corresponding to the interferometric baselines of approximately 4.5 and 4.9 km, respectively), and the correlation coefficients due to the thermal noise decorrelation are almost 0.909. Therefore, on the basis of (6.11), correlation coefficients due to the rotation-induced decorrelation in the two data acquisition methods are 0.523 and 0.996, respectively.

The rotation-induced decorrelations can be derived from the spatial spectral shifts of the InSAR pairs in azimuth. In Fig. 6.9, the azimuth spatial spectral shifts are 0.0153 m^{-1} and 3.40×10^{-7} m^{-1} for the ZDC data acquisition method and the OMRD data acquisition method, respectively. Because their spatial spectral bandwidth in azimuth are nearly 0.0326 m^{-1} (corresponding to the azimuth spectral bandwidth of about 51.28 Hz) in two cases, the azimuth spatial spectral shift of the InSAR pair in the ZDC data acquisition is large compared with the spatial spectral bandwidth in azimuth, giving rise to the obvious rotation-induced decorrelation. In contrast, the InSAR pair of the OMRD data acquisition method has the almost negligible azimuth spatial spectral shift and the rotation-induced decorrelation. Therefore, by the proposed method, the generated GEO InSAR pair has the maximal azimuth coherence. In the proposed simulation, the averaged total coherence coefficient of the GEO InSAR pair increases more than 0.4 compared with that in the ZDC data acquisition method. Because of the improvement of the averaged total correlation coefficient in the proposed case (from 0.475 to 0.891) after the OMRD

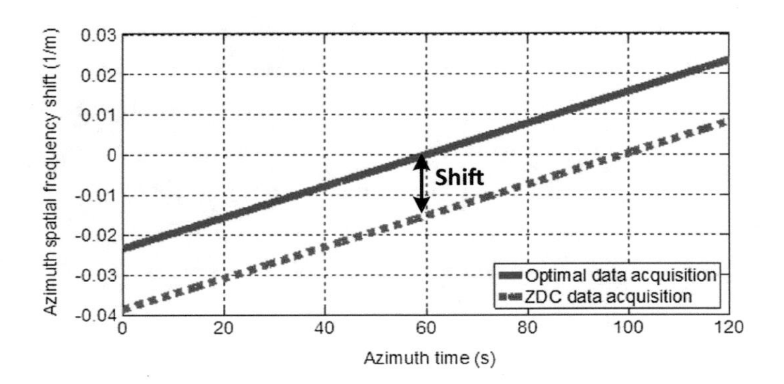

Fig. 6.9 Azimuth spatial spectral shifts corresponding to two data acquisition methods

data acquisition method is applied, the standard deviation of the phase noise reduces about 0.65 rads based on (6.25) and (6.26) in a single-look image pair case. Considering the GEO SAR parameters in Table 6.6 (the interferometric baseline is about 5 km in the proposed case) and (6.30), we can obtain the improvement of the height accuracy under the proposed case is more than 40 m, which suggests our OMRD data acquisition method has a much better performance.

Similarly, the OMRD data acquisition method can also raise the deformation retrieval accuracy in GEO D-InSAR processing by improving the coherence of the InSAR pairs and reducing the phase noise level in the corresponding differential interferogram. As for the D-InSAR data with a 1 day interval, Fig. 6.10 shows that the accuracy of the deformation retrieval has been improved significantly if the OMRD data acquisition method is applied in D-InSAR data acquisition.

Moreover, ignoring the spatial variation of the scene, the Doppler frequencies in different data acquisition methods are given in Fig. 6.11. By using the ZDC data acquisition method, the InSAR pair has no Doppler centroid shifts while the InSAR pair has the Doppler centroid shifts of nearly 20 Hz by using the OMRD data acquisition method. Combined with the previous analysis of the rotation-induced decorrelation of the two methods, it can be concluded that no Doppler centroid shifts does not mean the maximal overlapped azimuth spatial spectra and no rotation-induced decorrelation under the impacts of un-parallel repeated tracks. Thus, the classical common band filtering in the frequency domain cannot eliminate the rotation-induced decorrelation. Nevertheless, a proper processing in the wavenumber domain, which is similar to the common band filtering, will be an alternative method of the OMRD data acquisition method to improve the azimuth correlation.

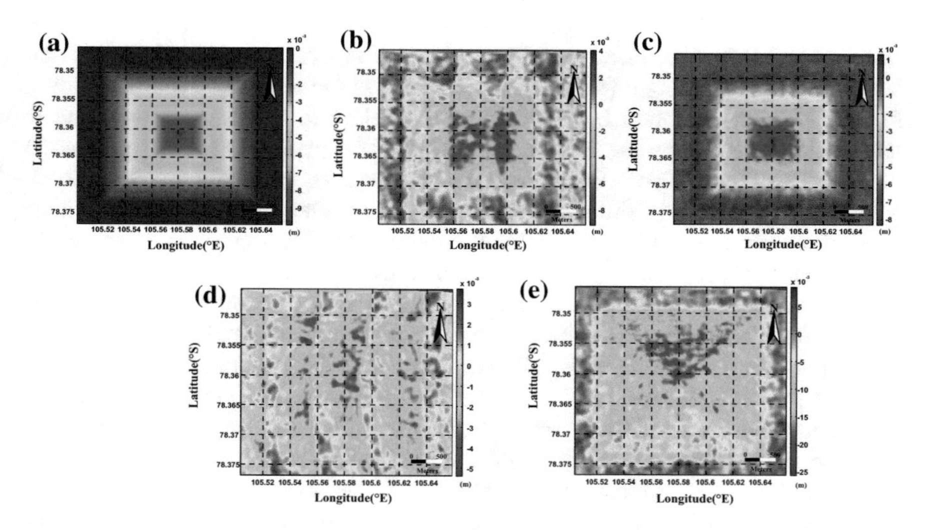

Fig. 6.10 **a** Reference deformation map; **b** obtained deformation map (ZDC data acquisition method); **c** obtained deformation map (OMRD data acquisition method); **d** deformation error map (ZDC data acquisition method) (root mean square error (RMSE) = 0.0016 m); **e** deformation error map (OMRD data acquisition method) (RMSE = 0.0009 m)

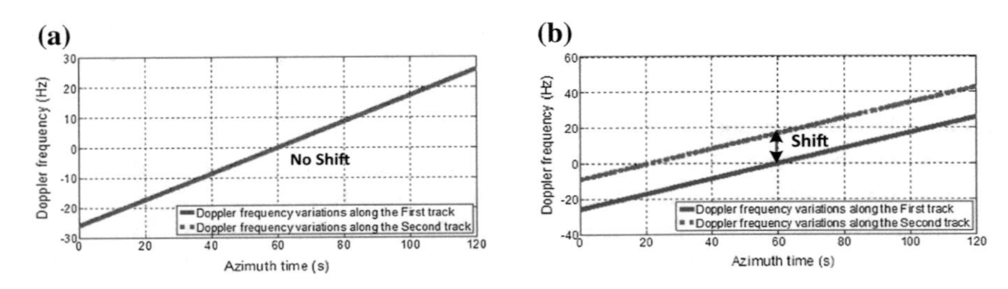

Fig. 6.11 Doppler frequencies in two data acquisition methods: **a** ZDC data acquisition method; **b** OMRD data acquisition method

6.3.3.2 Verification of the GEO InSAR Height Retrieval Model

In the simulations for the verification of the height retrieval model, the OMRD data acquisition method is utilized for the data acquisition. After obtaining the GEO InSAR pairs by the OMRD data acquisition at perigee and above the equator, the normal interferometry processing including co-registration and flat Earth removal is conducted until the height retrieval. Then, DEMs are generated based on the conventional model and the new model separately. The reference DEMs, the generated DEMs based on the conventional height retrieval model and the proposed model at perigee and above the equator are shown in Fig. 6.12. Some evaluation parameters of DEMs are given in Table 6.8.

At perigee, Fig. 6.12c has only a 26 m peak height, about ten times smaller compared with that in Fig. 6.12a. This is mainly because the repeated GEO InSAR tracks are un-parallel, thus the perpendicular baseline length (48.1 km) used in the conventional height retrieval model has a large bias with respect to the true interferometric baseline length (4.9 km). Meanwhile, as the retrieved height is utilized for the iteration operation in localization, the height retrieval errors will lead to localization errors, showing that the peak point of the pyramid moving to the north of the scene about 0.005°. In Fig. 6.12d, the geometric positioning error is intolerant at the satellite position above the equator. After the geocoding, latitude localization errors of more than 10° and longitude localization errors of about 30° emerge and the retrieved height information is distorted into a line-like area. These effects are caused by the failure of the convergence of the iteration in the broadside imaging localization equations because there exists no zero-Doppler frequencies for the targets along the GEO SAR full apertures above the equator.

Since the GEO InSAR height retrieval model ensures the correct expression of localization equations and the interferometric baseline length, Fig. 6.12e, f are almost the same as Fig. 6.12a, b both in height and localization. Their height error maps are shown in Fig. 6.13 and the related RMSEs of the height are 2.12 and 0.99 m, respectively. Generally, the height retrieval accuracy highly relates to the length of the interferometric baseline [5]. As the length of the interferometric baseline of the InSAR pair at the satellite position above the equator is 22 km, which is larger than that at perigee (4.9 km), the height retrieval accuracy of the GEO InSAR pair at the satellite position above the equator is better than that at perigee.

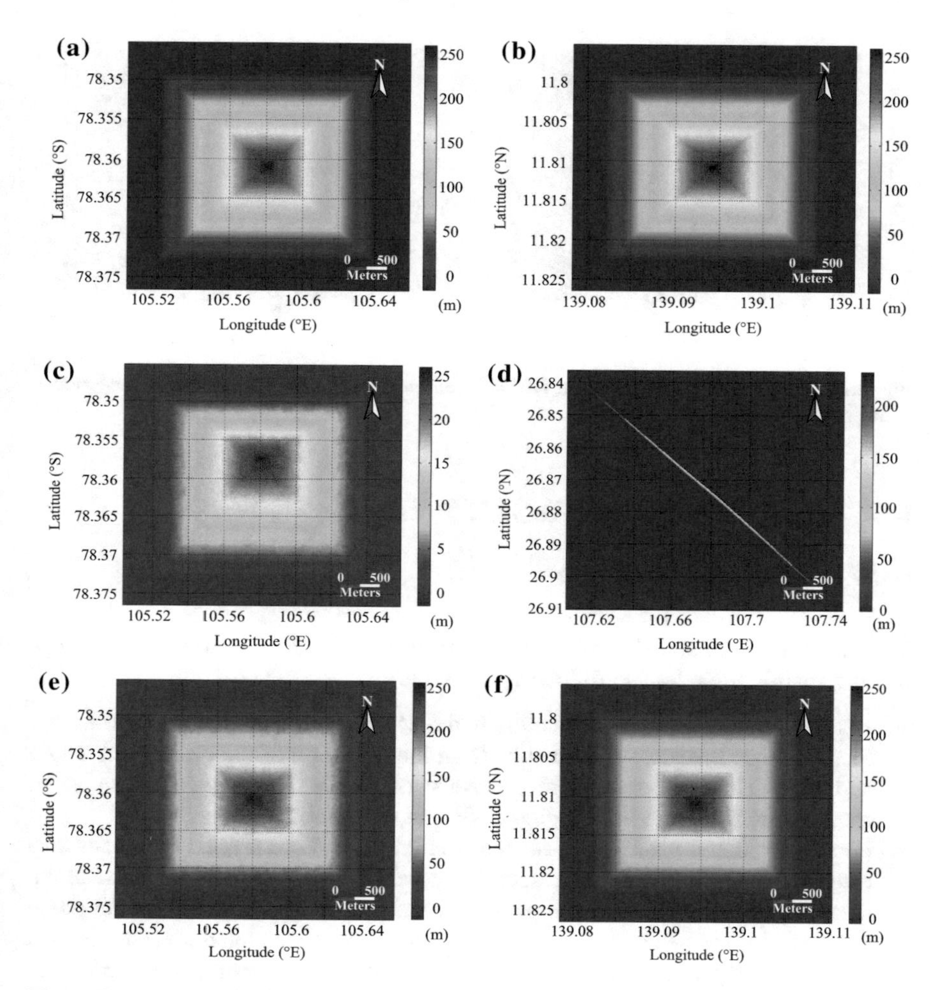

Fig. 6.12 Reference DEMs at perigee (**a**) and above the equator (**b**); generated DEM based on the conventional model at perigee (**c**) and above the equator (**d**); generated DEM based on the proposed model at perigee (**e**) and above the equator (**f**)

Table 6.8 Errors analysis of the DEMs

DEMs	Positions	Peak height (m)	RMS (m)
Reference	Perigee	260	–
	Above the equator	260	–
Conventional model	Perigee	26	76.49
	Above the equator	233	–
Proposed model	Perigee	254	2.12
	Above the equator	254	0.99

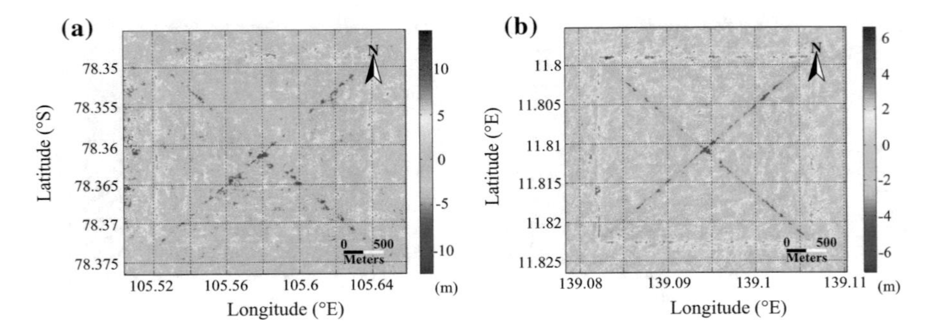

Fig. 6.13 Error maps of the generated DEMs based on the GEO InSAR height retrieval model. **a** At perigee; **b** above the equator

6.4 GEO InSAR and D-InSAR Processing

6.4.1 SAR Interferometry

Similar to traditional LEO SAR interferometry, GEO SAR interferometry processing mainly includes the following steps [20].

- Coregistration: there are dislocation and distortion in range and azimuth in two focused SAR images, which results from the deviation of the orbits, look angle and time. Thus, the consistent geometry of two SAR images contributes to acquire interferogram with higher SNR and coherence.
- Interferogram and flat-earth phase: the interferogram is constructed by a pointwise conjugate multiplication of corresponding pixels in both datasets. The phase of the interferogram includes flat-earth phase and topographic phase. The flat-earth phase is useless for the height retrieval and result in overcrowded interferometric fringes which seriously hinder the phase unwrapping. Thus, the flat-earth phase needs to be removed to obtain the interferogram with sparse fringes.
- Correlation coefficient map: The two focused SAR images after coregistration are utilized to generate correlation coefficient map. The map can be used to judge the quality of the interferogram and assist the phase unwrapping with the proper unwrapping path and weight.
- Phase filter: interferogram has low SNR because of the temporal decorrelation, baseline decorrelation, thermal decorrelation and processing induced decorrelation, which reduces the phase accuracy seriously and even makes phase unwrapping failed. Thus, interferometric phase noise must be remove by the effective filter to hence the quality of the interferogram.
- Phase unwrapping: the phase in interferogram is the 2π-modulus of the absolute phase, which means the two-dimensional relative phase signal is limited to the $[-\pi, \pi)$. In order to obtain the absolute phase, phase in each pixel need to be added with integer times of π. This processing is called phase unwrapping.

The phase unwrapping is one of the main technique in InSAR and it impacts the accuracy of digital elevation model (DEM) directly.

- Phase to height conversion and geocoding: after obtaining the absolute phase including the terrain information, the unwrapped interferometric phase can be converted into topographic height by adopting the GEO SAR models. Then, the image is transformed into WGS84 coordinate by interpolation into evenly grids to acquire DEM. By this step, the coordinate of the scene is shown by longitude, latitude and height.

The flow of the whole processing is shown in Fig. 6.14.

GEO InSAR simulation is based on the parameters in Table 6.9. The nadir-point trajectory of GEO SAR can be seen in Fig. 6.15. And the simulation results are shown in Fig. 6.16. Comparing the retrieval results with the setting scene, the RMSE of the whole scene is only 5.5778 m, which proves both the feasibility and the high accuracy of GEO InSAR.

6.4.2 *Differential Interferometry*

The algorithm of spaceborne differential interferometry includes two-pass differential interferometry and three-pass differential interferometry. The concept of two-pass differential interferometry is to remove topographic phase based on

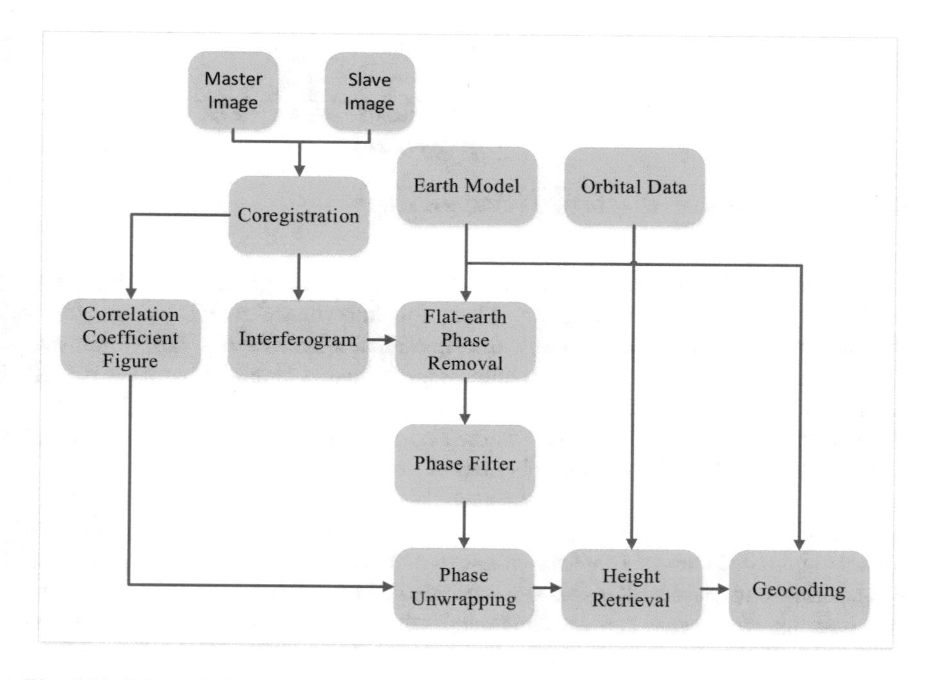

Fig. 6.14 Schematic block diagram of an interferometric processor

Table 6.9 GEO SAR system parameters

Orbit height	33,679 km (perigee)
PRF	150 Hz
Integration time	500 s
Bandwidth	20 MHz
Resolution	20 m
Spatial baseline	11.092 km
Elevation of the highest point	260 m
Setting elevation	260 m (highest point), 0 m (flat region)

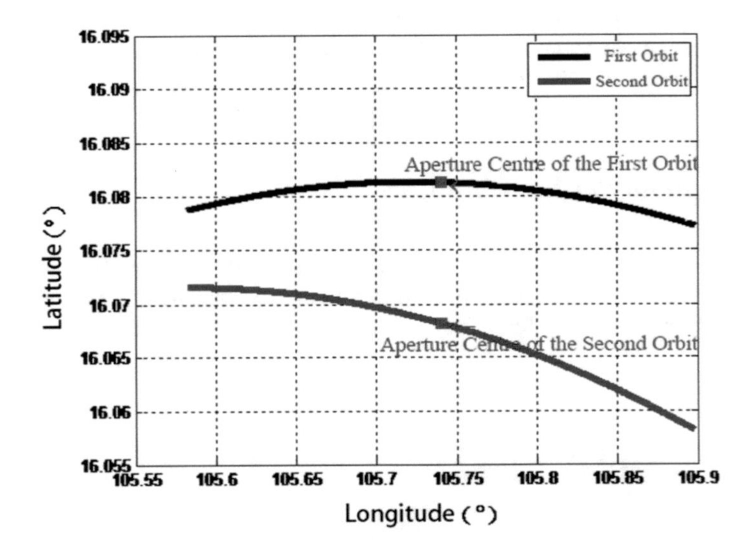

Fig. 6.15 Nadir-point trajectory in GEO InSAR repeat orbit

external known DEM. First, it uses two SAR images before and after the deformation to obtain the interferogram. Then, the known accurate DEM and orbit parameters are utilized to generate the simulated interferogram. The simulated interferogram is subtracted from the interferogram that is obtained by the two SAR images and the different phase caused by the deformation is acquired. The differential information can be directly obtained by the phase processing for the slight change comparing to the terrain information.

The advantage of two-pass interferometry is that the external DEM can assist to remove atmospheric impact in the two SAR images (it is extremely important to obtain simulated interferometric phase). On this basis, the coregistration, pre-filtering and phase filter can be continued. Also, it is worth noting that the two-pass interferometry can be conducted without flat-earth phase removal and phase unwrapping because the interferogram with accurate flat-earth phase can be

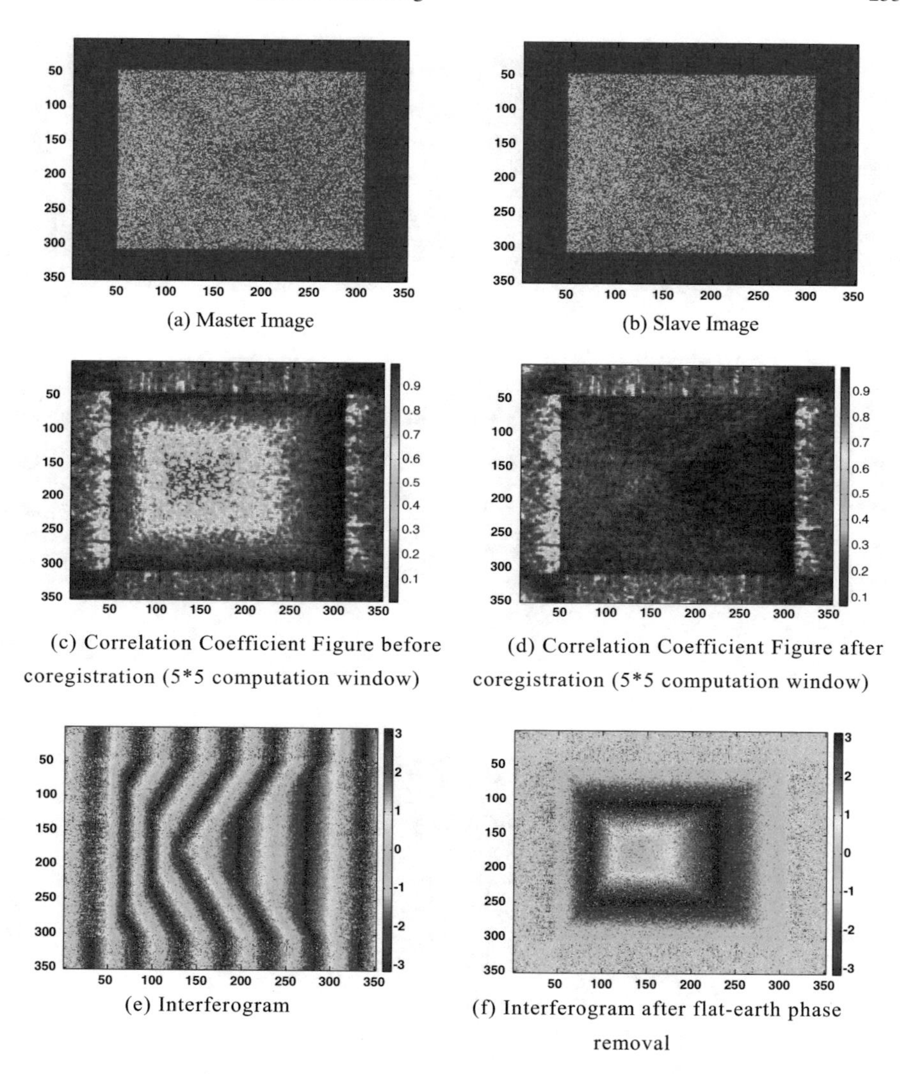

(a) Master Image

(b) Slave Image

(c) Correlation Coefficient Figure before coregistration (5*5 computation window)

(d) Correlation Coefficient Figure after coregistration (5*5 computation window)

(e) Interferogram

(f) Interferogram after flat-earth phase removal

Fig. 6.16 Results of InSAR

simulated by the external DEM and the change scope of deformation phase is during 2π. The processing flow of two-pass interferometry is shown in Fig. 6.17.

Three-pass interferometry needs three SAR images, one of which is acquired before the deformation and is taken as main image. The rest of images is regarded as slave images. One is acquired before the deformation and the other is acquired after the deformation. The three SAR images is utilized to generate topographic interferogram and deformation interferogram. In fact, the topographic interferometric phase needs to unwrap and multiply a factor, than minuses the deformation interferometric phase which has removed flat-earth phase. If the phase is wrapped,

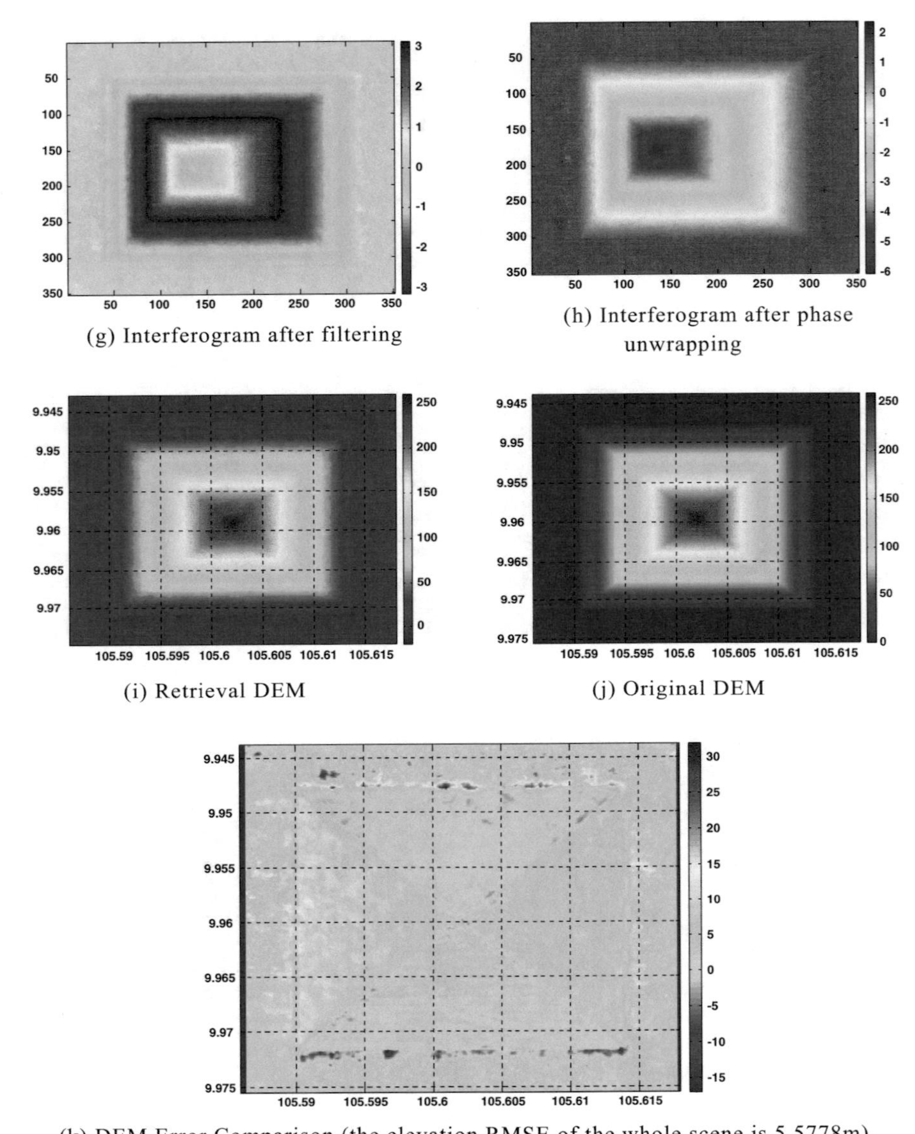

(g) Interferogram after filtering

(h) Interferogram after phase unwrapping

(i) Retrieval DEM

(j) Original DEM

(k) DEM Error Comparison (the elevation RMSE of the whole scene is 5.5778m)

Fig. 6.16 (continued)

the error will be generated. The processing flow of three-pass interferometry is shown in Fig. 6.18.

Here, the three-pass interferometry is taken as an example to conduct the simulation. Parameters of the simulation are shown in Table 6.10. And the nadir-point trajectory of GEO SAR can be seen in Fig. 6.19. The results of GEO SAR differential

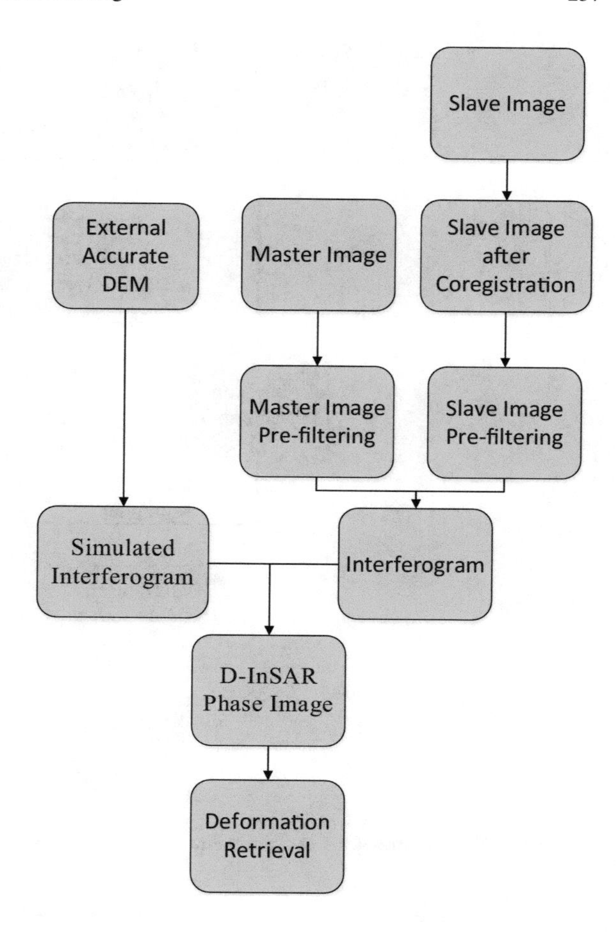

Fig. 6.17 Flow of two-pass interferometry

interferometry are shown in Fig. 6.20. Comparing with the deformation in setting scene, the RMSE of absolute deformation in the scene in vertical is only 2.9 mm, which shows high precision of deformation retrieval and proves that GEO D-InSAR has the ability of accurate deformation measurement.

6.4.3 Performance Analysis

6.4.3.1 Analysis of Decorrelation Factors

As similar to LEO SAR, the coherence coefficient is also an important aspect that needs to be studied in GEO InSAR and D-InSAR.

The coherence coefficient is a measurement of the stability of interferometric phase between the two radar complex images obtained by the InSAR system, which

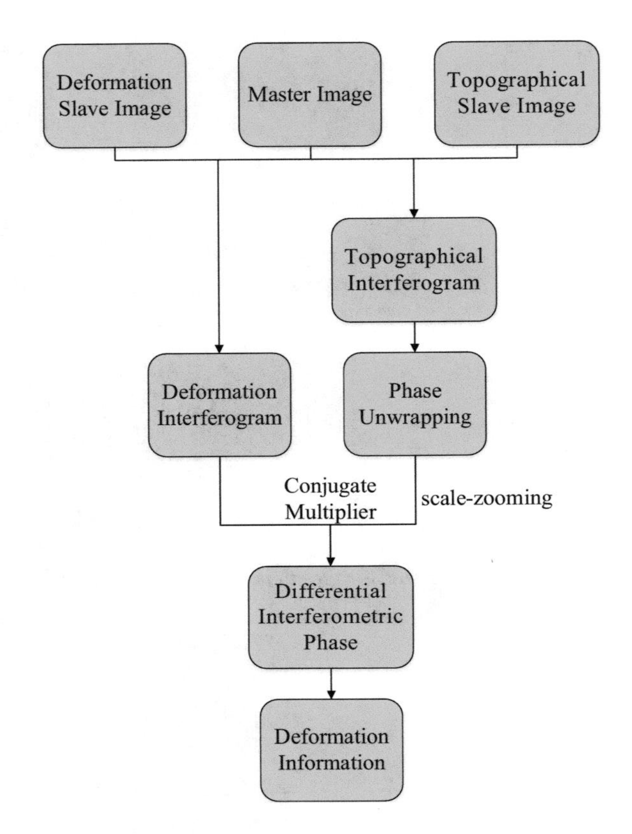

Fig. 6.18 Flow of three-pass interferometry

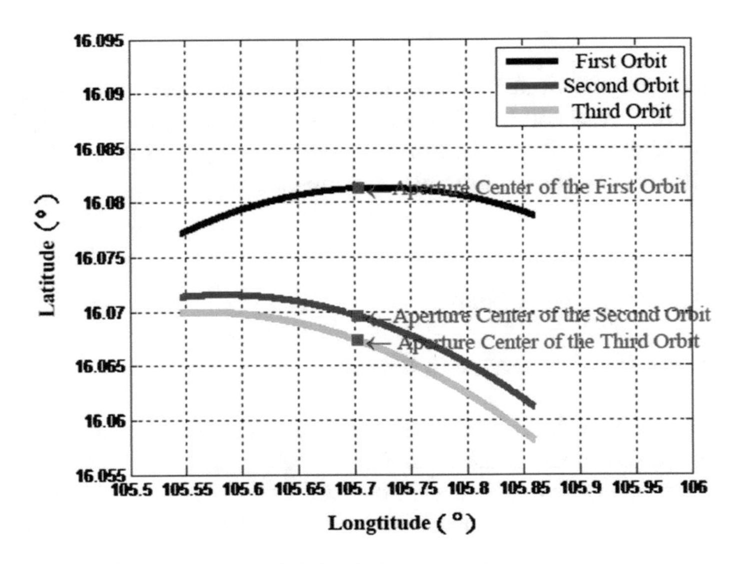

Fig. 6.19 Nadir-point trajectory in GEO DInSAR repeat orbit

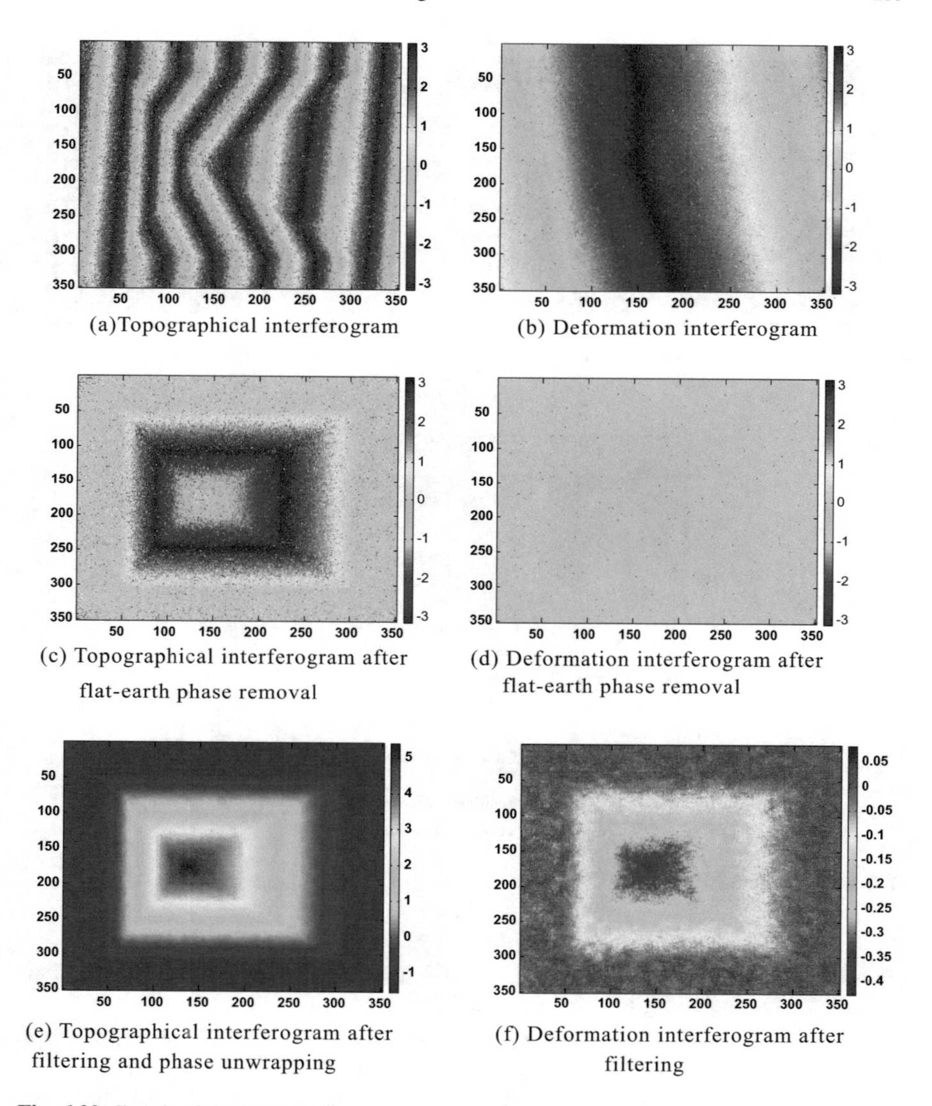

(a)Topographical interferogram

(b) Deformation interferogram

(c) Topographical interferogram after flat-earth phase removal

(d) Deformation interferogram after flat-earth phase removal

(e) Topographical interferogram after filtering and phase unwrapping

(f) Deformation interferogram after filtering

Fig. 6.20 Result of GEO DInSAR

is directly related to the accuracy of the interferometric phase. If the coherence coefficient is close to 1, it indicates that the two images have better correlation. However, the coherence coefficient is usually reduced due to the geometry, the surface scattering characteristics and the random thermal noise when acquiring the master and slave images.

There are many factors that affect the coherence of two SAR images. Generally, the decorrelation factors in GEO SAR are similar to those in LEO SAR, which are consisted of the following aspects.

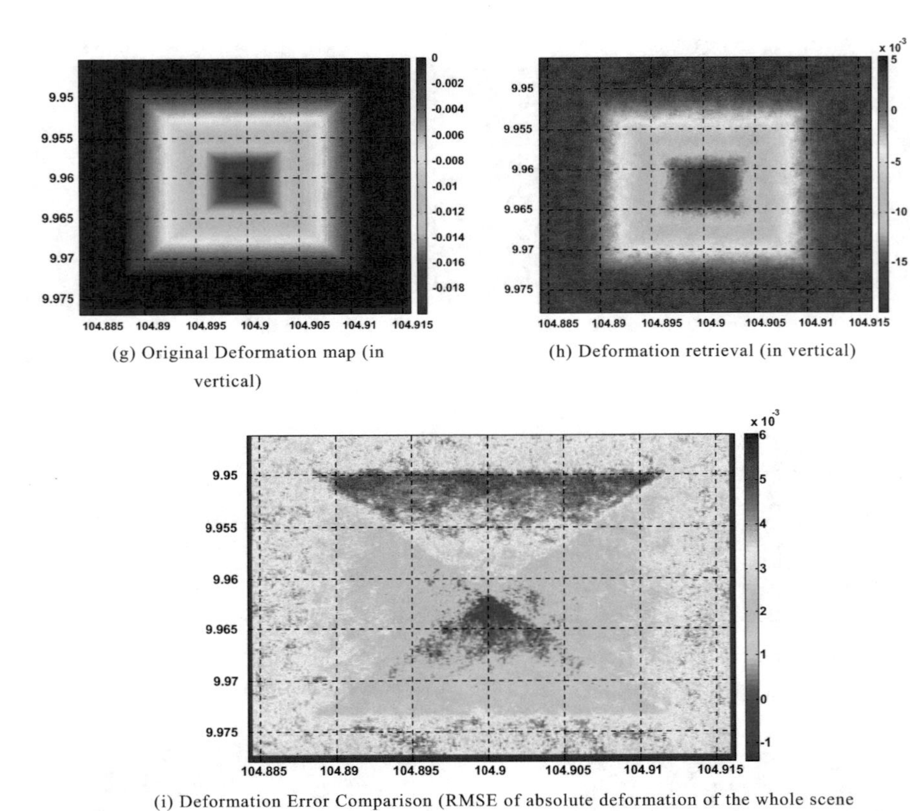

(g) Original Deformation map (in vertical)

(h) Deformation retrieval (in vertical)

(i) Deformation Error Comparison (RMSE of absolute deformation of the whole scene in vertical is 2.9mm)

Fig. 6.20 (continued)

- Thermal noise decorrelation $\gamma_{thermal}$. It's mainly affected by the system characteristics, including antenna gain and features;
- Spatial decorrelation. It is caused by a difference in angle of view (including spatial baseline decorrelation $\gamma_{spatial}$ and rotation decorrelation $\gamma_{rotation}$);
- Temporal decorrelation $\gamma_{temporal}$. It is the decorrelation caused by the change of surface scattering characteristics during the images acquisition.
- Data processing decorrelation $\gamma_{processing}$. It's the decorrelation introduced by the image co-registration and resampling.

The total coherence coefficient is the product of the each coherent term:

$$\gamma = \gamma_{thermal} \times \gamma_{spatial} \times \gamma_{rotation} \times \gamma_{temporal} \times \gamma_{processing} \tag{6.31}$$

Table 6.10 GEO D-InSAR system parameters

Orbit height	33,679 km (perigee)
PRF	150 Hz
Integration time	500 s
Bandwidth	20 MHz
Resolution	20 m
Spatial baseline	10.275 km (elevation) 2.065 km (deformation)
Setting deformation	20 mm (the maximal) 0 mm (the minimal)

6.4.3.2 Rotation Decorrelation

In contrast to LEO SAR, the non-parallel trajectory of GEO SAR caused by the orbital perturbation is very serious for the repeat-pass GEO SAR interferometry processing. This leads to the same target having different scattering characteristics in the line of sight direction when acquiring different image, which introduces the rotation decorrelation. In other words, illuminating the same area from two different angles results in the signal decorrelation. In a resolution cell as shown in Fig. 6.21, the polar coordinate position of each scattering center is rotated from (δ, ϕ) to $(\delta, \phi + d\phi)$. The formulas for converting polar coordinates into Cartesian coordinates are $x = \delta \cos \phi$ and $y = \delta \sin \phi$. Assuming that the distance from the antenna to one of the scattering points is $r + \delta \sin \theta \sin \phi_1$ before rotation, the distance after rotating a small angle $d\phi = \phi_1 - \phi_2$ is $r + \delta \sin \theta \sin \phi_2$. Due to the small angle of rotation, the distance from the antenna to the scattering center and the corresponding phases have changed slightly, so their coherence has also changed.

Assuming that the radar signals acquired by the two antennas are s_1 and s_2, they represent the echo signals before and after the rotation of the resolution cell, respectively. Therefore, the cross-correlation of the two signals is obtained as follows

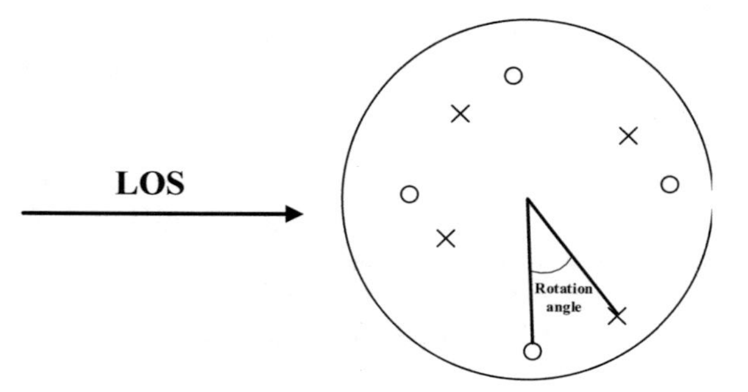

Fig. 6.21 Geometric schematic of rotation decorrelation (The circles represent scatterers in the first observation and the forks is the scatterers in the second observation)

$$
\begin{aligned}
s_1 s_2^* = & \iint \iint f(x - x_0, y - y_0) f^*(x' - x_0, y' - y_0) \\
& \exp\left\{ -j\frac{4\pi}{\lambda}\delta \sin\theta(\sin\phi_1 - \sin\phi_2) \right\} W(x, y) \\
& W^*(x', y') dx dy dx' dy'
\end{aligned} \tag{6.32}
$$

After taking the statistical average for the scatterers in the resolution cell, we can get the following formula:

$$
\langle s_1 s_2^* \rangle = \sigma_0 \iint \exp\left\{ -j\frac{4\pi}{\lambda} x \sin\theta d\phi \right\} |W(x, y)|^2 dx dy \tag{6.33}
$$

Through the Fourier transform of the formula, we can obtain the decorrelation expression caused by the corresponding rotation in the azimuth pulse response.

$$
\gamma_{rotation} = 1 - \frac{2\sin\theta |d\phi| R_x}{\lambda} \tag{6.34}
$$

6.4.4 GEO InSAR Baseline Analysis

Repeat-pass InSAR/DInSAR refers to observing the same area from a near-parallel trajectory with a slightly different angle of view, and then it uses the imaging mechanism and geometric relations to retrieve the elevation or deformation information. Baseline refers to the spatial distance between the two satellites when imaging the same target. The baseline length directly affects the measurement performance of the InSAR/DInSAR system.

The nadir track of GEO SAR is a closed graph, such as small "figure-8" and small "figure-O". Under the influence of perturbation, the orbital period of GEO SAR gradually increases, which leads to the decrease of the longitude of ascending node. And the nadir track of GEO SAR shows the phenomenon of 'westward'. After a period of time, the orbital period of GEO SAR gradually decreases, leading to the increase of longitude of ascending node, and the nadir track shows the phenomenon of 'eastward'. The consequence of perturbation is that the repeat GEO SAR orbit tracks are not parallel at the most southern and northern ends at different cycles, and there may exist crossings. In other words, it is difficult to ensure that the repeat tracks are approximately parallel to each other.

We use the HPOP model in STK to simulate the GEO SAR orbit under various perturbations. The baseline simulation results for different orbit configurations are given, as shown in Fig. 6.22, respectively.

In order to compare the results of baseline simulations more clearly, Table 6.11 gives the detailed baselines at typical locations under different orbits.

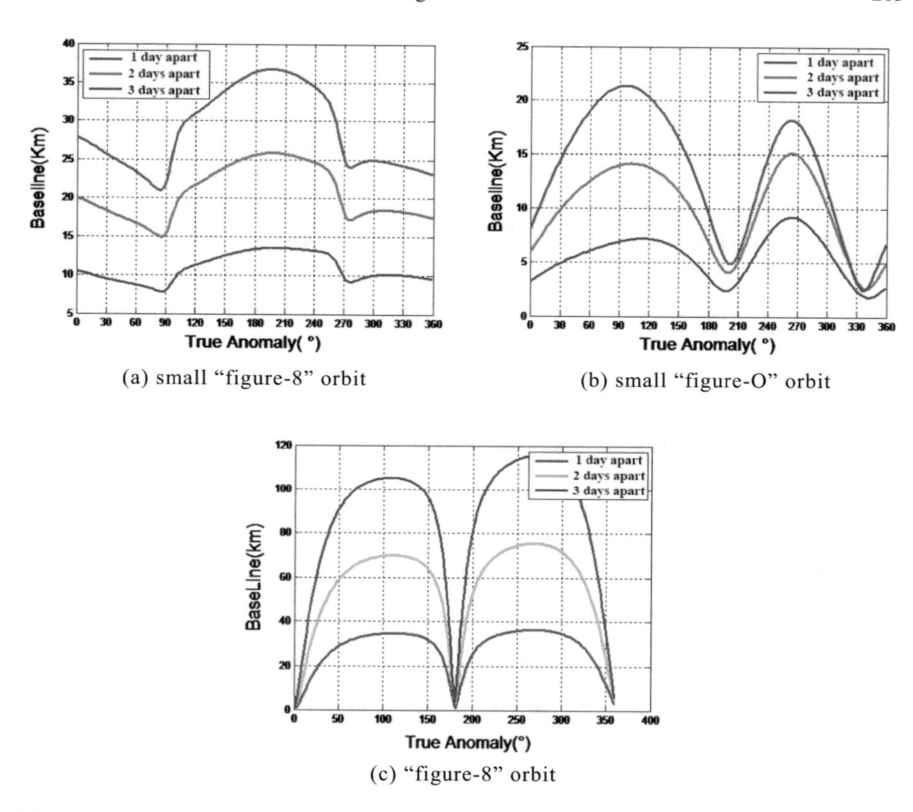

(a) small "figure-8" orbit

(b) small "figure-O" orbit

(c) "figure-8" orbit

Fig. 6.22 Baseline simulation results of GEO SAR repeat tracks

Table 6.11 Baseline of GEO SAR at different locations

Position		1st day	2nd day	3rd day
Small "figure-8" orbit	Southernmost (km)	7.97	15.38	21.96
	Equator (km)	10.49	20.05	27.92
	Northernmost (km)	9.89	18.29	25.31
Small "figure-O" orbit	Southernmost (km)	3.22	5.93	8.05
	Equator (km)	6.88	14.00	21.30
	Northernmost (km)	3.58	6.74	9.30
"figure-8" orbit	Southernmost (km)	3.06	5.02	5.95
	Equator (km)	36.28	75.28	115.50
	Northernmost (km)	2.24	2.96	2.90

It can be seen from Fig. 6.22 and Table 6.11 that the interferometric baselines of the 3 GEO SAR orbit configurations. The baseline of the small "figure-8" orbit is greater than the baseline of the small "figure-O" orbit. The "figure-8" orbit has the greatest baseline near the equator and the smaller baseline at both ends. In addition, the baseline increases by times as the time interval increases.

In InSAR processing, spatial baseline will affect accuracies of the height retrieval and the interferogram coherence. The selection of the baseline length is a compromise between the coherence and the height measurement accuracy. In GEO SAR, as slant ranges arise up to the level of 40,000 km, the baseline in GEO InSAR is relatively large compared with LEO SAR. Nevertheless, a long baseline will make the produced InSAR pair have a small overlapped ground spectrum. When the baseline length is larger than the critical baseline, a totally decorrelation will occur. The critical baseline $B_{\perp,c}$ is expressed as

$$B_{\perp,c} = \frac{\lambda R}{2R_y \cos^2 \theta} \tag{6.35}$$

where λ is the wavelength, R is the slant range, R_y is the range resolution and θ is the incidence angle. Thus in GEO InSAR, the long slant range will result in the large baseline.

If only the decorrelations of the spatial baseline and the thermal noise are considered, the optimized perpendicular baseline can be expressed as [6]

$$B_{\perp,opt} = \left\{1 - \gamma_{opt}\left(1 + SNR^{-1}\right)\right\} \frac{\lambda R}{2\rho_a \cos \theta} \tag{6.36}$$

where the optimized correlation coefficient $\gamma_{opt} = 0.6180334 - 1.717SNR^{-1}$. Assuming that the SNR is 10 dB, the azimuth resolution ρ_a is 20 m, and the geocentric-satellite distance is 42,000 km, the simulation of the optimized perpendicular baseline of GEO SAR interferometry is presented in Fig. 6.23. The optimized perpendicular baseline lies between 113 and 235 km, which are dependent on the incident angles.

6.4.5 Analysis of Measurement Accuracy

Measurement accuracy is an important indicator of all kinds of system design and the product quality evaluation. This section gives the analysis of the main factors

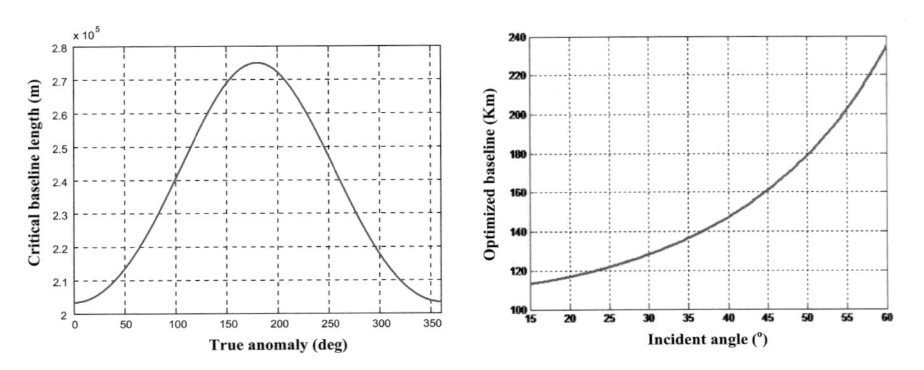

Fig. 6.23 Simulation of the critical and optimized perpendicular baseline of GEO InSAR

that affect the measurement accuracy of InSAR/DInSAR. And then we give simulations of the accuracy of elevation and deformation measurement.

6.4.5.1 Elevation Measurement Accuracy

According to the interferometry processing basics, the measurement accuracy of the target elevation is related to the following factors:

$$\sigma_h^2 = \left(\frac{\partial h}{\partial H}\right)^2 \sigma_H^2 + \left(\frac{\partial h}{\partial R}\right)^2 \sigma_R^2 + \left(\frac{\partial h}{\partial \alpha}\right)^2 \sigma_\alpha^2 + \left(\frac{\partial h}{\partial B}\right)^2 \sigma_B^2 + \left(\frac{\partial h}{\partial \varphi}\right)^2 \sigma_\varphi^2 \tag{6.37}$$

where h is the elevation of the target; H is the height of the satellite to the ground; R is slant range from target to satellite; α is the angle between the baseline and the horizontal direction; B is the length of baseline; φ is interferometric phase; $\sigma_h, \sigma_H, \sigma_R, \sigma_\alpha, \sigma_B$ and σ_φ are the measurement accuracy of the corresponding parameters respectively.

The above formula gives the absolute accuracy of elevation measurement. Because interferometry is a technique to measure the relative elevation, the absolute accuracy does not reflect the real performance of the system. Therefore, using relative accuracy of elevation measurement is more appropriate. According to the above analysis, the satellite height error, the slant range error and the angle error have the same effect on each point in the scene because of the small scene approximation. Comparatively, the phase error is a random error caused by thermal noise and other decorrelation factors. The baseline error is due to the existence of orbital misalignment errors, so the baseline error for each pixel is also a random error. Therefore the phase error and the baseline error are the main sources of error that affect the relative elevation measurement accuracy. We can draw the following formula:

$$\sigma_h^2 = \left(\frac{\partial h}{\partial B}\right)^2 \sigma_B^2 + \left(\frac{\partial h}{\partial \varphi}\right)^2 \sigma_\varphi^2 \tag{6.38}$$

The Influence of Interferometric Phase Error

The interferometric phase error is mainly affected by the coherence coefficient between the images and the number of looks. As the coherence coefficient and the number of looks increase, the phase error decreases. According to the theoretical derivation of the relationship between the target elevation and the interferometric phase, the elevation measurement error caused by the interferometric phase error can be expressed as:

$$\sigma_h^2 = \left| \frac{\lambda R \sin\theta}{4\pi B_\perp} \right|^2 \sigma_\varphi^2 \tag{6.39}$$

In the case of the coherence coefficient is 0.8 and the number of looks is 8, the interferometric phase error is 11.81°. The relationship between the elevation measurement error and the vertical baseline length is shown in Fig. 6.24. It can be seen that when the interferometric phase error is constant, the elevation measurement accuracy is improved with the increase of the vertical baseline length.

The Influence of Orbit Determination Error

Satellite orbit determination accuracy is defined as the accuracy in the directions of radial, along track and across track in the satellite coordinate system. In the InSAR/DInSAR measurement, the accuracy of the vertical baseline is related to the measurement accuracy, and the length of the vertical baseline is only related to the orbit determination accuracy in both directions of radial and across track.

It is assumed that the orbit determination errors of reference orbit and repeat orbit obey the Gaussian distribution of 0-mean and σ_R^2-variance in the direction of radial, and similarly the errors also obey the Gaussian distribution of 0-mean and σ_A^2-variance in the direction of across track. Also, we assume that the orbit determination errors of the two orbits are irrelevant. So the baseline errors in the radial and across track directions can be expressed as:

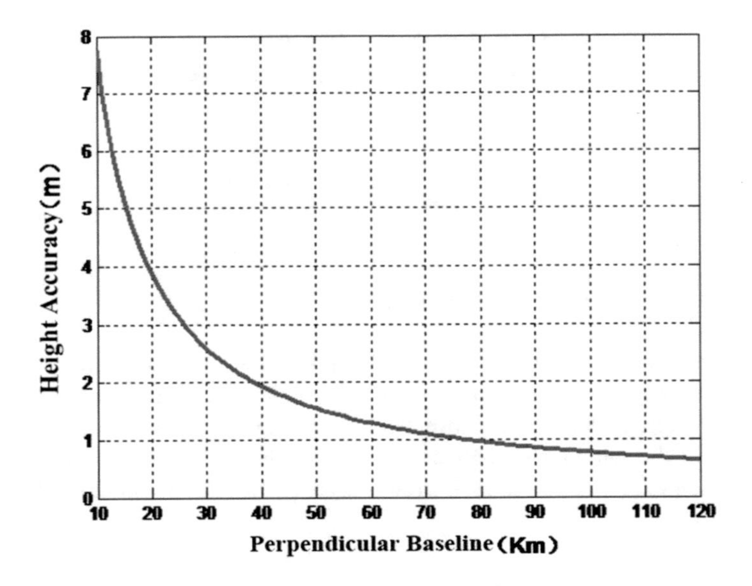

Fig. 6.24 Influences of interferometric phase error on elevation measurement accuracy

$$\sigma_{B_v} = \sqrt{\sigma^2_{Radial,1} + \sigma^2_{Radial,2}} = \sqrt{2}\sigma_R \tag{6.40}$$

$$\sigma_{B_h} = \sqrt{\sigma^2_{acrosstrack,1} + \sigma^2_{acrosstrack,2}} = \sqrt{2}\sigma_A \tag{6.41}$$

The components of the baseline in parallel and perpendicular to the direction of the line of sight have a transformation relationship with the components of the baseline in the radial and across track directions:

$$\begin{bmatrix} B_\perp \\ B_\parallel \end{bmatrix} = \begin{bmatrix} \cos\theta & \sin\theta \\ -\sin\theta & \cos\theta \end{bmatrix} \begin{bmatrix} B_h \\ B_v \end{bmatrix} \tag{6.42}$$

where θ is down-looking angle. According to the properties of the Gaussian distribution, $B_\perp$ and $B_\parallel$ also obey the Gaussian distribution, and the expression of the distribution can be described as:

$$\begin{aligned} B_\perp &\sim N(0, 2\sigma_A^2 \cos^2\theta + 2\sigma_R^2 \sin^2\theta) \\ B_\parallel &\sim N(0, 2\sigma_A^2 \sin^2\theta + 2\sigma_R^2 \cos^2\theta) \end{aligned} \tag{6.43}$$

Assuming that the orbit determination accuracy is exactly the same in the radial and across track directions, the distributions of the parallel and vertical baselines satisfy:

$$\begin{aligned} B_\perp &\sim N(0, 2\sigma^2) \\ B_\parallel &\sim N(0, 2\sigma^2) \end{aligned} \tag{6.44}$$

From the analysis of interferometric signal processing, the measurement error introduced by the baseline error comes from two steps, namely, the removal of the flat ground effect and the elevation inversion. In the case of the change of the down-looking angle is not very large, the interferometric phase difference of any two scatterers can be decomposed into two parts: the flat ground phase part and the elevation phase part. So we can get the following formula

$$\varphi = \varphi_{flat} + \varphi_z = -\frac{4\pi B_\perp \Delta R}{\lambda R \tan\alpha} - \frac{4\pi B_\perp q}{\lambda R \sin\alpha} \tag{6.45}$$

where ΔR is difference of slant range between the two scatterers; q is difference of elevation between the two scatterers; R is the slant range from satellites to the ground; α is incident angle. Therefore, the elevation measurement error introduced by the baseline error can be expressed as:

$$\sigma_h^2 = \left| \frac{\Delta R \cos\alpha}{B_\perp} + \frac{q}{B_\perp} \right|^2 \sigma_{B_\perp}^2 \tag{6.46}$$

Simulation

We used the typical GEO SAR parameters for the simulation and analysis of the elevation measurement accuracy. The system parameters are shown in Table 6.12. The simulation results are shown in Fig. 6.25. It can be seen from the figure that the length of vertical baseline must be longer than 16.3 km if the elevation measurement accuracy of 10 m is needed.

Thus, based on the analysis of the GEO SAR repeat track baseline shown in Fig. 6.22, it can be concluded that the perturbed small "figure-8" orbit can form a larger baseline which is suitable for interferometric measurements; the repeat track baseline of small "figure-O" orbit is relatively short which is not suitable for interferometric measurements; "figure-8" orbit has a better overall performance except for positions at the perigee and the apogee.

Table 6.12 System parameters for GEO SAR simulation

Parameters	Values	Parameters	Values
Frequency	1.25 GHz	Scene size	$\leq$ 100 km × 100 km
Down-looking angle	4.65°	Variations in terrain height	$\leq$ 1 km
Incident angle	32.62°	Correlation coefficient	$\geq$ 0.8
Slant range	36,500 km	Number of looks	8
Orbit accuracy	2 m	Average Earth radius	6371 km
Geocentric distance	42,000 km	SRTM accuracy	16 m

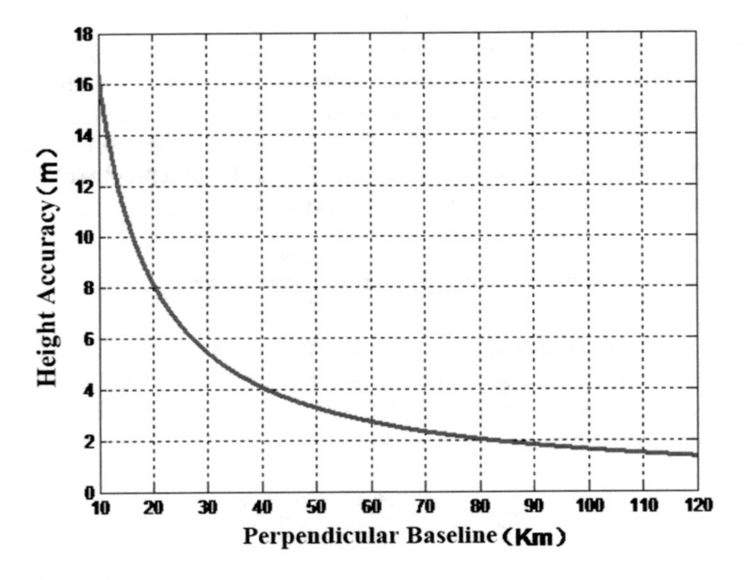

Fig. 6.25 Simulation results of elevation measurement accuracy

6.4.5.2 Deformation Measurement Accuracy

There are two ways to achieve deformation measurement for differential interferometry: the two-pass interferometry and the three-pass interferometry. The accuracy analysis of the two methods is different, and this section takes the two-pass interferometry as an example to carry on the accuracy analysis.

The measured output of the two-pass interferometry is

$$d = \frac{\lambda}{4\pi}(\varphi - \varphi_{simu}) \tag{6.47}$$

where φ is the measured interferometric phase; φ_{simu} is simulated interferometric phase according to the external DEM information and the orbit information. Therefore, the factors that affect the accuracy of differential interferometry are mainly interferometric phase error and simulated phase error, in which the simulated phase error is determined by the baseline error introduced by external DEM error and orbit error.

$$\Delta d^2 = \Delta d_{\varphi}^2 + \Delta d_{simu}^2 = \Delta d_{\varphi}^2 + \Delta d_{DEM}^2 + \Delta d_{track}^2 \tag{6.48}$$

In the case of the coherence coefficient is 0.8 and the number of look is 8, the interferometric phase error is 11.81°. According to the relationship between the differential interferometric phase and the deformation, the interferometric phase error introduces a deformation error of 3.9 mm.

The simulated phase error is caused by the external DEM error and the baseline error introduced by the orbital error. According to (6.48), the part affected by the external DEM error can be expressed as:

$$\Delta d_{DEM} = -\frac{B_{\perp}}{R \sin \alpha}\Delta q \tag{6.49}$$

where Δq is external DEM error. As can be seen from the expression, the longer the vertical baseline, the greater the effect of the external DEM error on the measurement accuracy. Therefore a shorter vertical baseline is more conducive to differential interferometry.

The error introduced by the vertical baseline error can be expressed as:

$$\Delta d_{Track} = -\frac{\Delta R}{R \tan \alpha}\sigma_{B_{\perp}} - \frac{q}{R \sin \alpha}\sigma_{B_{\perp}} \tag{6.50}$$

According to the simulation parameters listed in Table 6.12, the measurement accuracy of GEO SAR repeat track differential interferometry is simulated and the results are shown in Fig. 6.26. Based on the analysis of the GEO SAR repeat track baseline shown in Fig. 6.22, it can be concluded that the perturbed small “figure-8” orbit and perturbed “figure-8” orbit can form a larger repetitive orbital baseline

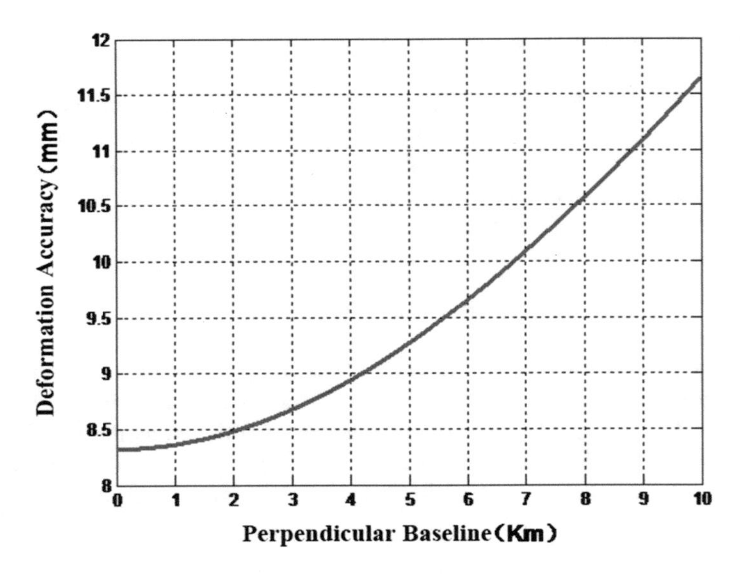

Fig. 6.26 Simulation of deformation measurement accuracy

which is not suitable for differential interferometry; the baseline formed by the perturbed small "figure-O" orbit meets the requirements of the simulation results, which is more suitable for differential interferometry.

6.5 Summary

GEO InSAR and D-InSAR are two of the most important applications in GEO SAR system. This chapter gives the corresponding basic concept, methodology, key technology and accuracy analysis of GEO InSAR and D-InSAR. The difference between LEO SAR and GEO SAR mainly lies in the geometrical relationship of data acquisition as the higher orbit in GEO SAR. Especially under the effects of complex perturbation forces in GEO SAR, the longitude of ascending node will oscillate within a certain range periodically. Furthermore, the nadir-point trajectories are not parallel in the positions of the perigee and the apogee, which results in the serious rotation decorrelation by ZDC data acquisition. Aiming at this problem, the proposed data acquisition method by minimizing the rotation decorrelation can remove the rotation decorrelation effectively and improve the performance of InSAR and D-InSAR. Meanwhile, the high-orbit perturbation can also generate longer perpendicular baseline as several kilometers or dozens of kilometers. However, the critical baseline of GEO SAR can reach several hundreds of kilometers which are difficult to reach for the repeat-track perpendicular baseline and thus will not impact the procession of GEO InSAR. GEO SAR always works in the squint looking mode.

Therefore, the height model under GEO SAR squint geometry is proposed to realize accurate height retrieval and geocoding. Based on the baseline selection and height retrieval model, the data processing and performance analysis are also presented which are similar to those in LEO SAR.

Aiming at different GEO SAR orbit configurations (such as different orbit inclination and eccentricity), this chapter also analyzes their differences in InSAR and D-InSAR. In different orbit configurations, small 'Fig. 8' orbit can form large perpendicular baseline which is suitable for InSAR, instead of D-InSAR. While the perpendicular baseline of 'figure O' orbit is relatively small and not suitable for InSAR but good for D-InSAR. The performance of big 'figure-8' orbit in InSAR and D-InSAR is moderate.

References

1. Ferretti A, Montiguarnieri A, Prati C et al (2007) InSAR principles—guidelines for SAR interferometry processing and interpretation. J Financ Stab 10(10):156–162
2. Goldstein RM, Zebker HA (1987) Interferometric radar measurement of ocean surface currents. Nature 328(6132):707–709
3. Orwig LP, Held DN (1992) Interferometric ocean surface mapping and moving object relocation with a Norden systems Ku-band SAR. Geoscience and remote sensing symposium, 1992. IGARSS'92. International. IEEE, pp 1598–1600
4. Long T, Dong X, Hu C et al (2011) A new method of zero-doppler centroid control in GEO SAR. IEEE Geosci Remote Sens Lett 8(3):512–516
5. Hu C, Li X, Long T et al (2003) GEO SAR interferometry: theory and feasibility study. Radar conference 2013, IET International. IET, pp 1–5
6. Guarnieri A M, Tebaldini S, Rocca F et al (2012) GEMINI: geosynchronous SAR for earth monitoring by interferometry and imaging. 137(60):210–213
7. Bruno D, Hobbs SE, Ottavianelli G (2006) Geosynchronous synthetic aperture radar: concept design, properties and possible applications. Acta Astronaut 59(1–5):149–156
8. Wertz JR, Larson WJ (1999) Space mission analysis and design. Kluwer Academic Publishers, Boston, USA
9. Marshall SR, Patrick RC (1997) Satellite tool kit user's manual: intuitive satellite systems analysis software design to assist in visualizing and analyzing complex relationships in space systems. Analytic Graphics, Incorporated, USA
10. Hu C, Long T, Zeng T, Liu FF, Liu ZP (2011) The accurate focusing and resolution analysis method in geosynchronous SAR. IEEE Trans Geosci Remote Sens 49:3548–3563
11. Wu X, Zhang SS, Xiao B (2012) An advanced range equation for geosynchronous SAR image formation. In: Proceedings of IET international conference on radar systems, Glasgow, UK, pp 1–4
12. Hu C, Liu ZP, Long T (2013) An improved focusing method for geosynchronous SAR. Adv Space Res 51:1773–1783
13. Bamler R, Hartl P (1998) Synthetic aperture radar interferometry. Inverse Prob 14:R1–R54
14. Ferretti A, Monti Guarnieri A, Prati C, Rocca F, Massonnet D (2007) InSAR principles: guideline for SAR interferometry processing and interpretation. ESA Publications, The Netherlands, pp B-11–B-55
15. Zebker HA, Villasenor J (1992) Decorrelation in interferometric radar echoes. IEEE Trans Geosci Remote Sens 30:950–959

16. Hu C, Li Y, Dong X et al (2015) Optimal data acquisition and height retrieval in repeat-track geosynchronous SAR interferometry. Remote Sens 7(10):13367–13389
17. Hanssen RF (2001) Radar interferometry: data interpretation and error analysis. Kluwer Academic Publishers, New York, USA, pp 23–60
18. Curlander JC (1983) Location of spaceborne SAR imagery. IEEE Trans Geosci Remote Sens GE-20, 359–364
19. Bruno D, Hobbs SE, Ottavianelli G (2006) Geosynchronous synthetic aperture radar: concept design. Prop Possible Appl Acta Aston 59:149–156
20. Bamler R (1998) Synthetic aperture radar interferometry. Inverse Prob 14(4):12–13

Chapter 7
Three Dimensional Deformation Retrieval in GEO D-InSAR

Abstract GEO SAR has characteristics of short revisit time of less than one day, extended coverage area with even larger than 1000 km and long coverage time of several hours for the scene of interest, and thus can provide data of a certain region of interest with lots of view angles. Consequently, employing GEO SAR for three-dimensional (3D) deformation retrieval can effectively address the drawbacks in LEO SAR cases, which are the lack of available data and the limited deformation retrieval accuracy. In this chapter, we first give some brief explanation about the reason why we should conduct 3D deformation retrieval instead of the simple one-dimensional (1D) line-of-sight (LOS) deformation measurement. Then, we focus on the GEO SAR 3D deformation retrieval by multi-angle measurement. To obtain the optimal accuracy, we consider the reasonable criterion to evaluate the 3D deformation measurement accuracy and implement it for optimal sub-aperture selection in 3D deformation retrieval.

7.1 Limitation of 1D Deformation Measurement

Although one dimensional (1D) deformation measurement of GEO SAR differential interferometry (GEO D-InSAR) has good deformation measurement accuracy, only a single direction (line-of-sight: LOS) deformation cannot be accurate to fully evaluate surface deformations [1–4]. As shown in Fig. 7.1, in the case of 1D deformation measurement, as the flight direction of the satellite is perpendicular to the north–south direction, the north–south direction deformation (the deformation marked as the green line) has a small projection component on the measurement direction. Resultantly, the LOS deformation measurement cannot measure the deformation in the north–south direction. In this case, even there is a serious disaster (earthquake, volcanic eruptions, etc.) with a large deformation in the north–south direction, it cannot be observed through the conventional D-InSAR system.

T. Long et al., *Geosynchronous SAR: System and Signal Processing*,
https://doi.org/10.1007/978-981-10-7254-3_7

Therefore, we need to study three dimensional (3D) deformation measurement through multi-angle data, and retrieve the complete deformation field information. Figure 7.1 is the schematic diagram of the 3D deformation measurement by SAR system. In the case of observation at three angles, although the deformation in the north–south direction has no projection component on the second observation angle, it has a large projection component on the first observation angle and can be measured. Therefore, due to the angle difference of the multi-angle observation, 3D deformation field will be projected upward at least one observation angle, and there will not be any problem in which dimension the deformation cannot be measured.

7.2 State of Art of 3D Deformation Retrieval in Spaceborne SAR

3D deformations measurement is very important as deformation usually occurs in multi-direction when natural disasters happen, Huang et al. [3] pointed that earthquake faults included both vertical and horizontal deformations based on the research of obviously moved Longmen Shan fault in 2008 Mw 7.9 China Wenchuan earthquake. The lack of the 3D deformation information will not only impede the accurately modelling and evaluating towards the disaster area (e.g. faults), but even any deformation cannot be detected when the deformation occurs along the flying direction.

To address the issue, the technologies of the 3D deformation retrieval have been developed for nearly 20 years. In 1999, Michel et al. [5] proposed a 3D deformation retrieval method based on the combination of azimuthal offsets (AZO) and cross-heading tracks D-InSAR data. Nevertheless, because the along-track deformation accuracy in AZO is only 1/10–1/30 of the spatial resolution, it only has the advantages in detecting large deformations (e.g. glacier movements). In 2004,

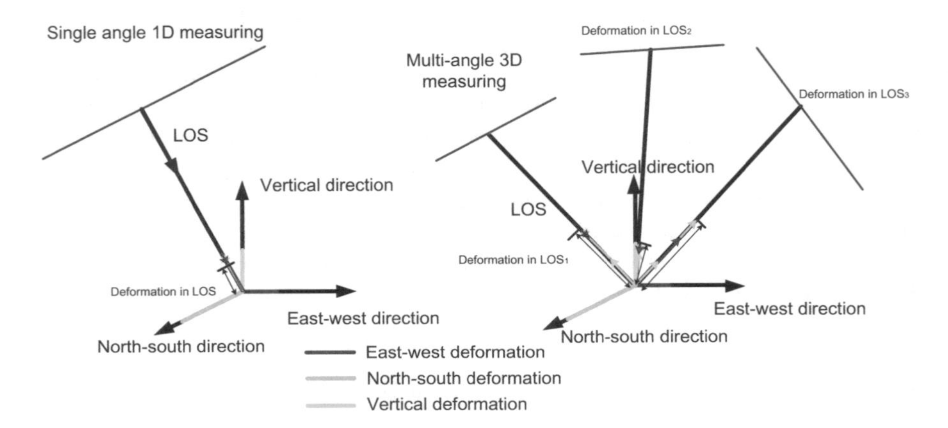

Fig. 7.1 Schematic diagram of the 1D and 3D deformation measurements by SAR system

Wright demonstrated a 3D deformation retrieval method based on multi-angle SAR data (more than three angles) [6]. However, restricted by the geometry configuration (e.g. only running in the polar orbit), the current running SAR satellites are hard to provide the effective multi-angle data [7]. Gudmundsson and Samsonov [8, 9] proposed the 3D deformation retrieval methods by fusing global positioning system (GPS) data and D-InSAR data, and the processing accuracy could be improved to the millimeter-level. However, the method is restrained by the numbers of GPS stations and the accuracy of the GPS data, which cannot improve the 3D deformation retrieval accuracy essentially. In order to improve the along-track deformation retrieval accuracy, Bechor et al. utilized multiple aperture interferometry (MAI) to obtain the more accurate along-track deformation by separating the full-aperture of the SAR data into the forward- and backward-looking sub-apertures and differentiating the corresponding interferograms [10]. MAI processing realizes the two-dimension (2D) (LOS and along-track) deformation retrieval rather than the simple 1D (LOS) deformation retrieval. 3D deformation can be obtained by combining the cross-heading tracks data after MAI processing based on weighted least squares solution [11]. Since the phase information is utilized, the deformation retrieval standard deviation error is improved by a factor 2 in the along-track by MAI compared with AZO [12–16].

Although 3D deformation retrieval can be retrieved, lots of problems exist in the real measurements for disaster evaluations: (1) a conventional low earth orbit (LEO) SAR has a long revisit time, from several days to dozens of days, which cannot satisfy the high temporal sampling rate requirement for disaster area observations; (2) because of the short coverage time of the LEO SAR systems, common areas of the cross-heading tracks SAR images used for 3D deformation retrieval are small. (3) 3D deformation retrieval accuracy of the LEO SAR MAI is still unsatisfactory, which is only about dozens of centimeters.

An effective method to address the shortcomings is to realize 3D deformation retrieval based on Geosynchronous SAR (GEO SAR) data [17, 18]. A GEO SAR runs in a geosynchronous orbit of 36,000 km height [19]. It has a short revisit time (<1 day, depending on different orbit design) and the long continuously coverage capability for the area of interest due to its high orbit [20, 21]. Therefore, GEO SAR MAI will increase the temporal sampling rate for 3D deformation retrieval. In addition, the data acquired within a long coverage time of GEO SAR towards the target could be effective to substitute the ascending (descending) data used in the 3D deformation retrieval, which can solve the problem of small common areas of the cross-heading tracks data.

3D deformation retrieval by GEO SAR has been conceived firstly by national aeronautic and space administration (NASA) in its global earthquake satellite system (GESS) project [22]. The expected 3D displacement accuracy can reach to a few millimeters after 24–36 h due to lots of GEO SAR images achieved in abundant view angles. Moreover, based on the study of the ideal GEO circular trajectory SAR (CSAR) system [23], it is validated that 3D deformation retrieval can be realized by using at least three sub-apertures measurements. However, some special issues in GEO SAR 3D deformation retrieval are necessary to be studied, such as the optimal sub-aperture selection method in GEO SAR 3D deformation retrieval and the accuracy analysis with the different conditions of orbits and the observation regions.

7.3 Multi-angle Measuring in GEO SAR: 3D Deformation Retrieval

7.3.1 Method and Accuracy Analysis

This section first explains the basic principle of GEO SAR 3D deformation retrieval by multi-angle measuring. The GEO SAR 3D deformation retrieval by multi-angle measuring can be realized by selecting at least three angles (sub-apertures) for interferometry processing, which are then be utilized for resolving the 3D deformation information of the whole scene. Basic sketch map of GEO SAR 3D deformation retrieval by multi-angle measuring is shown in Fig. 7.2. $\mathbf{P_0}$ is the position of the target. $\mathbf{S_1}$, $\mathbf{S_2}$, and $\mathbf{S_3}$ represent three phase centers of the sub-apertures obtained in different angles. L_{a1}, L_{a2} and L_{a3} are the corresponding sub-apertures in the three angles. $\widehat{\mathbf{E}}$ is the unit vector of the east–west direction, $\widehat{\mathbf{N}}$ is the unit vector of the north–south direction, $\widehat{\mathbf{H}}$ is the unit vector of the vertical direction. The black "figure 8" curve represents the nadir-point trajectory of the GEO SAR satellite.

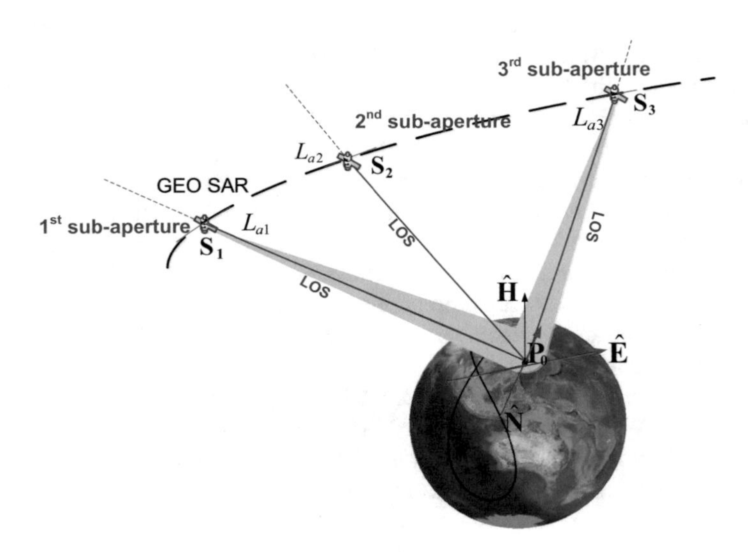

Fig. 7.2 Basic sketch map of GEO SAR 3D deformation retrieval by multi-angle measuring

It can be seen from Fig. 7.2 that the LOS deformation in each angle can be obtained respectively. We can get the estimation of the 3D deformation of the scene through the joint solution of the three angles.

Assuming the interferometric phases of angles L_{a1}, L_{a2} and L_{a3} are Φ_{s1}, Φ_{s2} and Φ_{s3}, the relationship between the multi-angle interferometric phases and the 3-D deformation is expressed as

$$\mathbf{\Phi} = \mathbf{\Theta}\mathbf{d} \tag{7.1}$$

where $\mathbf{\Phi} = (\Phi_{s1}, \Phi_{s2}, \Phi_{s3})^{\mathrm{T}}$ is the vector of the measured phases, $\mathbf{d} = (d_n, d_e, d_v)^{\mathrm{T}}$ is the 3D deformation vector in the local coordinate and the coefficient matrix $\mathbf{\Theta}$ is expressed as

$$\begin{aligned} \mathbf{\Theta} &= \begin{bmatrix} \frac{\partial \Phi_{s1}}{\partial d_n} & \frac{\partial \Phi_{s1}}{\partial d_e} & \frac{\partial \Phi_{s1}}{\partial d_v} \\ \frac{\partial \Phi_{s2}}{\partial d_n} & \frac{\partial \Phi_{s2}}{\partial d_e} & \frac{\partial \Phi_{s2}}{\partial d_v} \\ \frac{\partial \Phi_{s3}}{\partial d_n} & \frac{\partial \Phi_{s3}}{\partial d_e} & \frac{\partial \Phi_{s3}}{\partial d_v} \end{bmatrix} \\ &= \begin{bmatrix} -\frac{4\pi}{\lambda}\left\langle \widehat{\mathbf{R}}_{\mathbf{P_0,S_1}}, \widehat{\mathbf{N}} \right\rangle & -\frac{4\pi}{\lambda}\left\langle \widehat{\mathbf{R}}_{\mathbf{P_0,S_1}}, \widehat{\mathbf{E}} \right\rangle & -\frac{4\pi}{\lambda}\left\langle \widehat{\mathbf{R}}_{\mathbf{P_0,S_1}}, \widehat{\mathbf{H}} \right\rangle \\ -\frac{4\pi}{\lambda}\left\langle \widehat{\mathbf{R}}_{\mathbf{P_0,S_2}}, \widehat{\mathbf{N}} \right\rangle & -\frac{4\pi}{\lambda}\left\langle \widehat{\mathbf{R}}_{\mathbf{P_0,S_2}}, \widehat{\mathbf{E}} \right\rangle & -\frac{4\pi}{\lambda}\left\langle \widehat{\mathbf{R}}_{\mathbf{P_0,S_2}}, \widehat{\mathbf{H}} \right\rangle \\ -\frac{4\pi}{\lambda}\left\langle \widehat{\mathbf{R}}_{\mathbf{P_0,S_3}}, \widehat{\mathbf{N}} \right\rangle & -\frac{4\pi}{\lambda}\left\langle \widehat{\mathbf{R}}_{\mathbf{P_0,S_3}}, \widehat{\mathbf{E}} \right\rangle & -\frac{4\pi}{\lambda}\left\langle \widehat{\mathbf{R}}_{\mathbf{P_0,S_3}}, \widehat{\mathbf{H}} \right\rangle \end{bmatrix} \end{aligned} \tag{7.2}$$

Resolving (7.2), the estimation of $\mathbf{d}$ is given as

$$\widehat{\mathbf{d}} = \left(\mathbf{\Theta}^{\mathrm{T}}\mathbf{C_\Phi}\mathbf{\Theta}\right)^{-1}\mathbf{\Theta}^{\mathrm{T}}\mathbf{C_\Phi}\mathbf{\Phi} \tag{7.3}$$

where $\mathbf{C_\Phi}$ is the covariance matrix of the measured phase.

According to (7.3), the phase sensitivity of the deformation measuring direction is related only to the coefficient matrix during the GEO SAR multi-angle processing. The smaller the angle between the measured angle (LOS) and a component of the 3D deformation orthogonality basis, the more sensitive this angle is to the deformation along the component of the 3D deformation orthogonality basis, independent of the deformation value. The estimated covariance matrix of the deformation $\mathbf{C}_{\hat{\mathbf{d}}}$ is expressed as

$$\mathbf{C}_{\hat{\mathbf{d}}} = \left(\mathbf{\Theta}^{\mathrm{T}}\mathbf{C_\Phi}\mathbf{\Theta}\right)^{-1} \tag{7.4}$$

Especially, we have

$$\mathbf{E_\Phi} = diag(\mathbf{C_\Phi}) = \left(\sigma^2_{\Phi_{s1}}, \sigma^2_{\Phi_{s2}}, \sigma^2_{\Phi_{s3}}\right)^{\mathrm{T}} \tag{7.5}$$

where $\mathbf{E}_{\Phi}$ is the variance vector of the phase, $diag(\cdot)$ is the operation for extracting the diagonal elements of the matrix.

Assuming the interferograms of different angles and the deformations in different directions are not correlated, the variance vector of the deformation can be expressed as

$$\mathbf{E}_{\hat{\mathbf{d}}} = diag(\mathbf{C}_{\hat{\mathbf{d}}}) = \left(\sigma_{d_n}^2, \sigma_{d_e}^2, \sigma_{d_v}^2\right)^{\mathrm{T}} \tag{7.6}$$

From (7.6), the smaller the phase noise of the interferogram is, the higher the 3D deformation retrieval accuracy is obtained. In the meantime, $\mathbf{C}_{\hat{\mathbf{d}}}$ relates to $\boldsymbol{\Theta}$. Thus, 3D deformation retrieval accuracy also depends on the orbital configuration and the geographical location of the observed area, which will be discussed in the next section.

7.3.2 Optimal Multi-angle Data Selection

In general, the accuracy of the GEO SAR multi-angle 3D deformation retrieval is seriously affected by the coefficient matrix, that is, the retrieval accuracy relates to the spatial position of the satellite and the scene. Observation scenarios are generally determined by the actual application requirements, which cannot be arbitrarily changed. Therefore, it is necessary to study the optimal data acquisition angles (the positions of sub-apertures) within orbit interval for observing the scene to obtain the best 3D deformation retrieval performance. However, in practice, due to the complex geometric configuration of GEO SAR orbits, we often cannot find the appropriate combination of the data in three angles to minimize the deformation measurement errors in three dimensions simultaneously. Therefore, we need to consider the deformation retrieval accuracies in three dimensions holistically to select the optimal multi-angle data for 3D deformation retrieval.

Here, we introduce the concept of geometric dilution of precision (GDOP) in global navigation satellite system (GNSS) to evaluate the accuracy of the combination of the data obtained in different angles for the 3D deformation retrieval. Since there is no clock timing bias in SAR, in our analysis, GDOP will be determined by the position dilution of precision (PDOP). The PDOP is mainly characterized by the spatial configuration of different angles for the 3D deformation retrieval is good or bad. A poorer spatial configuration will result in a higher PDOP value.

According to the definition of PDOP, assuming the phase variance of interferograms in different angles is same (σ_e^2), we define the measurement accuracy coefficient of 3D deformation as

$$PDOP_m = \sqrt{\frac{16\pi^2\Lambda}{\lambda^2\sigma_e^2}} \tag{7.7}$$

where

$$\Lambda = \sqrt{\mathbf{E}_{\hat{\mathbf{d}}}^{\mathbf{T}} \mathbf{E}_{\hat{\mathbf{d}}}} \tag{7.8}$$

is the square root of the sum of the variances in three dimensions, representing the holistic 3D deformation measurement accuracy.

Considering (7.7), (7.8) can be simplified as

$$PDOP_m = \sqrt{\frac{16\pi^2}{\lambda^2} tr\left[\left(\mathbf{\Theta}^{\mathbf{T}}\mathbf{\Theta}\right)^{-1}\right]} \tag{7.9}$$

where $tr(\cdot)$ represents the trace of a matrix.

By (7.9), it is more intuitive to find that $PDOP_m$ is directly associated with the coefficient matrix. As a small $PDOP_m$ will be helpful for improving the 3D deformation retrieval accuracy, to find the optimal multi-angle positions to obtain the best holistic 3D deformation measurement accuracy, we need to consider

$$\begin{aligned} &\min_{\mathbf{S_1},\mathbf{S_2},\mathbf{S_3}} \{PDOP_m(\mathbf{S_1}, \mathbf{S_2}, \mathbf{S_3})\} \\ &s.t. \ \{\mathbf{S_1}, \mathbf{S_2}, \mathbf{S_3}\} \in \mathbf{S_a} \end{aligned} \tag{7.10}$$

where $\mathbf{S_a}$ represents the GEO SAR trajectory part which can observe the scene.

The analytical solution of (7.10) is difficult to directly obtain due to the complex trajectory of GEO SAR. At the same time, because of the long observation time and large observation area of GEO SAR system, it is not of high efficiency to obtain the optimal multi-angle data and 3D deformation accuracy of each ground observation position by searching algorithm along $\mathbf{S_a}$. To solve this problem, the searching algorithm can be firstly used to obtain the solutions of (7.10) for a number of sample ground observation points. Then, the neural networks can be constructed to obtain the results of the whole observation region of interest. Neural networks are very good at calculating the problem with a complex mathematical model [24]. After learning many times, the model parameters adaptability, the data calculation accuracy and the convergence rate can suit for the practical applications [24]. Therefore, based on the basic principles of neural networks, our model is set as

$$\mathbf{y}_i = f_i(\mathbf{w}_i^{\mathbf{T}}\mathbf{x}_i - \theta_{i,j}) \tag{7.11}$$

where i represents the ith layer of the neural network, $\mathbf{x}_i$ is the input of the neuron in the ith layer, $\mathbf{y}_i$ is the output of the neuron in the ith layer, $\mathbf{w}_i$ is the weight of the neuron in the ith layer, $\theta_{i,j}$ is the threshold of the jth neuron in the ith layer, f_i is the activation function of the neuron in the ith layer, which is usually the nonlinear sigmoid function shown as

$$f(x) = \frac{1}{1 + \mathrm{e}^{-x}} \tag{7.12}$$

When training samples, the input samples have two characteristics: longitude *lon* and latitude *lat*. The outputs are the minimal $PDOP_{m_o}$ and the corresponding true anomalies of the multi-angle positions A_{S_1}, A_{S_2} and A_{S_3}. Thus, the input vector $\mathbf{x}_1$ and the output vector $\mathbf{y}_e$ can be expressed as

$$\begin{cases} \mathbf{x}_1 = (lat, lon)^{\mathrm{T}} \\ \mathbf{y}_e = (PDOP_{m_o}, A_{S_1}, A_{S_2}, A_{S_3})^{\mathrm{T}} \end{cases} \tag{7.13}$$

Based on the above analysis, the complete 3D deformation retrieval based on GEO SAR multi-angle processing is depicted in Fig. 7.3, which is summarized as:

Step 1: based on the region of interest, GEO SAR orbit and the steering capability of the GEO SAR system (the look angle variation scope and the ground squint angle scope), calculating the required look angle and the long coverage GEO SAR orbit part which can observe the target region;
Step 2: using searching algorithm with the objective function of to determine the optimal sub-aperture locations for multi-angle data acquisition, which can minimize the PODP value by the selected sub-apertures;
Step 3: realizing the optimal OMRD InSAR pair data acquisition and completing imaging and interferometry processing; compensating the ionospheric disturbance phase by using the time-spatial ionospheric phase screen model;

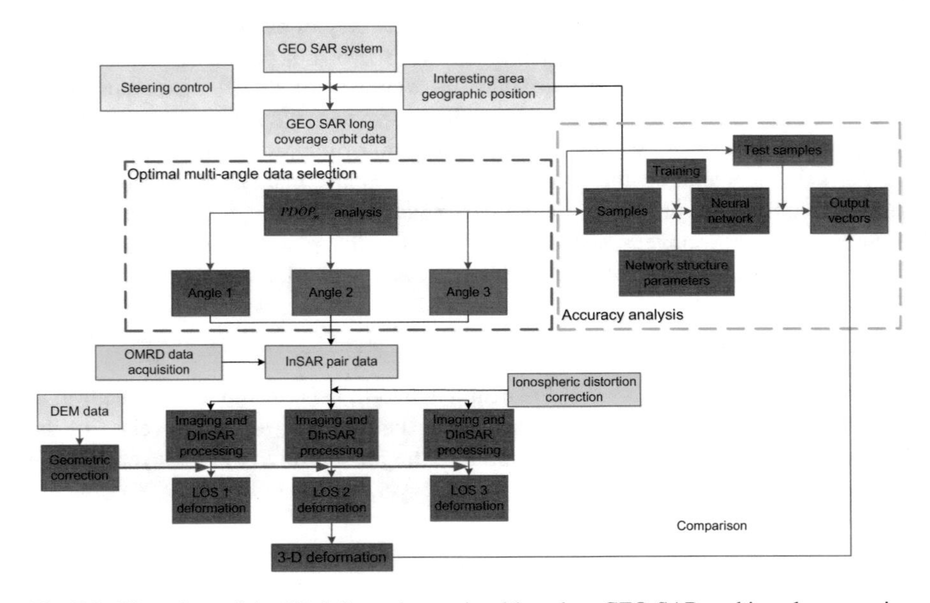

Fig. 7.3 Flow chart of the 3D deformation retrieval based on GEO SAR multi-angle processing

Step 4: co-registering the interferograms obtained at different angles by using external DEM data;
Step 5: generating the LOS deformations at different angles and realizing 3D deformation retrieval of the scene;
Step 6: utilizing PDOP analysis to calculate the 3D deformation retrieval accuracy of the samples in the scene, and obtaining the 3D deformation retrieval accuracy of the whole scene by neural network, which will be compared with the simulated 3D deformation retrieval results.

7.3.3 Simulation and Discussion

In the simulation, we use the GEO SAR orbit with a large "figure 8" nadir-point trajectory. The orbit position at apogee is selected as an instance for the following processing. The geographic position of the center of the scene is (40°N, 105°E). The orbit and system parameters of GEO SAR are given in Table 7.1.

7.3.3.1 Optimal Multi-angle Data Selection

The selection of the multi-angle data directly affects the accuracy of GEO SAR 3D deformation retrieval. Therefore, we first simulate and discuss the optimal multi-angle data selection for GEO SAR 3D deformation retrieval. The antenna has the ability to be continuously steered, and only has the right-look ability. The look angle ranges from 1° to 8° and the ground squint angle of GEO SAR is limited within ±60°. In this case, the GEO SAR can keep observing the scene [scene center locates at (40°N, 105°E)] within a true anomaly scope of [116.4°, 243.4°]. When all the positions of the three angles used for 3D deformation retrieval are changed, the whole optimization space is a spatial topology, which is very complicated. In order to simplify the calculation, in the following simulation analysis, we assume that the position of the first aperture is known at a position in the full aperture (true anomaly is 174.4°), which is obtained at the ground 0° squint angle and 2° look angle. Then, the 3D deformation retrieval performance is investigated by selecting the other two angles (sub-apertures) along the orbit scope of covering the scene. Using (7.4), we obtain the 3D deformation retrieval accuracies of different positions of the two angles, which are shown in Fig. 7.4.

Table 7.1 GEO SAR orbit and system parameters

Items	Value	Items	Value
Semi-major axis (km)	42164.170	True anomaly (°)	0
Inclination (°)	53	Eccentricity	0.07
Argument of perigee (°)	270	Right ascension of ascending node (°)	265
Wavelength (m)	0.24	Bandwidth (MHz)	18

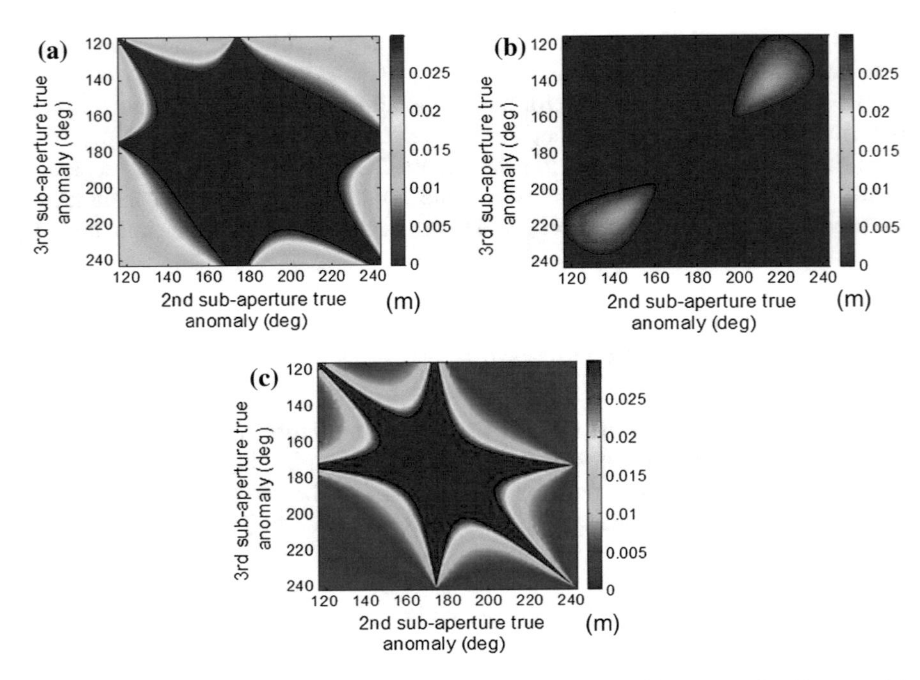

Fig. 7.4 3D deformation retrieval accuracies of different positions of the two angles (the area with the accuracy worse than 0.03 m has been excluded and shows the darkest). **a** North–south direction; **b** East–west direction; **c** Vertical direction

From Fig. 7.4, it can be first observed that it will introduce a large error in the 3D deformation retrieval when two or more angles have the same or similar positions (corresponding to the dark-colored region in Fig. 7.4). When the positions of at least two angles are accordance, theoretically, 3D deformation retrieval is impossible. The bias between the positions of the angles will be helpful to improve the 3D deformation retrieval accuracy. This also means the coefficient matrices are not correlated. In addition, according to Fig. 7.4, the deformation accuracies in different directions are largely difference. The deformation measurement accuracy in vertical direction is higher than those of the other two directions by an order of magnitude. The deformation measurement accuracy in the east–west direction is doubled than that in the north–south direction. This phenomenon is mainly due to the different deformation measurement sensitivities of the deformation measurements in three directions. The results of the deformation measurement sensitivity analysis in different directions are shown in Fig. 7.5. The deformation measurement sensitivity in the vertical direction is the highest among the three directions. Moreover, in the most part of the orbit, the deformation measurement sensitivity in the north–south direction improved about twice than that in the east–west direction.

As the optimal positions of the deformation measurement accuracy in different directions are different, we should determine the optimal multi-angle positions for GEO SAR 3D deformation measurement based on the method of PDOP analysis.

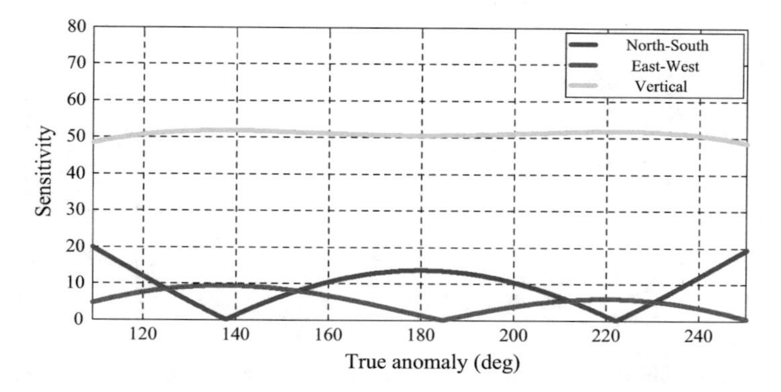

Fig. 7.5 Deformation measurement sensitivity analysis in different directions

The PDOP analysis of the selection of other two multi-angle positions is shown in Fig. 7.6. As shown in Fig. 7.6, when the true anomalies of the three angles are (132.1°, 174.4°, 222.4°) (marked as cross star), the holistic 3D deformation retrieval accuracy is the optimal, which has only a PDOP of 6.2. Under such multi-angle data selection, the 3D deformation measurement accuracies are 0.009 m (north–south direction), 0.021 m (east–west direction) and 0.003 m (vertical direction). Since the east–west direction has the worst deformation measurement accuracy, its accuracy has the greatest effect on the PDOP. Comparing Figs. 7.4b and 7.6, it can be found that the location of the multi-angle positions determined optimally by PDOP is closer to the position with the best deformation measurement accuracy in the east–west direction. Finally, the multi-angle data selected by optimal PDOP makes the deformation measurement accuracy in the three dimensions all less than 2.1 cm.

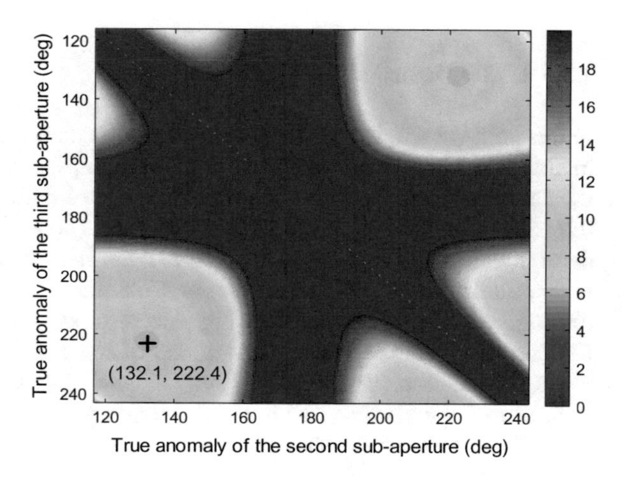

Fig. 7.6 PDOP analysis of the selection of other two multi-angle positions (assuming the correlation coefficient of a single interferogram is 0.95)

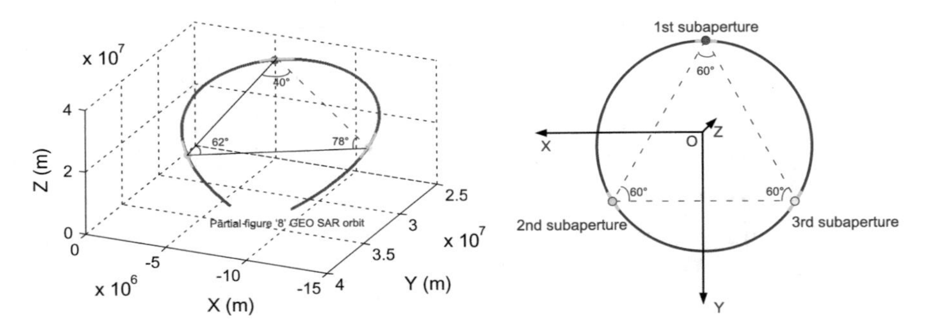

Fig. 7.7 Relationship of the angles between the link lines of the three phase centers of the multi-angle data. **a** inclined GEO SAR; **b** GEO CSAR

According to the optimal multi-angle positions, we study the relationship of the angles between the link lines of the three phase centers of the multi-angle data, which is shown in Fig. 7.7a. The angles between the three link lines are 62°, 40°, and 78°, respectively (clockwise). Compared with the angles between the three link lines of optimal multi-angle positions in CSAR (Fig. 7.7b), which are all 60°, there is an obvious difference, which derives from the following two aspects:

(1) the inclined "figure 8" GEO SAR orbit is affected by the earth rotation and orbit perturbation, which make trajectory very curved, like a 'collar' edge. It is not a standard circle, which is pear-like or elliptical;
(2) the steering capability of the actual GEO SAR satellite system is limited and the system cannot stare at the target all the time. Thus, only part of the orbit (part of space angle) can be used for multi-angle retrieval.
Therefore, as the influence of the aforementioned factors, the spatial configuration of the optimal multi-angle data is different from that in GEO CSAR.

7.3.3.2 3D Deformation Retrieval

In this part, we will exhibit the results of the 3D deformation retrieval. The scene of interest is 7 km × 7 km. There is a 260 m high pyramid in the scene. We generate the deformation in the pyramid area. The vertical deformation is 2 cm, decreasing linearly from the peak to the bottom. There are about 0.26 m deformation and no deformation in the east–west direction and north–south direction, respectively. The integration time is 100 s. The temporal baseline is 1 day. In the following processing, we only consider the single look processing and the Goldstein phase filtering coefficient is 0.1. The positions of the multi-angle data used in the following simulation are shown in Fig. 7.8. A–E represent the five positions of the utilized multi-angle data and the corresponding true anomalies are also marked in Fig. 7.8. The covered yellow line along the nadir-point trajectory represents the orbit scope which can observe the target.

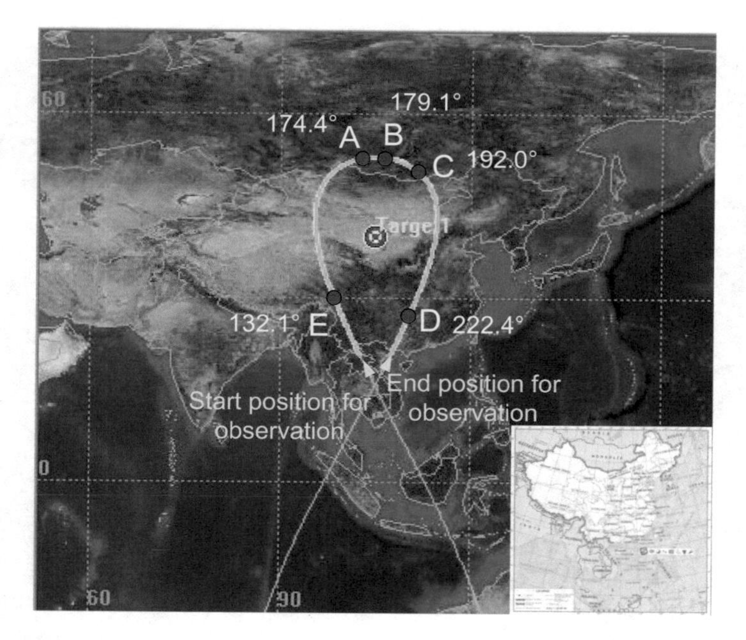

Fig. 7.8 Positions of the multi-angle data used in the following simulation

Two simulation groups are conducted and their results are compared. We first use the multi-angle data with the sub-apertures of A, B and C (true anomalies are 174.4°, 179.1° and 192.0°) for the 3D deformation retrieval, which has a theoretical PDOP value of 106 (assuming the correlation coefficient of a single interferogram is 0.95). In the comparison group, the optimal multi-angle data, including the sub-apertures of A, D, and E (true anomalies are 132.1°, 174.4° and 222.4°), is selected (theoretical PDOP = 6.2 for this group). The interferograms of the sub-apertures all have the high coherence coefficients above 0.8 [0.91 (A), 0.91 (B), 0.83 (C), 0.91 (D), and 0.89 (E)]. The original set 3D deformation and the 3D deformation retrieval results are shown in Fig. 7.9. The corresponding error analysis is provided in Table 7.2. Because of the non-optimal sub-apertures, the deformation measurement accuracy in the north–south is worse in the first group, which is more than 20 cm. In addition, its east–west deformation measurement accuracy (ten-centimeter level) and vertical deformation measurement accuracy (centimeter-level) are not very high as well. In comparison, when the optimal sub-apertures are selected, all the deformation measurement accuracies in the three dimensions are improved significantly. Both the east–west and north–south deformation measurement accuracies are centimeter-level, and the vertical deformation measurement even reaches to millimeter-level. Therefore, the simulation results suggest the importance of the optimal sub-aperture selection.

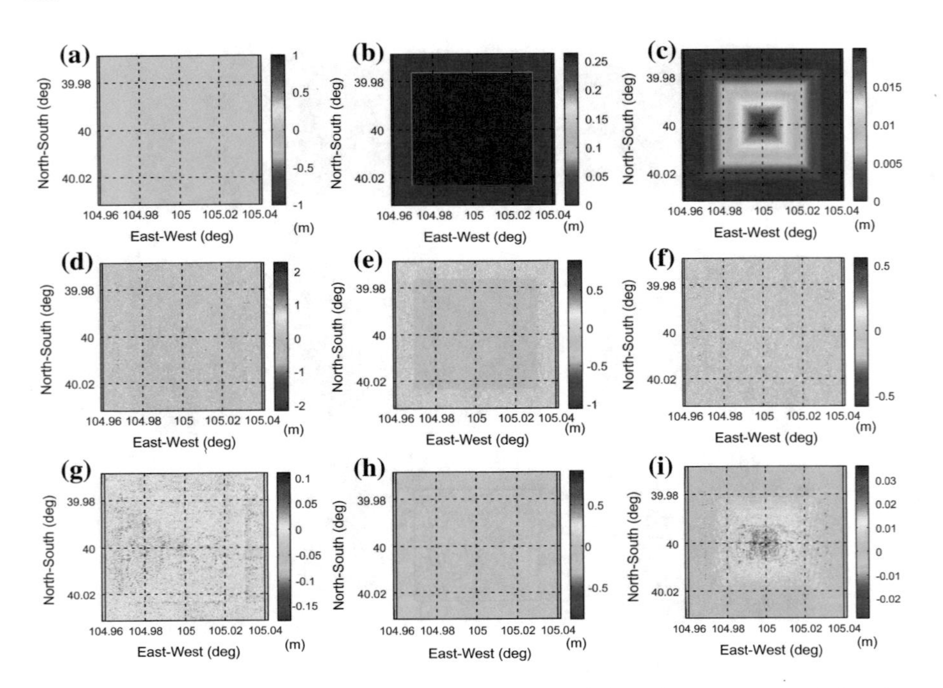

Fig. 7.9 Results of the 3D deformation retrieval. Original deformation **a** north–south direction (N–S); **b** east–west direction (E–W); **c** vertical direction (VD). The first group (A + B + C): **d** N–S; **e** E–W; **f** VD. The second group (A + D + E): **g** N–S; **h** E–W; **i** VD

Table 7.2 3D deformation retrieval accuracy evaluation (RMSE analysis)

Group no.	North–south direction (cm)	East–west direction (cm)	Vertical direction (cm)
1	25.1	10.0	6.6
2	1.72	6.27	0.35

7.3.3.3 Influence Analysis of Orbit and Geographical Position of Scene

Three classical GEO SAR orbits are used in this part for simulation analysis the impacts of orbit and geographical position of scene on 3D retrieval accuracy. Although they all run in geosynchronous orbit height, but have different orbit elements (e.g. eccentricity and inclination). The prime orbit elements of the three classical GEO SAR orbits are listed in Table 7.3 and the nadir-point trajectories are given in Fig. 7.10a. Large “figure 8” and “figure O” are set to work in right-look for obtaining good coverage performance for China. In contrast, small “figure 8” is designed to work in left-look to obtain good coverage performance for China.

In the simulation, we use the neural network method described in Sect. 7.3.3 for the analysis. We selected 44 samples (43 in China and 1 in Singapore) in the

Table 7.3 Prime orbit elements of the classical GEO SAR orbit configurations

Nadir-point trajectory	Large "figure 8"	Small "figure 8"	"figure O"
Eccentricity	0.07	0.05	0.1
Inclination (°)	53	0	16

following simulation, which are shown in Fig. 7.10b. The training set constitutes 39 training samples and test set has 5 samples. We have constructed a three-layer Back Propagation (BP) neural network. Each layer of the neural network includes 10 neurons. Sigmoid function is selected as the activation function. To reduce computation burden, when conducting the 3D deformation accuracy analysis, we assume that the position of the first angle is known, which locates at the start of the orbit scope for observing the target. Then, the optimal positions of the other two angles are analyzed, that is the output vector $\mathbf{y}_e = (PDOP_{m_o}, A_{S_2}, A_{S_3})^{\mathrm{T}}$. Holistic 3D deformation retrieval accuracies Λ under the three orbits and the different geographical positions and the coverage maps for China are given in Fig. 7.11.

Comparing the results of Fig. 7.11a, c and e, it can be seen that within the 3 types of orbits, the large "figure 8" orbit has the best holistic 3D deformation retrieval accuracy, which reaches to centimeter-lever. The holistic 3D deformation retrieval accuracy even reaches to millimeter-level in Qinghai-Tibet Plateau of China; small "figure 8" has the worst holistic 3D deformation retrieval accuracy, which is close to meter-level. The 3D deformation retrieval accuracy in the east part of China is about 10 cm, which is better than that in the west area of China; "figure O" has a moderate 3D deformation retrieval accuracy, which ranges from several centimeters to dozens of centimeters. 3D deformation retrieval accuracy in the south is better than that in the north in the "figure O" case. Various 3D deformation retrieval accuracies derive from different orbit configurations. According to Fig. 7.10a, the observation angle scope of the large "figure 8" is the greatest for observing China, which results in a small correlation of each row in the observation

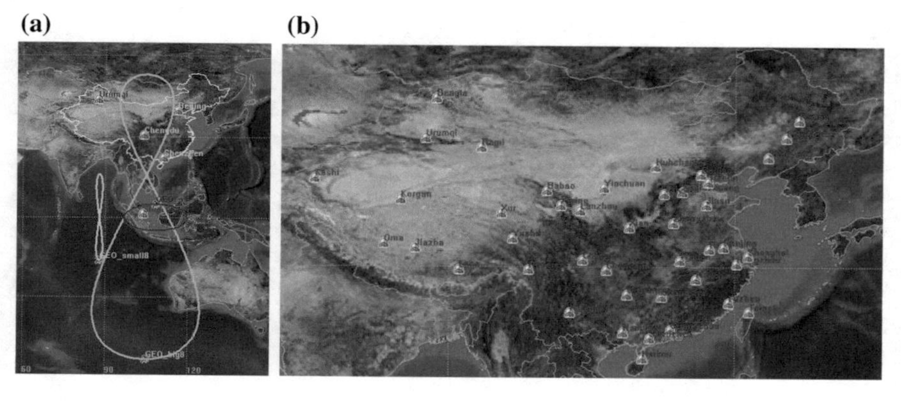

Fig. 7.10 Nadir-point trajectories of the three GEO SAR orbits (**a**) and the distribution of the samples (**b**)

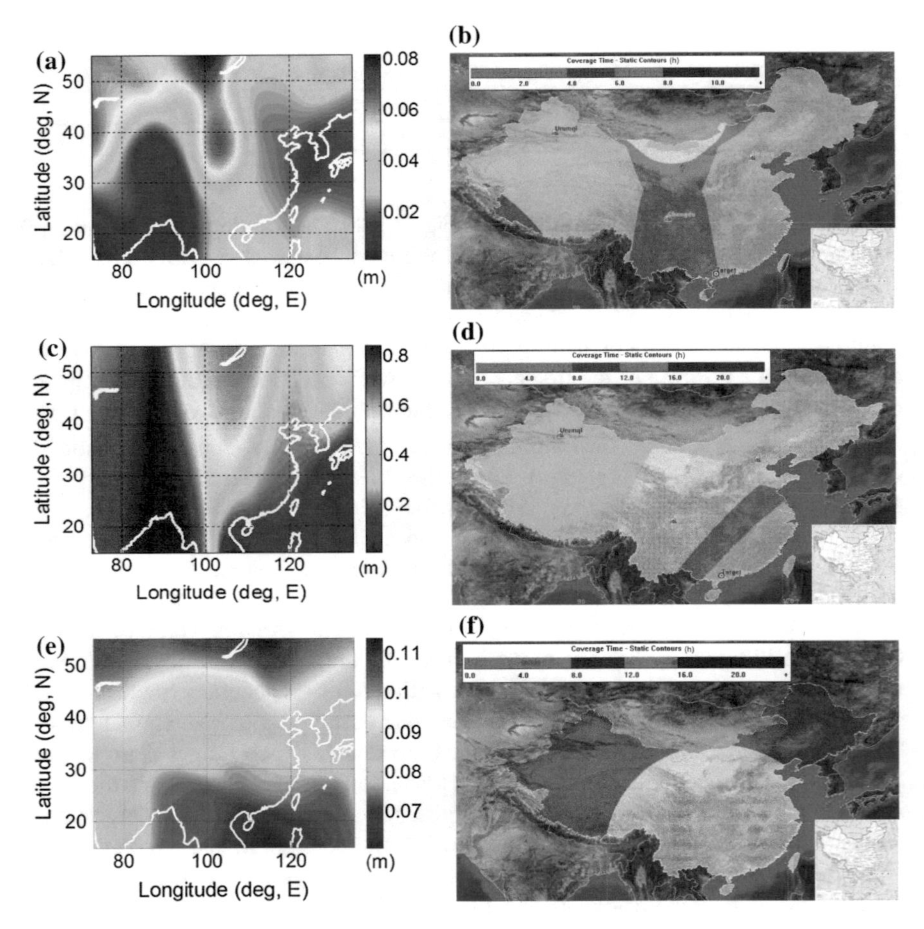

Fig. 7.11 Holistic 3D deformation retrieval accuracies under the three orbits and the different geographical positions **a** large "figure 8"; **c** small "figure 8"; **e** "figure O". Coverage maps for China **b** large "figure 8"; **d** small "figure 8"; **f** "figure O"

matrix. Thus, its 3D deformation retrieval accuracy is the best. In contrast, as the observation angle scope of the small "figure 8" is very small, its 3D deformation retrieval accuracy is poor. Analyzing Fig. 7.11a, b, c, d, e and f, it can be found that the 3D deformation retrieval accuracy of different orbits and the coverage of the corresponding orbit for China have a positive correlation, this is, an area with a longer coverage corresponds to a better 3D deformation retrieval accuracy. As a longer coverage gives rise to a larger observation angle scope for the scene, 3D deformation retrieval accuracy will be improved.

7.4 Summary

In this chapter, we introduce the 3D deformation retrieval method by the multi-aperture processing method. We explain the detailed 3D deformation retrieval methods and the accuracies. Multi-aperture processing method is a more traditional method to obtain 3D deformation. With the method, 3D deformation can be obtained directly after the InSAR data obtaining of three sub-apertures. Due to the more flexible selection of the sub-apertures, the 3D deformation retrieval accuracy by multi-aperture processing method is much higher than MAI method. Under the single-look processing, the accuracy can be as lower as centimeter-level and even millimeter-level.

The optimal GEO SAR sub-aperture selection method for 3D deformation retrieval is needed. Based on the optimal configuration of the sub-apertures for the minimum PDOP, the proper optimal data selection method is proposed. The simulation results show the proposed optimal method can greatly improve the 3D deformation retrieval accuracy.

Moreover, the orbital configurations and locations of scene have the large impacts on the GEO SAR 3D deformation retrieval accuracy. We analyze the impact in the multi-angle processing case. In the best case of the analyzed orbits, "figure 8", it has the best holistic 3D deformation retrieval accuracy, which reaches to centimeter-lever. In contrast, as for the small "figure 8" orbit, its 3D deformation retrieval accuracy is close to meter-level in the west part of China.

References

1. Massonnet D, Rossi M, Carmona C, Ardagna F, Peltzer G, Feigl K, Rabaute T (1993) The displacement field of the landers earthquake mapped by radar interferometry. Nature 364:138–142
2. Ferretti A, Guarnieri AM, Prati C, Rocca F, Massonnet D (2007) InSAR principles. ESA Publications, The Netherlands
3. Huang R, Fan X (2013) The landslide story. Nat Geosci 6:325–326
4. Li YH, Hu C, Long T (2015) A novel SAR interferometry processing method in high resolution spotlight SAR. J Electromag Waves Appl 29(13):1786–1803
5. Michel R, Avouac JP, Taboury J (1999) Measuring ground displacements from SAR amplitude images: application to the landers earthquake. Geophys Res Lett 26(7):875–878
6. Wright TJ, Parsons BE, Lu Z (2004) Toward mapping surface deformation in three dimensions using InSAR. Geophys Res Lett 31(1):L01607
7. Ansari H, Zan FD, Parizzi A, Eineder M, Goel K, Adam N (2016) Measuring 3-D surface motion with future SAR systems based on reflector antennae. IEEE Geosci Remote Sens Lett 13(2):272–276
8. Gudmundsson S, Sigmundsson F (2002) Three-dimensional surface motion maps estimated from combined interferometric synthetic aperture radar and GPS data. J Geophys Res 107 (B10):2250
9. Samsonov S, Tiampo K (2006) Analytical optimization of a DInSAR and GPS dataset for derivation of three-dimensional surface motion. IEEE Geosci Remote Sens Lett 3(1):107–111

10. Bechor NBD, Zebker HA (2006) Measuring two-dimensional movements using a single InSAR pair. Geophys Res Lett 33(16):L16311
11. Jung HS, Lu Z, Shepherd A, Wright T (2015) Simulation of the SuperSAR multi-azimuth synthetic aperture radar imaging system for precise measurement of three-dimensional earth surface displacement. IEEE Trans Geosci Remote Sens 53(11):6196–6206
12. Jung HS, Won JS, Kim SW (2009) An improvement of the performance of multiple-aperture SAR interferometry (MAI). IEEE Trans Geosci Remote Sens 47(8):2859–2869
13. Jo MJ, Jung HS, Won JS (2015) Detecting the source location of recent summit inflation via three-dimensional InSAR observation of Kilauea Volcano. Remote Sens 7(11):14386–14402
14. Jung HS, Lee DT, Lu Z, Won JS (2013) Ionospheric correction of SAR interferograms by multiple-aperture interferometry. IEEE Trans Geosci Remote Sens 51(5):3191–3199
15. Jo MJ, Jung HS, Won JS, Poland MP, Miklius A, Lu Z (2014) Measurement of slow-moving along-track displacement from an efficient multiple—aperture SAR interferometry (MAI) stacking. IEEE Trans Geosci Remote Sens 52(6):3421–3427
16. Hu J, Li ZW, Ding XL, Zhu JJ, Zhang L, Sun Q (2012) 3D coseismic displacement of 2010 Darfield, New Zealand Earthquake estimated from multi-aperture InSAR and D-InSAR measurement. J Geophys Res 11:1029–1041
17. Hu C, Li Y, Dong X, Wang R, Cui C (2017a) Optimal 3D deformation measuring in inclined geosynchronous orbit SAR differential interferometry. Sci China Inf Sci 60(6)
18. Hu C, Li Y, Dong X, Wang R, Cui C, Zhang B (2017b) Three-dimensional deformation retrieval in geosynchronous SAR by multiple-aperture interferometry processing: Theory and Performance Analysis. IEEE Trans Geosci Remote Sens 55(11):6150–6169
19. Tomiyasu K (1978) Synthetic aperture radar in geosynchronous orbit. In: Proceedings of antennas and propagation society international symposium. College Park MD, USA, pp 42–45
20. Prati C, Rocca F, Giancola D, Monti A (1976) Passive geosynchronous SAR system reusing backscattered digital audio broadcasting signals. IEEE Trans Geosci Remote Sens 36: 1973–1976
21. Monti-Guarnieri A, Bombaci O, Catalano TF, Germani C, Koppel C, Rocca F, Wadge G (2015) ARGOS: a fractioned geosynchronous SAR. Acta Astronaut. https://doi.org/10.1016/j.actaastro.2015.11.022
22. NASA, JPL, Global Earthquake Satellite System, a 20 year plan to enable earthquake prediction, 2003, (http://www.jpl.nasa.gov), 20 May 2005
23. Kou L, Wang X, Xiang M, Zhu M (2012) Interferometric estimation of three-dimensional surface deformation using geosynchronous circular SAR. IEEE Trans Aerosp Electron Syst 48(2):1619–1635
24. Stramondo S, Del Frate F, Picchiani M, Schiavon G (2011) Seismic source quantitative parameters retrieval from InSAR data and neural networks. IEEE Trans Geosci Remote Sens 49(1):96–104

Printed in the United States
By Bookmasters